Helge Ingolf Geisler

Finite Element Method (FEM) Model and Performance Analysis of Solid Oxide Fuel Cells

Schriften des Instituts für Angewandte Materialien –
Werkstoffe der Elektrotechnik
Karlsruher Institut für Technologie

Band 36

Eine Übersicht aller bisher in dieser Schriftenreihe
erschienenen Bände finden Sie am Ende des Buchs.

Finite Element Method (FEM) Model and Performance Analysis of Solid Oxide Fuel Cells

by
Helge Ingolf Geisler

KIT Scientific Publishing

Karlsruher Institut für Technologie
Institut für Angewandte Materialien – Werkstoffe der Elektrotechnik

Finite Element Method (FEM) Model and Performance Analysis of Solid Oxide Fuel Cells

Zur Erlangung des akademischen Grades eines Doktor-Ingenieurs von der KIT-Fakultät für Elektrotechnik und Informationstechnik des Karlsruher Instituts für Technologie (KIT) genehmigte Dissertation

von Dipl.-Ing. Helge Ingolf Geisler, geb. in Wiesbaden

Tag der mündlichen Prüfung: 14. Dezember 2018
Hauptreferentin: Prof. Dr.-Ing. Ellen Ivers-Tiffée
Korreferent: Prof. Dr.-Ing. Thomas Schulenberg

Impressum

KIT Scientific Publishing

Karlsruher Institut für Technologie (KIT)
KIT Scientific Publishing
Straße am Forum 2
D-76131 Karlsruhe

KIT Scientific Publishing is a registered trademark of Karlsruhe Institute of Technology.
Reprint using the book cover is not allowed.

www.ksp.kit.edu

Print on Demand 2019 – Gedruckt auf FSC-zertifiziertem Papier

ISSN 2365-8029
ISBN 978-3-7315-0895-3
DOI 10.5445/KSP/1000090508

Finite Element Method (FEM) Model and Performance Analysis of Solid Oxide Fuel Cells

Zur Erlangung des akademischen Grades eines

DOKTOR-INGENIEURS

von der KIT-Fakultät für

Elektrotechnik und Informationstechnik
des Karlsruher Instituts für Technologie (KIT)

genehmigte

DISSERTATION

von

Dipl.-Ing. Helge Ingolf Geisler
geb. in Wiesbaden (Hessen)

Tag der mündlichen Prüfung:	14.12.2018
Hauptreferentin:	Prof. Dr.-Ing. Ellen Ivers-Tiffée
Korreferent:	Prof. Dr.-Ing. Thomas Schulenberg

Acknowledgements

This dissertation is the result of my research at the *Institut für Angewandte Materialien - Werkstoffe der Elektrotechnik (IAM-WET)*. First and foremost I would like to express my sincere gratitude to Prof. Dr.-Ing. Ellen Ivers-Tiffée, who gave me the opportunity to accomplish this work by supervising my diploma thesis and subsequently accepting me as a scientist at her institute. I am truly grateful for the support and confidence she has placed in me, for the excellent working environment and high quality equipment supplied in her institute, and for the experience I have gained at various conferences at very special locations abroad. The time spent under her guidance has greatly encouraged my professional and personal development and the knowledge gained will accompany and support me for the rest of my life.

Second, I would like to thank Prof. Dr.-Ing. Thomas Schulenberg from the Institut für Kern- und Energietechnik (IKET) for taking on the role as second reviewer of my dissertation, even though its scientific content is somewhat distant from his current work. His careful evaluation of each written page and the critical review in several discussions have significantly increased the quality of this work, and I am proud to have received his approval in the end.

My position as a scientist was largely funded by the EU project METSAPP, which not only offered diversified tasks, but also helped me to understand more about the bigger picture in fuel cell development. I really appreciate the time spent at the different locations in the EU during the project meetings, where I made personal contacts with wonderful people from other research institutions and industrial partners. As representatives of all project members, I would like to thank Peter Blennow and Henrik Lund Frandsen from the Technical University of Denmark (DTU) and Richard Schauperl from AVL for the scientific exchange and, probably even more importantly, for the time we spend together during various off-work activities.

The foundation of this work has been laid by my diploma thesis advisor and former colleague, Alexander Kromp. His role as an advisor to this day and his encouragement made the insane idea of even starting this work possible. Further, I am indebted to my group leader André Weber for the scientific discussions, for constantly sharing his expertise and the overall trusting cooperation. Also, I had the pleasure to supervise several student works in collaboration with colleagues and on my own. In this context I would like to thank especially Sebastian Dierickx, who is now my colleague and whose scrutinizing mentality often made me think twice.

Apart from the intriguing scientific challenge, it was above all the pleasant and friendly working atmosphere at the institute that convinced me to give up my private life to a large extent and pursue my dissertation. I would like to thank all institute employees for their always reliable support in technical and organizational matters. Alongside the ever freely shared scientific advice of my colleagues, the sometimes escalating distractions during and after work, especially at conferences, were more than welcome and made life worthwhile. Most responsible characters to thank in this context are Jörg Illig, Dino Klotz, Michael Schönleber, Jan Hayd, Michael Kornely, André Leonie, Daniel Manka and Andreas Häffelin. However, a special thanks goes to Jochen Joos and Julian Szász, with whom I shared alternately the office, numerous lunch breaks and sports activities after work and during the holidays. I sincerely hope, we have a long way to go together!

Cheers to all my fellow students and friends, especially to Johannes Sartiono, Andreas Gobell and Arkadiusz Dmytrow, for reminding me that life is beautiful, even when it is sometimes difficult to understand.

Last but not least I want to thank my parents and my sister for their unconditional love and support at all times. Their trust that I was going my way made this work possible in the first place. Finally, I would like to express my deepest and most heartfelt gratitude to my love Leonie and to our wonderful son Jona, who have endured me in these difficult times and who have created for me the necessary distance to worldly affairs. Now that this work is finished, I am looking forward to our next step as one of many to come.

Karlsruhe,
December 2018

HELGE GEISLER

Zusammenfassung

In der heutigen, modernen Gesellschaft ist die zuverlässige Versorgung mit Energie eine Grundvoraussetzung für das Gemeinwohl. Daraus leitet sich eine moralische Verantwortung ab, den immer weiter wachsenden Bedarf möglichst klimaneutral zu decken. Die dezentrale Energieproduktion durch erneuerbare Quellen (Wind und Sonne) erfordert allerdings eine Neuausrichtung der ehemals zentralistisch organisierten Energieversorgungsinfrastruktur. Zusätzlich besteht ein erhöhter Bedarf an Speicherkapazität, um die volatile Natur der erneuerbaren Energieerzeugung auszugleichen. In diesem Kontext rückt die Brennstoffzellentechnologie in letzter Zeit verstärkt in den Fokus, da sie eine saubere und effiziente Umwandlung von chemisch gespeicherter Energie direkt in elektrische Energie ermöglicht. Zusätzlich bietet sie gleichzeitig die Möglichkeit, durch den Elektrolysebetrieb elektrische Energie in Wasserstoff zu speichern. Brennstoffzellen haben daher das Potential, eine Schlüsselposition in der modernen Energieversorgung einzunehmen. Die Hochtemperatur Festoxid-Brennstoffzelle (SOFC, engl. solid oxide fuel cell) spielt dabei eine besondere Rolle: Aufgrund der erhöhten Betriebstemperatur können in der katalytisch aktiven Brenngaselektrode (Anode) nicht nur reiner Wasserstoff, sondern auch kohlenwasserstoffhaltige Brenngase direkt verwendet werden. Dadurch ist die SOFC im Gegensatz zu anderen Brennstoffzelltypen nicht direkt abhängig von einer flächendeckenden Wasserstoffversorgung. Diese Brenngasflexibilität ermöglicht ein breites Einsatzspektrum: Als stationäre Kombi-Kraftwerke zur Bereitstellung von Wärme und Energie in Ein- und Mehrfamilienhäusern, als Hilfskrafterzeuger (APU, engl. auxiliry power unit) in mobilen und Inselanwendungen, sowie in Großindustriellen Anlagen und Energieerzeuger/-Speicher Einheiten.

Das primäre Entwicklungsziel in der SOFC Forschung liegt neben der Verbesserung der Langzeitstabilität und generellen Kostensenkung im Transfer der bereits hervorragenden Leistungsfähigkeit von Einzelzellen in die technisch relevante Anwendung. Dies erfordert eine serielle Verschaltung der Einzelzellen zu einem Stack mit Hilfe von Interkonnektoren. Dadurch werden im Betrieb die zellinternen Verlustprozesse durch eine nicht ideale Kontaktierung der Luftelektrode (Kathode) negativ beeinflusst und zusätzliche Verlustprozesse verursacht.

Das Ziel der vorliegenden Arbeit ist es, ein detailliertes Modell basierend auf der Finite-Elemente-Methode (FEM) zu entwickeln, mit dessen Hilfe die Leistung der SOFC im Betrieb mit Wasserstoff sowie mit kohlenwasserstoffhaltigem Brenngas zuverlässig vorhergesagt werden kann. Die FEM ermöglicht dabei, Verlustmechanismen ortsaufgelöst zu berücksichtigen. Darauf basierend wurde eine numerische Untersuchung von relevanten Kathodenmaterial- und Interkonnektor Geometrie Parametern durchgeführt, um deren Einfluss auf die Leistung systematisch zu analysieren und somit das potentielle Optimierungspotential für die kommerzielle Anwendung aufzuzeigen.

Experimentelle Messungen zur Parameterbestimmung und Modell Validierung wurden an anodengestützten Zellen (ASCs, engl. anode supported cells), bestehend aus Ni/YSZ Zermet-Anode, YSZ-Elektrolyt, GDC-Diffusionsbarriere und LSCF Kathode durchgeführt. Die aktive Elektrodenfläche der am Forschungszentrum Jülich hergestellten ASCs betrug 1 cm^2 sowie 16 cm^2. Als Messmethode kam die elektrochemische Impedanzspektroskopie (EIS), die Aufnahme von Strom/Spannungskennlinien (C/Vs) sowie gaschromatographische Analytik zum Einsatz.

Die wichtigsten Informationen zur Modellerstellung und die damit gewonnenen Ergebnisse dieser Arbeit werden im Folgenden zusammengefasst.

Modellentwicklung - Kapitel 4

Das Modell wurde mit Hilfe der kommerziellen FEM Software COMSOL Multiphysics v4.3b und MATLAB entwickelt. Dabei wurde größtenteils auf die vorgegebenen Modellgleichungen verzichtet und eigene Ansätze verwendet. Dies bietet die Möglichkeit, das Modell auch in anderen Entwicklungsumgebungen zu implementieren. Die symmetrische Anordnung von Gaskanälen und Kontaktstegen des planaren Interkonnektor Flowfield Designs ermöglicht eine Reduzierung der erforderlichen Modellgeometrie auf eine Wiederholeinheit (RPU, engl. repeat unit), und damit einhergehend eine starke Reduzierung des erforderlichen numerischen Aufwands (Section 4.2). In der Arbeit wird unterschieden zwischen einer Modellgeometrie mit *idealer Kontaktierung* (Laborbedingungen, Section 2.5.2) für die Modell Validierung und einer Geometrie mit *Stack Kontaktierung* (planares F-Design, Section 2.5.3) für betriebsrelevante Parameterstudien. Die folgenden physikalischen Prozesse werden in der Modellierung berücksichtigt: (i) Elektronischer/ionischer Ladungstransport in den porösen Elektroden, in der Elektrolyt/Interdiffusionsschicht (Section 4.3.1) sowie in der Materialphase der metallischen Kontaktnetzen und im Interkonnektor, (ii) Multikomponenten Gastransport in den Gaskanälen und im Porenraum der semi-porösen Kontaktnetze und der Elektroden (Section 4.3.2), (iii) elektrochemische Ladungstransferreaktionen (Section 4.3.3) und (iv) Gasumsatzreaktionen auf Grund von Reformierungsreaktionen in der Anode (Section 4.3.4). Es wurden isotherme Bedingungen angenommen, da im Prüfstand unter Last der maximal zu erwartende Temperaturgradient $< 5\ ^{\circ}C$ betrug und nur ein sehr geringer Einfluss auf die auftretenden Verlustmechanismen zu erwarten wäre. Im Stackbetrieb bilden sich zwar größere Temperaturgradienten entlang das Gaskanals, diese können aber in einem 2D-Querschnitt durch den Kanal, (2D-RPU Geometrie) als konstant verteilt angenommen werden.

Besondere Sorgfalt galt bei der Auswahl und Implementierung des verwendeten Gastransportmodells (Section 4.3.2). Es hat sich gezeigt, dass die Aufteilung in konvektive und diffusive Transportflussberechnung eine strikte Einhaltung des verwendeten Einheitssystems (massenbasiert oder molbasiert) von entscheidender Bedeutung für eine physikalisch korrekte Berechnung ist (Section 4.3.2.1, Figure 6.28). Mehrere Ansätze aus der Literatur wurden als Modell für den porösen Stofftransport implementiert, um deren Anwendbarkeit zu vergleichen: (i) Das Fick'sche Modell (FM, Section 4.3.2.3), (ii) das Dusty-Gas Modell (DGM, Section 4.3.2.3) und das Mean Pore Transport Modell (MPTM, Section 4.3.2.3). Der Vergleich zwischen Gastransportverluste der Anode (R_{1A}), bestimmt aus Impedanzmessungen, und simulierten Verlusten zeigte, dass das MPTM die genauesten Ergebnisse erzielt (Section 6.5). Wurde das FM im massenbasiertem Einheitssystem implementiert (FM-mass), war eine isobare Modellierung unzureichend (Figure 6.27). Basierend auf den erzielten Ergebnissen dieser Arbeit wird die Verwendung des molbasierten DGM oder MPTM empfohlen, wobei die Implementierung des DGM im geringen Maße weniger Aufwand erforderte. Ab einer Anzahl von sechs oder mehr Komponenten im Brenngas war der numerische Aufwand auch für das DGM zu groß und es wurde auf das FM zurückgegriffen.

Der elektrochemische Ladungstransfer in der Anode wurde mit dem Butler-Volmer Ansatz realisiert (BVM, Section 4.3.3.1). Dabei wurde die Elektrooxidation von H_2 als dominierende Ladungstransferreaktion im Reformat Betrieb implementiert. An der Kathode wurde neben dem BVM ein erweiterter Modellansatz für elektronisch-ionisch mischleitende Kathoden (MIEC, engl. mixed-electronic-ionic conductive) implementiert. Dadurch konnte die Eindringtiefe der Ladungstransferreaktion in die Elektrode berücksichtigt und alternative MIEC-Materialien im Modell implementiert werden. Dies war eine wichtige Eigenschaft in der numerischen Optimierung der Kathode; eine gleichwertige Implementierung für die Anode wäre eine sinnvolle Modellerweiterung.

Jeder sinnvolle Betriebspunkt kann mit Hilfe des Modells berechnet und die dabei auftretenden Verlustmechanismen getrennt voneinander bestimmt werden. Als zusätzliche Eigenschaft kann der differentielle Widerstand einzelner Verlustprozesse angebenen werden, was einen direkten Vergleich mit Messdaten bestimmt aus Impedanzmessungen ermöglicht (Section 4.4.2). Die Berechnung aller auftretenden Verlustprozesse und die daraus resultierende Zellspannung und Stromdichte ist gekoppelt an die Berechnung der Aktivierungsüberspannung der Ladungstransferreaktion $\eta_{\mathrm{act,an/cat}}$ (Section 4.4). An der Schnittstelle zwischen Elektrode und Elektrolyt werden die elektrischen und ionischen Potenzialverteilungen mit dem jeweiligen (partialdruckabhängigen) Eletrodenpotenzial nach Kirchhoff's 2. Gesetz bestimmt. Die Beachtung der entsprechenden Vorzeichen auf Grund der Stromflussrichtung ist von entscheidender Bedeutung für eine korrekte Überspannungsberechnung.

Material und Modellparameter - Kapitel 5

Die poröse Mikrostruktur der Elektroden wurde im Modell nicht ortsaufgelöst, sondern homogenisiert und daher als isotrop verteilt betrachtet (Section 5.1.5). Dies ermöglichte die Berücksichtigung eines weitaus größeren Zellabschnitts im Vergleich zu einem ortsaufgelösten Modell. Die daraus resultierende Verwendung von effektiven Parametern berücksichtigt den Einfluss der Mikrostruktur auf die Transport- und Reaktionsprozesse. Die poröse Mikrostruktur der Elektroden der verwendeten Zelltypen wurde mit hochauflösenden Tomographie Methoden numerisch rekonstruiert (Section 2.6). Diese wurden im Anschluss mit Hilfe geeigneter mathematischer Methoden analysiert und charakteristische Mikrostrukturparameter für die Modellierung bestimmt (Section 5.1). Zusätzlich wurde ein Mikrostrukturfaktor für die Anode aus der Auswertung von experimentell aufgenommenen Impedanzspektren bestimmt (Section 5.1.7). Als weiteres Feature dieser Arbeit wurden die Ergebnisse eines in der Arbeitsgruppe entwickelten Kathoden-Mikrostrukturgenerator implementiert, wodurch die charakteristischen Abhängigkeiten bei Variation der Porosität im Modell erhalten blieb.

Für die Bestimmung des mittleren Porendurchmessers wurden rekonstruierte Elektroden ausgewertet, einmal mit Hilfe des Arithmetisches Mittels und in einem zweiten Ansatz auch gemittelt, aber skalierend bezogen auf das jeweilige Porenvolumen (Section 5.1.6). Geeignete analytische Ansätze wurden zur Berechnung der Permeabilität für der Anode mit Messdaten aus Jülich (Anode, Section 5.1.8) und für die Kathode mit numerischen Werten, welche an künstlich erstellen Strukturen bestimmt wurden (Section 5.1.8). Für beide Elektroden zeigte der Kozeny-Carman Ansatz mit Oberflächenabhängigkeit die beste Übereinstimmung. Die Parameter der Kontaktnetze wurden aus den Herstellerangaben abgeleitet (Section 5.4). Weitere charakteristische Materialparameter der Elektroden sowie die effektive Leitfähigkeit der Elektrolyt/Interdiffusions-Schicht wurden zum Teil aus anderen (am IAM-WET durchgeführten) Arbeiten übernommen, in denen der gleiche Zelltyp behandelt wurde bzw. sie wurden für diese Arbeit mit Hilfe der Ersatzschaltbildauswertung von aufgenommenen Impedanzspektren neu bestimmt (Sections 5.2 and 5.3). Die Reaktionskinetik der Wasser-Gas-Shiftreaktion (WGS) wurde anhand einer kombinierten experimentellen, numerischen Methode bestimmt und ist darüber hinaus übertragbar auf andere Zelltypen, deren aktive Ni-Oberfläche bekannt ist (Section 5.5). Kontaktwiderstände für die Modellierung der Stackebene wurden vom Forschungszentrum Jülich gemessen und freundlicher Weise zur Verfügung gestellt (Section 5.6).

Modell Validierung und Ansatzanalyse - Kapitel 6

In der durchgeführten Sensitivitätsanalyse stellten sich die Aktivierungsenergien der Elektrodenreaktionen und die der ionischen Leitfähigkeit der kombinierten Elektrolytschicht als sensitivste Parameter heraus (Section 6.1). Außerdem wurde festgestellt, dass der Einfluss bestimmter Parameter erst unter besonderen Betriebsbedingungen an Einfluss gewann. Daher war es wichtig, die Modellvorhersagen nicht nur unter Standardbedingungen zu testen, sondern auch in Bereichen, wo ein Transportprozess limitierenden Charakter entwickeln kann. Nur auf diese Weise konnte die physikalisch korrekte Reproduktion der Verluste durch die verwendeten Modellansätze und Parameter verlässlich überprüft werden.

Für die Validierung wurden aufgenommene Strom/Spannungskennlinien (C/Vs), aus Impedanzmessungen bestimmte Widerstände einzelner Verlustprozesse sowie mit Hilfe von Gasanalytik bestimmte Konzentrationsverteilungen entlang des Brenngaskanals herangezogen. Zusammengefasst lag die relative Abweichung zwischen simulierter und gemessener Zellleistung im Wasserstoffbetrieb unter $\sim$ 2.5 %. Insbesondere die hohe Genauigkeit der reproduzierten limitierenden Stromdichte bei niedrigem $pH_{2,\text{an}}$ (Section 6.3) demonstrierte die Genauigkeit der bestimmten Anoden Mikrostrukturparameter und des gewählten Stofftransportmodells (DGM-c, Section 6.5). Zusätzlich zeigte das Modell eine sehr gute Genauigkeit in der Reproduktion aufgenommener Messdaten im Elektrolysemodus (Figure 6.9). Für die Zukunft ist eine erweiterte Modell Validierung diesbezüglich geplant. Im Betrieb mit Reformat als Brenngas wurde bei niedrigeren Temperaturen eine leicht erhöhte Abweichung zwischen simulierter und gemessener Leistung festgestellt. Die gemessenen Stofftransportwiderstände und Strom/Spannungskurven bei variierte Brenngasausnutzung, sowie die Konzentrationsverteilungen unter OCV und Last konnten allerdings mit ausreichender Genauigkeit vom Modell reproduziert werden (Section 6.3). Dies spricht dafür, dass die Abweichung bei niedriger Temperatur auf die Verwendung des isothermen Ansatz zurückzuführen ist und nicht auf das verwendete Gastransportmodell. Ein um die Energiegleichung erweitertes Modell ist momentan in Arbeit und sollte in naher Zukunft zur Verfügung stehen.

In der Standardkonfiguration konnte das Modell mit der 2D-RPU Geometrie gemessene Daten bei $pO_{2,\mathrm{cat}} > 0.1$ atm (technisch relevant) mit guter Genauigkeit widergeben, zeigte aber eine steigende Abweichung hin zu geringeren $pO_{2,\mathrm{cat}}$-Werten (Section 6.4). Ein vereinfachter Ansatz zur Berücksichtigung des Gasumsatzes bei steigender Last zeigte bereits eine Verbesserung der simulierten Ergebnisse (Section 6.5). Die Anpassung der Kontaktnetzparameter, begründet durch eine temperaturabhängige Veränderung mit der Zeit, zeigte eine weitere Verbesserung (Section 6.5) und die Verwendung der 3D-RPU Geometrie lieferte schlussendlich hervorragende Ergebnisse (Section 6.5). Dies führte aber zu einer stark gesteigerten Simulationszeit (Faktor 84 bis 1050, Table 6.5).

Eine detaillierte Analyse der Einflüsse verschiedener Mikrostrukturparameter ergab die Nichtanwendbarkeit der arithmetisch gemittelten Porendurchmesser (Section 6.5). Darüber hinaus wurde eine Abweichung für $\tau_{\mathrm{por,an}}$ festgestellt, die auf eine inhomogene Porengrößenverteilung im Anodensubstrat und dem daraus resultierenden Fehler bei der 3D-Elektrodenrekonstruktion zurückgeführt wurde.

Des weiteren hat sich gezeigt, dass die Gastransportmodellkombination aus FM-mass + Darcy-Gleichung nicht geeignet war, die SOFC Stackebenen Leistung zuverlässig vorherzusagen. Für den Fall einer O_2-Verarmung erfolgte eine fehlerhafte Druckberechnung, hervorgerufen durch die erforderliche Randbedingung der Darcy-Gleichung. Dadurch wurden die erforderlichen Zwangsbedingungen in den Bilanzgleichungen nicht mehr erfüllt und die Simulation führte zu einem Fehler; der molbasierte DGM-Ansatz war hingegen fehlerfrei und prognostizierte eine physikalisch plausible Lösung (Section 7.1.2.1).

Ergebnisse der numerischen SOFC Performance Analyse - Kapitel 7

Basierend auf der 2D-RPU Modelgeomtrie konnte anhand simulierter Überspannungs-, Stoff- und Stromverteilungen sowie der Gesamtleistung (Figures 7.3 to 7.6) gezeigt werden, dass die Leistung einer SOFC durch eine nicht ideale Kontaktierung im Stack negativ beeinflusst wird. Anhand der ortsaufgelösten Gasverteilung konnte gezeigt werden, dass limitierende, kathodische Gastransporteigenschaften mit wachsender Polarisation zu einer Gasverarmung in den Bereich unterhalb des Kontaktstegs führte und dadurch eine verminderte elektrochemische Aktivität in dem betroffenen Bereich hervorgerufen wurde. Dadurch stiegen nicht nur die kathodische Gastransportverluste, sondern alle übrigen Verlustprozesse ebenfalls (ausgenommen die anodischen Gastransportverluste). Die zusätzliche Querleitung in Kathode und Elektrolyt auf Grund der inhomogenen elektrochemischen Reaktionsverteilung führte zu einem Anstieg der ohmschen Verlust sowie der Anodenaktivierungsverluste. Der totale Leistungsverlust bei betriebsrelevanter Polarisation für die verwendete Kontaktierungsgeometrie betrug dabei $\sim$24 %, wobei ein Kontaktwiderstand auf Grund der Verwendung des Au-Flowfields vernachlässigt werden kann.

Ein Vergleich von simulierter Stackebenen Leistung mit dem 2D-RPU Modell (Figure 7.10) und am Forschungszentrum Jülich gemessener Stackebenen Leistung zeigte eine gute Reproduktion der gemessenen Werte (Abweichung $\sim$2 %) vom gemessenen Wert bei $T = 800$ °C. Die anwachsende Abweichung bei niedrigeren Temperaturen konnte mit einer leicht veränderten Zellproduktion erklärt werden.

Mit Hilfe der 2D-RPU Modellgeometrie wurde eine Parameterstudie durchgeführt, in der relevante Kathoden- und Flowfield Designparameter systematisch variiert wurden. Die folgenden Ergebnisse wurden dabei festgestellt:

- *Schichtdicke*: Um den Einfluss der O_2-Verarmung zu minimieren benötigte es eine Kathodenschichtdicke von $h_{\mathrm{cat}} > 100 - 300$ µm, abhängig von der gewählten Polarisation. Der Einfluss der O_2-Verarmung nahm hin zu niedrigeren Temperaturen ab, da die anwachsenden Aktivierungsverluste und ohmschen Verluste eine größere Stromdichte und damit verbundenen Gasumsatz verhinderten. Bei Standardbedingungen betrug der relative Leistungsverlust auf Grund der Stackkontaktierung $\sim 14.8\%$ (verglichen zur idealen Kontaktierung, Section 7.2.1).
- *Elektrische Leitfähigkeit*: Limitierte Querleitung und ein dadurch verursachter Anstieg der ohmschen Verluste in der Kathode waren zu vernachlässigen für Materialien mit $\sigma_{\mathrm{cat,eff}} > 100$ S/m (*e.g.* LSC). Für LSCF betrug der relative Verlust $\sim 4\%$ verglichen mit LSC bei $T = 800$ °C, wohingegen die Verluste für LSM im Vergleich bis zu $\sim 18\%$ betrugen. Bei niedrigeren Betriebstemperaturen war der Einfluss der Querleitung geringer.
- *Mikrostrukturparameter*: Eine ideale Mikrostruktur hing von der Betriebstemperatur und der Höhe der Polarisation ab. Eine erhöhte Porosität ($\varepsilon_{\mathrm{cat}} = 0.5$) verhinderte einen wachsenden Verlust auf Grund von Gasverarmung bei höheren Temperaturen. Bei niedrigeren Temperaturen hingegen verhinderte eine dichtere Mikrostruktur ($\varepsilon_{\mathrm{cat}} = 0.5$) eine bessere elektronische Querleitung (Section 7.2.3).
- *MIEC-Materialeigenschaften*: Die Verwendung von MIEC-Materialien mit erhöhter elektrochemischer Aktivität beeinflusste die SOFC Stackleistung entscheidend. Insbesondere ein erhöhter Sauerstoffionen Transportkoeffizient (D^{δ}) verhinderte eine Gasverarmung in schlecht versorgte Zonen unter den Stegen. Verglichen mit LSCF wurde für LSC oder BSCF als Kathodenmaterial ein Leistungszuwachs zwischen $9 - 18\%$ (bei 800 °C) vom Modell vorhergesagt. Bei $T = 600$ °C sogar bis zu 30%. Dabei war allerdings die elektronische Leitfähigkeit der Materialien zu beachten, welche bei BSCF entscheidend niedriger ausfällt. Daher wurden für BSCF im Vergleich zu LSC wesentlich höhere ohmsche Verluste vom Modell prognostiziert, weshalb LSC bei alleiniger Verwendung als Kathodenmaterial zu bevorzugen ist (Section 7.2.4).
- *Kathodenkontaktschicht (engl. cathode current collector, CCL)*: Ein fortschrittliches CCL-Design bot eine höhere Flexibilität bezüglich der Mikrostruktureigenschaften. Die beste Leistung wurde vorhergesagt für $h_{\mathrm{CCL}} > 200$ µm und $\varepsilon_{\mathrm{CCL}} = 0.6$ ($T = 800$ °C, Section 7.4.1).

- *Flowfield Parameter*: Die modellgestützten Ergebnisse zeigten, dass eine möglichst geringe RPU Dicke kombiniert mit dicken Stegen sich vorteilhaft auf die Stackleistung auswirkt. Das optimale Steg/Kanal Verhältnis verschob sich bei niedrigeren Temperaturen noch weiter hin zu breiteren Stegen, vor allem für dickere Kathodenschichten (Section 7.3.1). Dies liegt daran, dass ein optimales Flowfield Design hauptsächlich vom Kontaktwiderstand zwischen Steg und Elektrode abhängt (Section 7.3.2).

Anhand der oben genannten Erkenntnisse konnte gezeigt werden, dass die SOFC Stackebenen Leistung (und damit die des gesamten Stacks) zwischen $15 - 32$ % gesteigert werden kann, indem die Standard ASC mit LSCF Kathode mit einer optimierten Kontaktschicht (CCL) und Flowfield-Design kombiniert wird. Der Leistungszuwachs ist dabei abhängig von der gewählten Betriebstemperatur (größer bei höherer Betriebstemperatur, Table 7.3).

In einer abschließenden numerischen Studie wurde eine Flowfield Optimierung für ASCs mit alternativer MIEC Kathode durchgeführt. Im Modell wurde dabei die Leitfähigkeit einer metallischen Legierung als Schutzschicht auf dem Interkonnektor implementiert, welche extrem niedrige Kontaktwiderstände verspricht. Die Ergebnisse zeigten, dass eine Leistung von $P_{\mathrm{SL}} = 160$ W·cm^{-2} (bei $U_{\mathrm{cell}} = 0.8$ V und $T = 600$ °C) möglich ist, wenn BSCF als Kathodenmaterial mit einer LSC Kontaktschicht kombiniert wird und das Flowfield dementsprechend optimiert wurde. Eine weitere Leistungssteigerung könnte mit einem noch dünneren Elektrolyten oder alternativen Elektrolytmaterial mit besserer Leitfähigkeit erreicht werden, kombiniert mit einer hoch elektrochemisch aktiven, nanoskaligen Ni/8YSZ-Anode.

Die Ergebnisse dieser Dissertation ermöglichen eine modellbasierte Leistungsanalyse der SOFC in betriebsrelevanter Kontaktierung. Der Einfluss von Zell- und Kontaktgeometrieparametern auf die gesamte SOFC Stackleistung kann mit Hilfe des Modells gezielt untersucht und dadurch das Optimierungspotential der SOFC in Zukunft voll ausgenutzt werden.

Contents

1 Introduction **1**

2 Fundamentals **3**
2.1 Solid Oxide Fuel Cell (SOFC) Working Principle 3
2.2 Current/Voltage (C/V) Characteristic 5
2.3 Electrochemical Impedance Spectroscopy (EIS) 7
2.3.1 Technical Realization 9
2.3.2 Impedance Analysis 10
2.4 Technical Realization of SOFCs - From Single Cell to System 14
2.4.1 Single Cell Material and Design Requirements 14
2.4.2 Stack Design 19
2.4.3 SOFC-System Application 21
2.5 Contacting: Ideal and Stack 24
2.5.1 Requirements and Technical Realization 25
2.5.2 Laboratory Design - *Ideal Contacting* 25
2.5.3 Technically Relevant Design - *Stack Contacting* 26
2.6 3D Reconstruction of Porous Electrodes 32
2.7 Application of Numerical Simulation in Research and Development 35

3 Experimental Methods **39**
3.1 Cell Samples 40
3.2 Setup for Electrochemical Testing of SOFCs 42
3.2.1 Test Bench A - Impedance and Performance Measurements 42
3.2.2 Test Bench B - Performance and Gas Conversion Measurements 44
3.3 Cell Measurement Details 45

4 Description of Finite Element Method Model Framework for SOFCs **47**
4.1 Overview about Model Approaches in SOFC Research 48
4.2 Model Geometries 51
4.3 Loss Process Modeling 54
4.3.1 Electric and Ionic Charge Carrier Transport 54
4.3.2 Gaseous Species Transport 57
4.3.3 Electrochemical Charge Transfer Reactions 74
4.3.4 Chemical Reforming Reactions 81

4.4 Electrical Function Principle . . . 82
4.4.1 Cell Voltage, Current Density and Loss Calculation . . . 83
4.4.2 Differential Resistance Calculation . . . 86
4.4.3 Model-based Loss Process Quantification . . . 87

5 Model Framework Parametrization . . . 89
5.1 Porous Electrode Microstructure Properties . . . 89
5.1.1 Volume Fraction . . . 90
5.1.2 Tortuosity . . . 90
5.1.3 Stochastic Microstructure Generator . . . 92
5.1.4 Specific Surface Area . . . 93
5.1.5 Homogenization . . . 94
5.1.6 Mean Pore and Particle Size Diameter . . . 95
5.1.7 EIS-based Effective Microstructural Parameter . . . 97
5.1.8 Permeability . . . 100
5.2 Electrochemical Charge Transfer Kinetics . . . 107
5.2.1 Butler-Volmer Model (BVM) . . . 107
5.2.2 MIEC-Cathode Model . . . 108
5.3 Electronic and Ionic Conductivities . . . 112
5.3.1 Electronic Conductivity of Ni/8YSZ-Anodes . . . 112
5.3.2 Electronic Conductivity of various Cathode Materials . . . 112
5.3.3 Effective Ionic Conductivity of Electrolyte/Interdiffusion-Layer . . 114
5.4 Metallic Contact Mesh . . . 115
5.5 Chemical Reforming Reaction Kinetics . . . 116
5.6 Contact Resistance . . . 121

6 Validation and Modeling Approach Assessment . . . 123
6.1 Sensitivity Analysis . . . 124
6.2 Temperature Dependence . . . 128
6.3 Anode Gas Composition Dependence . . . 133
6.4 Cathode Oxidant Gas Composition Dependence . . . 141
6.5 Modeling Considerations and Accuracy Aspects . . . 148

7 Results of SOFC Stack Layer Performance Analysis . . . 167
7.1 SOFC Performance under Ideal and Stack Contacting Conditions . . . 167
7.1.1 DRT Analysis . . . 168
7.1.2 Performance and Loss Process Analysis . . . 170
7.1.3 Comparison of Model Predictions with Stack Measurements . . . 176
7.2 Influence of Cathode Parameters . . . 178
7.2.1 Layer Thickness . . . 179
7.2.2 Electronic Conductivity . . . 183
7.2.3 Microstructure Properties . . . 187
7.2.4 Charge Transfer Kinetics (MIEC-Material) . . . 190
7.2.5 Parameter Interdependencies . . . 195

7.3 Influence of Interconnector Parameters 196
7.3.1 Flowfield Design 197
7.3.2 Contact Resistance 205
7.4 Model-based Performance Optimization 209
7.4.1 Advanced Current Collector Layer (CCL) Design 209
7.4.2 Optimal Flowfield Design for LSCF Cathodes 212
7.4.3 Outlook on SOFC Stack layer Performance 214

8 Summary and Conclusions **217**

Appendix **225**
A Reformate Fueled SOFC Stack Layer Performance (2D-GCh Model) 225
B Parameters for Diffusion Coefficient Calculation 227
C Calculation of Dynamic Viscosity 228
D Measured Cells 229
E Supervised Diploma, Master and Bachelor Theses 230
F Publications 231
G Conference Contributions 232

Nomenclature **235**
Symbols 235
Acronyms 238
Subscripts 240

Bibliography **241**

1. Introduction

Modern society is founded on the reliability of electricity sources. Until recently our ever increasing demands have been supplied by a few hundred centralized power plants. Nowadays this is increasingly replaced by thousands of decentralized, mainly renewable power generators. While this is a positive development in the light of recent nuclear incidents (such as Fukushima in 2011) or global warming [1], the main renewable power sources, wind and solar power, are unreliable contributors to the electric transmission grid. Germany pursues the political agenda "Energiewende", where the share of renewable energy to the energy supply is projected to be 35 % in 2020 and up to 80 % by 2050, whereof wind and solar power will account for 60 % of the gross energy consumption [2]. The increasing network connectivity of infrastructure installations and commercial services are also driving demand for “secured power”. Furthermore, wind power has the unfortunate disadvantage of being produced in remote locations, far away from the main power consumers. Distribution via land lines is often delayed due to political disputes or environmental protection considerations, or is not met with particular delight by local communities [3]. Fortunately, an elegant solution is supplied by the well-connected natural gas grid. Artificial reformate, methane or hydrogen can be produced by Power-to-Gas processes (PtG) and fed into the gas grid [4, 5].

For this, an efficient energy converter is required. *Solid Oxide Fuel Cell* (SOFC) technology offers a highly efficient energy converter. At its high operating temperature, the Ni contained in the SOFC's cermet anode is catalytically active and enables the use of pure hydrogen and various hydrocarbons as potential fuels. When fed with natural gas produced from renewable sources, it offers emission free energy conversion with an unchallenged efficiency up to 60 % . Possible applications range widely: From small portable units (0.01-0.1 kW) up to microCHP (Combined Heat and Power) units (1-3 kW) supplying electricity and heat for communities [6]. New emerging markets are battery charging stations or highly resilient power supplies for data centers (10-100 kW) and power plant application (> 1 MW) [7]. Furthermore, an SOFC can operate in reverse mode as a *Solid Oxide Electrolyser* (SOE). It then functions as a highly efficient PtG to produce hydrogen or (in co-electrolysis mode) an artificial reformate [5]. As a Power-to-Liquid (PtL) unit the SOFC/SOEC-technology can function as part of a liquid fuel production plant [8].

The state-of-the-art anode supported cell (ASC) design supplies single-cell, high-power output under ideal laboratory conditions. In recent years researchers have managed to unravel the individual loss processes contributing to the overall single cell resistance by means of impedance spectroscopy [9]. Aside from long-term stability, it is now one of the most demanding development goals in SOFC research to transfer the already excellent single cell performance from the laboratory into commercial applications in order to provide an economically attractive alternative for the highly competitive energy market.

Goals of this Thesis

To increase overall power output up to an application relevant level, a stack of single cells is constructed from individual, in series connected, stack layers. Each ASC is hosted between two interconnector (IC) plates. In the state-of-the-art planar stack design, the IC's alternating flowfield is made of contact ribs to ensure electron transport throughout the stack and gas channels to supply fuel and oxidant to each cell surface. Both functions are fundamental but nevertheless competing due to their simultaneous requirements under non-ideal conditions. A numerical approach is advantageous for cost reasons and flexibility. The finite element method (FEM) provides a powerful tool for this purpose and is capable of spatially resolving involved species distributions. This is the only method where the loss processes taken into account by the model are dependent on a selected geometry. The literature provides miscellaneous modeling approaches, however none sufficiently ascertain the complex coupling of occurring losses under non-ideal stack operating conditions. Furthermore, comprehensive sets of experimentally determined material parameters are rarely given, but all too often adopted from other literature sources with unknown thermal cell histories. Those (mostly inattentively performed) model validations often rely on fitting in order to match predicted and experimental data. The goal of this thesis is to provide a framework based on the FEM whereby above mentioned requirements are attended in detail and SOFC stack performance is predicted reliably for varied cell and contact designs.

Outline

After the introduction, basic knowledge of SOFC technology is given, followed by a description of the numerical and experimental methods used in this work. Then the necessary model geometries and adequate equations for the implementation of the relevant physical loss processes are described. The next chapter is dedicated to the acquisition of corresponding material parameters by experimental and numerical methods. A comparison of model predictions and experimentally recorded data under conditions that go far beyond standard operation shows in detail the reliability of the framework in the validation section. Then, various modeling aspects and their influence on the predicted results are discussed. In the last part of this work the framework is used for the analysis of influence and performance optimization potentials of different cathode and interconnector parameters under application-relevant conditions.

2. Fundamentals

2.1. Solid Oxide Fuel Cell (SOFC) Working Principle

An SOFC is a type of fuel cell. Similar to batteries and accumulators, the SOFC is an electrochemical energy conversion device, functioning on the principle of galvanic cells [10, 11]. It converts chemical energy (Gibbs free energy ΔG) stored in a gaseous fuel (e.g. hydrogen, H_2) into electric energy (and heat) through a chemical reaction with an oxidant (e.g. oxygen, O_2). The gaseous reactants are thereby supplied and removed constantly, which differentiates the SOFC (and any other fuel cell) from batteries or accumulators, where the amount of stored energy and thus the operation time is inherently limited. The underlying redox-reaction (combustion) is described by the following reaction equation:

$$H_2 + \frac{1}{2}O_2 \rightleftharpoons H_2O. \tag{2.1}$$

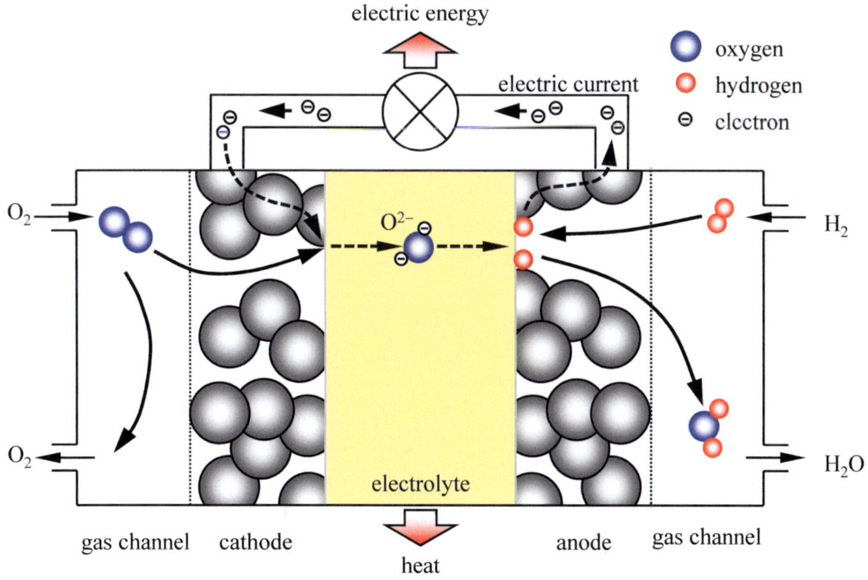

Figure 2.1.: Function principle of the SOFC (adapted from Ref. [12]).

However, no electric energy can be harvested from direct fuel combustion, as it only results in heat. In fuel cells, the direct reaction of fuel and oxidant is prevented by combining materials with special abilities and structure.

Figure 2.1 illustrates the simplified working principle and structure of an SOFC. Fuel and oxidant gas atmospheres are separated by a gas-tight ceramic layer (**electrolyte**), which becomes conductive for oxygen ion (O^{2-}) at higher temperatures (above $500°C$). The redox-reaction (Equation 2.1) is thereby split up into two spatially divided electrochemical partial reactions, which enables the incorporation and removal of O^{2-} from the dense solid oxide electrolyte. In order for the partial reaction to take place, porous and electron (e^-) conductive electrodes are placed on either side of the electrolyte.

At the **cathode (air electrode)**, oxygen (O_2) molecules from the gas phase dissociate, reduce to O^{2-} by accepting two e^- and are incorporated into the electrolyte phase, thus charging the cathode positively:

$$\frac{1}{2}O_{2(g)} + 2e^- \rightleftharpoons O^{2-}_{(el)}. \tag{2.2}$$

Due to a gradient in chemical potential, O^{2-} ions diffuse within the electrolyte from the cathode (high oxygen partial pressure $pO_2 = 0.01 \ldots 1$ atm) towards the **anode (fuel electrode)**, where there is a low $pO_2 = 10^{-27} \ldots 10^{-13}$ atm and the electrochemical fuel oxidation occurs. For hydrogen as fuel, H_2 reacts with the O^{2-} ions emerging from the electrolyte to water (H_2O), thereby releasing two e^- into the anode material and charging it therewith negatively:

$$O^{2-}_{(el)} + H_{2(g)} \rightleftharpoons H_2O_{(g)} + 2e^-. \tag{2.3}$$

The created electrons are repelled by the negatively charged anode and are drawn towards the positively charged cathode. As the electrolyte is not ideally electronically conductive, the electrons are forced to flow via the external circuit back to the cathode. If an external load is attached into the external circuit, electrical work is performed by the electrons. In the case of an open circuit, the electrons have no way of returning to the cathode. As a consequence, an electrical potential difference arises between both electrodes, inducing an electrical field force between both electrodes in opposite direction to the O^{2-} ion diffusion flux caused by the chemical potential difference. After some time however, the reversible electrode reactions (Equations 2.2 and 2.3) reach equilibrium when both chemical and electrical potential differences are of equal magnitude and the net flux of oxygen ions is impeded. The resulting equilibrium voltage is called theoretical Nernst voltage U_{th} after the German chemist Walter Nernst. Based on the Gibbs free energy ΔG of the overall cell reaction (Equation 2.1), Nernst formulated the following expression [13]:

$$U_{th} = -\frac{\Delta G}{zF} \tag{2.4}$$

Therein, z denotes the number of electrons transferred per mole of fuel, which is $z = 2$ for the SOFC and H_2 as fuel, and F the Faraday constant. Another way to predict the Nernst voltage is to use the oxygen partial pressures pO_2 of anode and cathode:

$$U_N = \frac{RT}{zF}\ln\frac{\sqrt{pO_{2,cat}}}{\sqrt{pO_{2,an}}} \tag{2.5}$$

wherein R denotes the universal gas constant in J/mol·K and T the absolute temperature in K. A more practical expression (as the oxygen partial pressure in the fuel gas ($pO_{2,\mathrm{an}}$) is difficult to determine) can be deduced by using the temperature dependence of ΔG with

$$\Delta G = \Delta G_0 + RT\ln K \tag{2.6}$$

and the equilibrium constant, K, of the overall cell reaction (Equation 2.1) as

$$K = \frac{\sqrt{pO_{2,\mathrm{cat}}} \cdot pH_{2,\mathrm{an}}}{pH_2O_{\mathrm{an}}}, \tag{2.7}$$

resulting in [11]

$$U_\mathrm{N} = -\frac{\Delta G_0(T)}{2F} + \frac{RT}{2F}\left(\frac{\sqrt{pO_{2,\mathrm{cat}}} \cdot pH_{2,\mathrm{an}}}{pH_2O_{\mathrm{an}}}\right). \tag{2.8}$$

Other common expressions for denoting U_N are open circuit voltage (OCV) or electromotive force (EMF), because it represents the driving force for the overall cell reaction (Equation 2.1). For SOFCs operated between $T = 600\ldots950$°C with hydrogen as fuel (+1% H_2O) and air as oxidant, Equation 2.8 yields values between 1.18 and 1.13 V. Please note, these given expressions are for an ideal fuel cell under open circuit conditions and do not take any losses into account. In a real cell, various loss mechanism exhibit overpotentials, thus lowering the theoretical Nernst voltage, U_th, to the operating voltage, U_cell. This will be discussed in the next section.

2.2. Current/Voltage (C/V) Characteristic

The operating voltage, U_cell, of an SOFC under load drops below the theoretical Nernst voltage, U_th. This voltage loss is due to the occurrence of irreversible processes, called overpotential or polarization. It is highly desirable to understand the origin and nature of these processes in order to advance the development of SOFCs. The most common way to characterize a single cell is the current/voltage (C/V) measurement. This technique will now be explained along with a description of the individual loss processes that occur within a single cell.

A C/V measurement is conducted by applying a defined steady-state current load, j, to a cell, then incrementally increasing the load and recording the resulting cell voltage, U_cell. Figure 2.2 qualitatively shows a C/V measurement with idealized overpotentials. As can be seen, the occurring polarizations are: (i) Parasitic losses, (ii) gas conversion, (iii) ohmic losses and (iv) polarization losses.

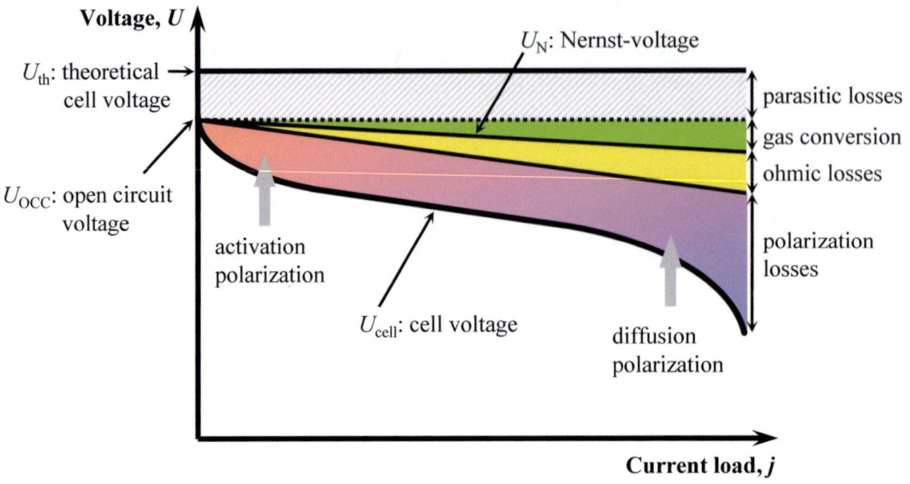

Figure 2.2.: Schematic plot of the C/V characteristic of an SOFC showing the different types of polarization (adapted from Refs. [11, 14]).

Parasitic losses occur even under open circuit conditions (OCC). They lower the theoretical Nernst voltage, U_{th}, to the open circuit voltage U_{OCC} or OCV (as predicted by Equation 2.4). The loss is caused by electron loss due to an undesired electric conductivity in the electrolyte and/or gas leakage in the system. Gas leakage may arise with a non-100% gas-tight electrolyte or cell sealants. In any case, gas leakage alters the electrode gas concentrations and thereby the chemical potential difference. Furthermore, an additional voltage loss may be attributed to the fuel gas composition, if it is not in chemical equilibrium.

Gas conversion describes the Nernst voltage, U_N, decrease with increasing current load, j. The electrochemical conversion of fuel and oxidant changes the gas mixture concentration in both electrodes and thereby the chemical potential difference between each (Equation 2.8). Gas conversion is therefore a measure for fuel utilization (f.u.). It should be pointed out that gas conversion is immanent during the cell operation and is actually not a real loss process. It can be offset, however, by raising the fuel and oxidant gas flow rates during the cell operation and thus excluding it from the analysis of the real loss processes.

Ohmic losses arise from non-ideal electronic and ionic transport in the electrodes and electrolyte. The greater part originates in the electrolyte, because the ionic conductivity is one order of magnitude smaller compared to the electronic conductivity in the electrodes for common electrolyte material (Section 5.3).

Polarization losses are the main reason for the characteristic non-linear shape of the C/V curve [15, 16] and have to be further distinguished according to their physical origin:
(a) *Activation polarization* relates to the non-ideal charge transfer reactions taking place at the electrochemically active sites (triple-phase boundary TPB) where the ionic, electronic and gas phases meet. The arising activation overpotential is attributed to the charge transfer reaction activation energies and are dominant for low current densities.

(b) *Diffusion polarization* or *concentration polarization* denotes the overpotential due to gas transport limitations in the porous electrode microstructure. Non-ideal gas transport properties cause a concentration gradient formation of the electrochemical reactions reactants (Eqs.2.3, 2.2), which is the dominant limiting factor at high current densities. This is especially relevant for anode supported cells (ASCs) with their thick support layer (300 ... 1500 μm) and therefore longer diffusion pathways.

In summary, the above described loss processes exhibit different dependencies on operating conditions (temperature, fuel and oxidant gas composition, total gas flow rate) and accordingly shape the course of measured C/V curves [9, 16]. The C/V measurement is the most common and application-related technique, but all loss processes occur simultaneously and therefore interfere with each other. A detailed process separation and subsequent analysis is therefore impossible and a different measurement method, such as the Electrochemical Impedance Spectroscopy (EIS), is required. The EIS and its analyzing method will be described in the next section.

As a last but important remark for this section, it has to be pointed out that above described loss processes hold for single cell operation with ideal electric contact and gas supply (Section 2.5.2). However, additional losses have to be taken into account for non-ideal conditions in practical applications (stack contacting, (Section 2.5.3). The matter will be discussed in more detail in Section 2.5, as it is essential for understanding the results of this thesis.

2.3. Electrochemical Impedance Spectroscopy (EIS)

Electrochemical Impedance Spectroscopy (EIS) is regarded as one of the most promising, non-destructive methods to unravel complex electrochemical systems, such as the SOFC, where several loss processes appear simultaneously. In the literature, several textbooks about impedance analysis and its application in electrochemical systems are available [18, 19, 20]. A comprehensive introduction about EIS in SOFC research is given in Ref. [21]. The following section is mainly based on Refs. [9, 17].

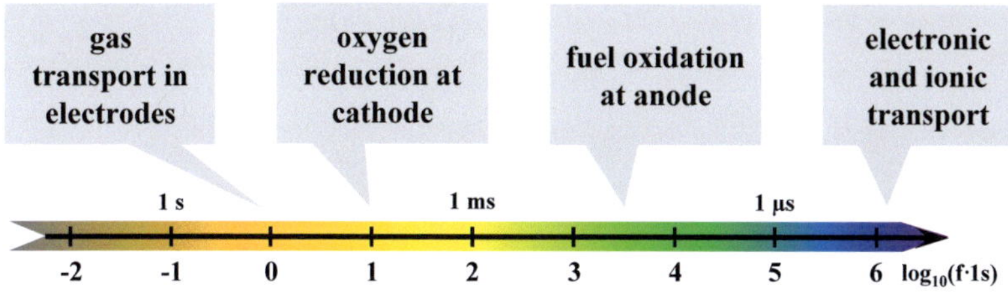

Figure 2.3.: Dynamics of a SOFC: Characteristic relaxation frequencies f of the individual electrochemical loss mechanisms occurring under operation of an anode supported SOFC (adapted from [17]).

In contrast to steady-state C/V measurements (Section 2.2), the transient EIS exploits the fact that the various occurring loss processes exhibit different time constants. The term *time constant* refers thereby to the relaxation time, τ_0, a process requires to return to its equilibrium state after the excitation by a perturbation signal, whereby $f_0 = 1/(2\pi\tau_0)$ denotes the corresponding relaxation frequency. Figure 2.3 illustrates qualitatively the various loss processes occurring in an SOFC with respect to their relaxation frequency.

The underlying theory of conducting an EIS measurement is to apply a periodic signal perturbation to an electrochemical system (i.e. an SOFC) at a defined stimulus frequency, perform a frequency sweep and record the systemic response signal for each frequency. The result is a measurable excitation of occurring processes due to their different relaxation frequencies.

The next two subsection are dedicated to the technical realization of EIS on SOFCs (single cells) and the subsequent impedance analysis methods.

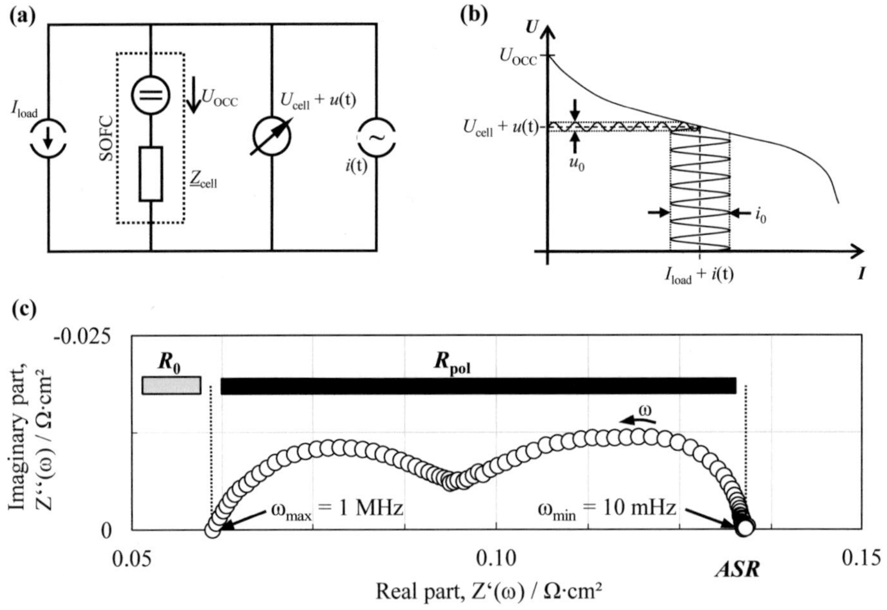

Figure 2.4.: (a) Electrical circuit of the measurement setup for the impedance measurement on a SOFC single cell with the internal impedance, $\underline{Z}_{cell}$. (b) Illustration of the impedance measurement with the corresponding C/V curve. A sinusoidal current of small amplitude, $i(t)$, is superimposed to a defined bias current, I_{load}, and the voltage response, $u(t)$, is measured [21, 22]. (c) Typical Nyquist-plot of a complex electrochemical impedance spectrum recorded on an anode supported SOFC single cell.

2.3.1. Technical Realization

The dynamic behavior (impedance) of an electrochemical system is measured by applying a sinusoidal perturbation and measuring the system's response signal, phase shift angle and amplitude. It is differentiated between the provided perturbation signal: *Galvanostatic* operation (current stimulus) and *potentiostatic* operation (voltage stimulus). Hence, the measured variables are current and voltage.

Figure 2.4a provides the basic experimental arrangement for impedance measurements on an SOFC single cell. In general, galvanostatic operation is preferred as it provides a lower signal to noise ratio (SNR) and therefore better measurement data quality [23]. On top of a fixed current bias, I_{load}, a sinusoidal current signal, $i(t) = i_0 \cdot sin(\omega t)$, is superposed with a fixed angular frequency, $\omega = 2\pi f \text{ s}^{-1}$. The signals current amplitude, i_0, is thereby determined a priori so that the voltage responds amplitude is below $u_0 <= 12$ mV, thus ensuring a linear operation [9, 24]. In the main EIS measurement the sinusoidal voltage respond, $u(t) = u_0(\omega) \cdot sin\left[\omega t + \varphi(\omega)\right]$, is recorded over the frequency sweep (Figure 2.4b) and the complex impedance, $\underline{Z}(\omega)$, can be calculated for each frequency, ω [9, 25]:

$$\underline{Z}(\omega) = \frac{u_0(\omega)}{i_0(\omega)} \cdot e^{j\varphi(\omega)} = |\underline{Z}(\omega)| = Z' + jZ'' \tag{2.9}$$

wherein φ denotes the frequency dependent phase shift angle between voltage and current signal and Z' and Z'' denote the real and imaginary part of the complex impedance $\underline{Z}(\omega)$, respectively. Hereby, the following three assumptions must hold [19, 25]:

- *Causality*: A system is causal, when the measured response signal at any point of time exclusively depends on the perturbation signal at this point of time and/or its evolution until this point of time.
- *Linearity*: The measured response is a linear function of the perturbation signal, i.e., the relation between output and input underlies the principles of superposition and amplification.
- *Time-invariance*: The output of a time-invariant system does not depend explicitly on time, i.e., the system response on a certain perturbation signal should be exactly the same for any shift of time.

The fulfillment of these criteria are verified in this work for each measured spectra by applying the Kramers-Kronig relation [19].

The resulting impedance values are usually plotted in the complex plane, also known as Nyquist-plot. Figure 2.4c shows an example of such a Nyquist-plot for a measured electrochemical impedance spectrum, recorded on a anode-supported SOFC single cell as used in the measurements for this thesis. The high frequency intercept at ω_{max} with the real axis corresponds to the purely ohmic resistance R_0 of the cell, whereas the low frequency intercept at ω_{max} is identical to the differential cell resistance, which can be obtained in a corresponding C/V measurement at the given operation point. The difference between the low and high frequency intercept is thereby called polarization resistance, R_{pol}, of the cell, whereby R_{pol} is the sum of all individual polarization resistances caused by the individual loss mechanisms. The sum of R_0 and R_{pol} is the *area specific resistance* ASR (or R_{ASR}) of the cell.

EIS measurements conducted on SOFCs offer clear separation of the cells ohmic and polarization resistances. However, a clear separation of occurring processes is still not possible due to the overlap in the spectrum. Hence, further analyzing methods have to be applied to the measured data in order to reveal the underlying dynamic processes. These methods are presented in the following subsection.

2.3.2. Impedance Analysis

It is common practice to employ an *equivalent circuit model* (ECM) in the analysis of measured electrochemical impedance spectra. An ECM is an electrical model composed of (usually) in series connected single impedance elements known from electrical engineering (i.e. resistor, capacitor and inductance) as well as other impedance functions, which have been specifically developed for this purpose. A comprehensive description of common equivalent circuit elements is given in Refs. [9, 23, 18].

Equivalent Circuit Model

In general, it is best to set up a physically motivated ECM, where each element represents a specific loss process occurring in the measured system in order to gather meaningful information. This requires knowledge of (i) the amount of physical processes contributing to the overall impedance and (ii) their physical origin or at least the assignment to a dominating physical mechanism. For both tasks, it is very helpful to employ the method called *Distribution of Relaxation Times* (DRT), which will be presented in the next paragraph. With an adequate ECM set up, a measured impedance can be accurately described by the sum of the individual ECM element impedances.

So, each element contains important information about the loss process it represents: Area specific resistance (ASR), characteristic time constant and degree of time constant dispersion. The last item accounts for the fact that real physical processes do not exhibit a single defined time constant due to spatially distributed physical properties.

In this work, a well-established ECM for anode-supported SOFCs with a mixed-ionic-electronic conductive (MIEC) cathode will be used to analyze the recorded impedance measurements for the modeling parameter determination (Section 2.3.2). The model was proposed by Leonide *et al.* [9, 27] and successfully applied for operation with H_2/H_2O or CO/CO_2 as fuel and air as oxidant [15, 23]. Furthermore, the model was extended and applied by Kromp *et al.* [17, 28, 29] for operation with reformate (H_2/H_2O/CO/CO_2/N_2) as fuel.

Figure 2.5a displays this extended ECM model. It shows the in series connected elements in their order of occurrence with respect to their relaxation frequencies from high to low frequencies. The elements account for the individual loss processes taking place in an ideally contacted SOFC, as described in Section 2.2:

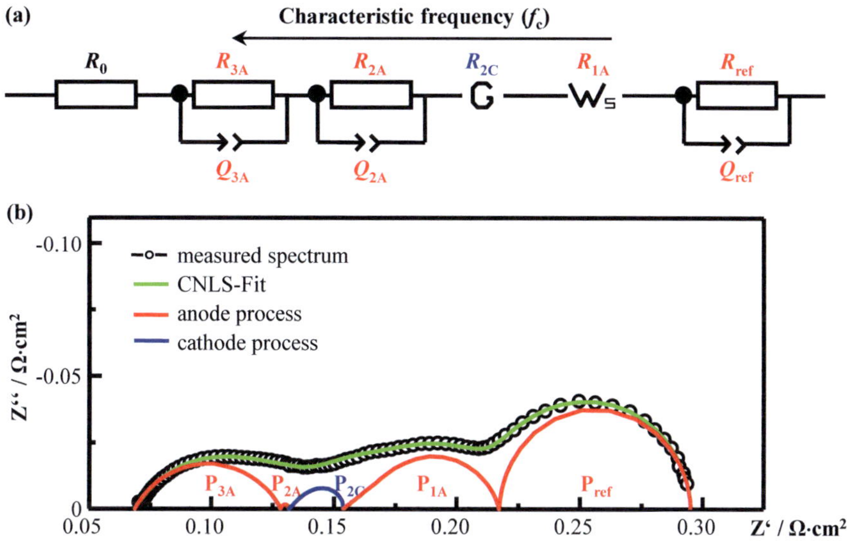

Figure 2.5.: (a) Extended equivalent circuit model used for the CNLS-fit of the impedance data measured under reformate operation of anode-supported SOFCs. (b) CNLS-fit of an impedance spectrum measured under $pH_{2,an} = 0.15$ atm, $pH_2O_{,an} = 0.05$ atm, $pCO_{an} = 0.15$ atm, $pCO_{2,an} = 0.05$ atm, balance: N_2; cathode: air; $T = 800°C$ (adapted from Ref. [26]).

(i) **Ohmic resistor - $\mathbf{R}_0$**: non-ideal ionic and electronic charge transport.

(ii) **Constant phase - RQ** (P_{3A} + P_{2A}): gas diffusion coupled with charge transfer and ionic transport in the anode functional layer (AFL) [27].

(iii) **Gerischer - G** (P_{2C}): oxygen surface kinetics and oxygen ion diffusifity in the mixed-ionic-electronic-conductive cathode [30, 31].

(iv) **Finite length Warburg - $\mathbf{W}_s$** (P_{1A}):
gas diffusion in the anode substrate [18, 29].

(v) **Constant phase - RQ** (P_{ref} / P_{1C}):

- reforming chemistry coupled with gas diffusion in the anode substrate [17, 29],
- gas diffusion in the cathode layer ($pO_{2,cat} << 0.21$ atm) [27].

Please note, P_{ref} only occurs in presence of hydrocarbons in the fuel gas (Section 6.3) and P_{1C} only contributes to the overall resistance notably for $pO_{2,cat} << 0.21$ atm and is otherwise accounted to P_{1A} [27, 9].

The ECM parameters are found by performing a *complex non-linear least-squares* (CNLS) fit algorithm. The CNLS-fit method is based in minimizing the sum of a quality criterion given by [18]:

$$S = \sum_{n=1}^{N} \left[w_{\text{Re},n} \left(Z'(\omega_n) - Z'_{\text{mod}}(\omega_n, \mathbf{a}) \right)^2 + w_{\text{Im},n} \left(Z''(\omega_n) - Z''_{\text{mod}}(\omega_n, \mathbf{a}) \right)^2 \right] \quad (2.10)$$

with the complex model fit function as

$$\underline{Z}_{\text{mod}}(\omega_n, \mathbf{a}) = Z'_{\text{mod}}(\omega_n, \mathbf{a}) + jZ''_{\text{mod}}(\omega_n, \mathbf{a}) \tag{2.11}$$

and the measured complex impedance as

$$\underline{Z}(\omega_n) = Z'(\omega_n) + jZ''(\omega_n). \tag{2.12}$$

In Equation 2.10, the vector $\mathbf{a}$ contains all free parameters of the fit problem, while the factors $w_{\text{Re},n}$ and $w_{\text{Im},n}$ are the weighting factors associated with the *n*-th data point. The fit quality is thereby given by the relative residuals. The CNLS-fit result and thereby determined single element impedance spectra are displayed in Figure 2.5b for the measured impedance and the operating conditions given in the caption.

Distribution of Relaxation Times (DRT)

The Distribution of Relaxation Times is a electrochemical impedance analysis method, which was developed at the Institute of Applied Materials (IAM-WET) [25]. It has been demonstrated by Leonide *et al.* [9], that applying the DRT is a powerful aid in the development of a physically meaningful ECM for SOFC single cells. In the following section this concept will be briefly introduced (the interested reader is referred to Refs. [9, 25]).

An impedance spectrum, $Z(\omega)$ (e.g. as displayed in Figure 2.4), measured under compliance with the mentioned criteria (causality, linearity and time invariance) can be represented by an ECM model with a sufficiently large number of in-series connected resistor-capacitor (RC) elements [18]. The RC-element is the parallel connection of a resistor and a capacitor and is one of the most basic equivalent circuit elements by which almost all technical processes can be described. Its impedance expression is given by $\underline{Z}_{\text{RC}}(\omega) = \frac{R}{1+j\omega\tau}$ with its characteristic relaxation time, $\tau = RC$. A detailed explanation of its application is given in Refs. [18, 23]. Now, the DRT method is based on the fact, that $Z(\omega)$ can be expressed by an infinite number of RC-elements with continuously increasing relaxation times, τ from 0 to ∞, with the following integral equation, containing the distribution function, $\gamma(\tau)$:

$$\underline{Z}(\omega_n) = R_0 + R_{\text{pol}} \int_0^\infty \frac{\gamma(\tau)}{1+j\omega\tau} \mathrm{d}\tau \tag{2.13}$$

with

$$\int_0^\infty \gamma(\tau)\mathrm{d}\tau = 1. \tag{2.14}$$

The general model given in Equation 2.13 describes the impedance, $\underline{Z}(\omega_n)$, by its ohmic resistance, R_0, and polarization resistance, R_{pol}, without any physical meaning but still reflecting the system's dynamics. The fraction $\frac{\gamma(\tau)}{1+j\omega\tau}\mathrm{d}\tau$ thus describes the specific part of R_{pol} with the relaxation times between the interval τ and $\tau + \mathrm{d}\tau$. However, in practice an infinite number of RC-elements is not feasible and $\tau(\omega)$ is approximated by a discrete function γ_n for N serial RC-elements with predefined, logarithmically distributed values for τ_n [23]:

$$\underline{Z}(\omega_n) = R_0 + R_{\text{pol}} \sum_{0}^{N} \frac{\gamma_n}{1 + j\omega\tau_n}. \tag{2.15}$$

Therein, γ_n weighs the contribution of the n-th RC-element with the relaxation time τ_n to the overall polarization resistance. The calculation of γ_n represents an inverse ill-posed problem [32, 33], which can be solved by approximating a solution numerically with the help of regularization. The regularization is based on the method proposed by Tikhonov [34, 35], which has been implemented at the Institute of Applied Materials (IAM-WET) in a Matlab-code, in accordance to Refs. [9, 17, 27, 36].

Figure 2.6 demonstrates the benefit of using the DRT in the impedance analysis. Therein displayed is a comparison between the imaginary part of the impedance shown in Figure 2.4 and the DRT (in red) calculated from it. Peaks are clearky distinguishable in the DRT whereas in the imaginary part the peaks overlap. Hence, the DRT offers a higher resolution of the systems dynamics. Each peak in the DRT represents a physical process with its characteristic relaxation frequency corresponding to the peak frequency, f_c. This is most valuable information for the initial values in the CNLS-fitting procedure, described in the previous paragraph. Furthermore, by analyzing a series of DRTs calculated from measured impedance data under varied operating parameters, the according process dependencies can be observed more accurately. Comparing these results with obtained CNLS-fit results increases the reliability tremendously.

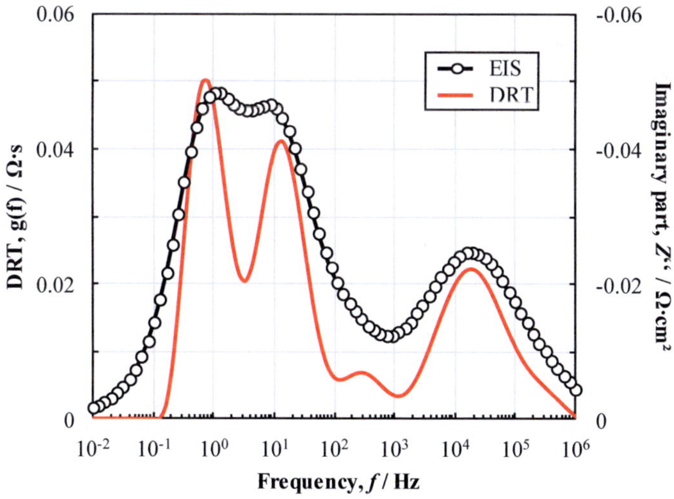

Figure 2.6.: Imaginary part of a measured impedance spectrum along with the corresponding Distribution of Relaxation Times (DRT, in red).

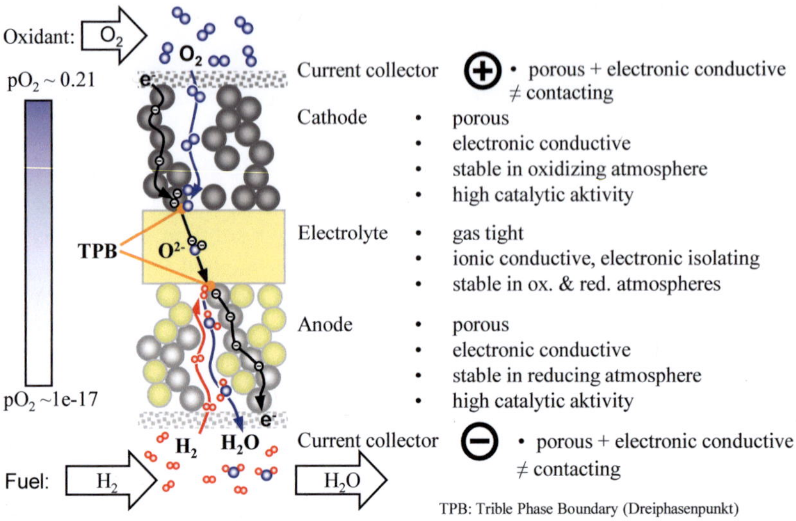

Figure 2.7.: Schematic display of physical processes occurring during the operation of an SOFC and the therefrom deduced technical requirements for the individual cell layers (adapted from [37]).

2.4. Technical Realization of SOFCs - From Single Cell to System

Extensive research on SOFC materials and their application has been carried out in the last three decades. Hence, this section will give only a basic overview, beginning with the requirements on the individual single cell components, followed by a concise description of state of the art stack designs, a brief subsection covering SOFC fuel flexibility and closing with an overview of various SOFC-system applications. Various comprehensive works can be found in Refs. [14, 38, 39, 40, 41, 42, 43, 44].

2.4.1. Single Cell Material and Design Requirements

A single fuel cell is an assembly of three layers and is generally called a *membrane electrode assembly* (MEA). The individual layers are electrolyte, anode and cathode: each must have different technical requirements to guaranty efficient operation. The governing goal is to create a material system which provides highly efficient half cell reactions (Equations 2.2 and 2.3) combined with a fast transport of the involved reactants (gaseous species, ions and electrons). Hence, the electrodes should have high catalytic activity, high ionic and electronic conductivity as well as high porosity. In contrast to the latter, the electrolyte is required to be as gas-tight as possible with a high ionic conductivity and an as low as possible electronic conductivity. Common requirements for all three components are the stability in reducing/oxidizing atmospheres at high temperatures (up to 1000°C), the chemical compatibility to the neighboring components and matching temperature expansion coefficients.

Figure 2.7 briefly summarizes the features and illustrates the SOFC working principle with a schematic display of occurring species transport processes.

Electrolyte

The electrolyte has the main purpose of separating the fuel and oxidant gas atmospheres to prevent direct fuel oxidation (combustion) so that the chemical potential between the electrodes can be exploited to the highest possible extent. Hence, the layer must be fabricated as a gas-tight solid. Furthermore, it should exhibit a significantly high ionic conductivity (σ_{io}) coupled with an as low as possible electronic conductivity (σ_{el}) in order to prevent an unprofitable exchange of charge carriers between the electrodes. Moreover, its chemical structure must be stable in both oxidizing and reducing atmosphere (cathode and anode, respectively), with a thermal expansion coefficient matching to the thermal characteristics of the electrodes.

A common electrolyte material used in SOFCs is stabilized zirconium oxide (ZrO_2) as it meets the above described requirements with a high proficiency. Its cubic phase is stable over the required temperature range, further increased by cation dopants. Substitution of ZrO^{4+}-cations with bi- or tri-valent cations from rare-earth or alkaline earth materials (e.g. Y, Sc, Ca, Mg, Nb, Sm, Yb) also improves the ionic conductivity (O^{2-} hopping-mechanism [44]) by creation of additional oxygen vacancies within the crystal lattice [38]. It has been found, that Ytrium (Y) and Scandium (Sc) meet the requirements best. The most commonly employed electrolyte material in SOFCs is ZrO_2 doped with 8 mol% Y_2O_3 (8YSZ), which is also the electrolyte material in the cells used in this work. It possesses an ionic conductivity $\sigma_{\text{ion,8YSZ}} = 5$ S/m at $T = 800°\text{C}$ [9, 38] and a thermal expansion coefficient $\alpha_{\text{8YSZ}} = 10.5 \cdot 10^{-6}$ K^{-1} in the temperature range of $30 \ldots 800°\text{C}$ [40, 45].

In early SOFC research, the electrolyte layer was the thickest layer in the MEA with $200 \ldots 500$ µm thickness, thereby giving the MEA structural stability and even its name (*electrolyte supported cells*, ESC). As a consequence, high operating temperatures were required due to the temperature dependent ionic conductivity (increased at higher temperature) to minimize the ohmic losses in the electrolyte. With increasingly optimized fabrication methods like thin film technology, it was possible to lower the electrolyte thickness and therewith related ohmic losses, enabling efficient SOFC operation at lower temperatures ($600 \ldots 800°\text{C}$). Nowadays, state-of-the-art cells (as used in this work) have a thickness of around 10 µm thickness. Recent research at Forschungszentrum Jülich has produced experimental cells with an electrolyte layer thickness of around 1 µm, paving the way to even lower operating temperatures in the near future. As a result of ever decreasing electrolyte thickness, one of the electrodes has to provide the support for MEA stability. Cell design has evolved the *anode supported cell* (ASC), which is also the cell type used in this work. The eponymous electrode (anode) will be briefly described in the next paragraph.

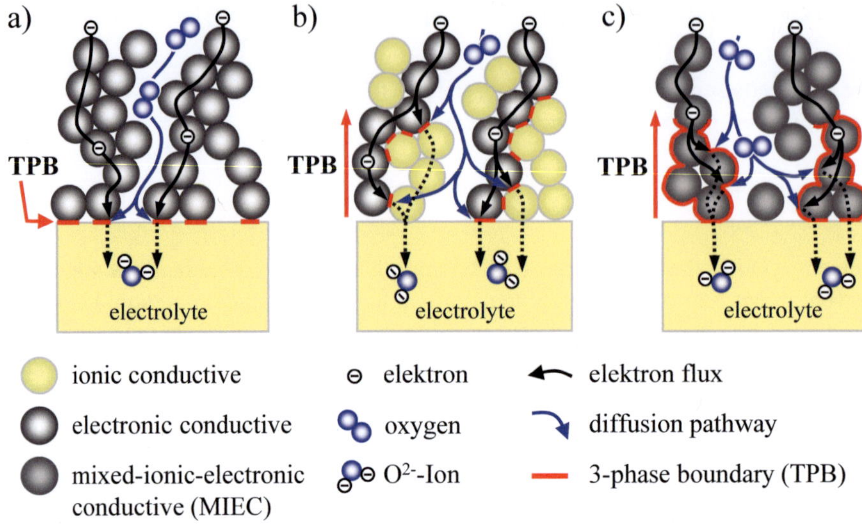

Figure 2.8.: Electrode working principles illustrated by means of the oxygen electrode (cathode): (a) Single-phase electronic conductive electrode (LSM), TPB only at electrode/electrolyte interface. (b) Composite electrode exhibits increased TPB length by two-phase mixed-conducting properties (LSM/YSZ). (c) Single-phase mixed-ionic-electronic conductive (MIEC) electrode with increased TPB length (LSCF, LSC, BSCF). (Adapted from [37]).

Anode

The main purpose of the anode is to provide conditions for the fuel oxidation (anode reaction Equation 2.3) to proceeds most efficiently. Hence, a high catalytic activity combined with a high electronic and ionic conductivity as well as high porosity is required to transport the reactants (i.e. electrons, ions, gaseous fuel components) to and away from the electrochemical reaction zone. Further, a compatible thermal expansion is required as well as chemical stability over a long time-period in highly reducing conditions. As mentioned at the end of the previous paragraph, the anode in a state-of-the-art cell design responsible for the structural support of the whole MEA. High sustainability against mechanical stress is demanded for ASC anodes. A double-layer cell design is the consequence with a highly porous but still resilient *anode support* (300 ... 1500 µm thick) and a thin (3 ... 30 µm) *anode functional layer* (AFL) providing the triple-phase boundary (TBP) for the electrochemical reaction.

The reducing conditions in the anode gas atmosphere ($pO_{2,\mathrm{a}} \leq 10^{-18}$ atm) call for the use of metals as preferred material class, which provide a high electronic conductivity. However, metals possess a high thermal expansion coefficient (compared to the common electrolyte materials), which has to be adjusted by mixing the metal with electrolyte material, this form a *cermet* [46]. This has an additional advantage as such a cermet-anode is also ionically conductive and the TPBs can expand spatially into the electrode volume. The catalytic activity of such a composite-electrode is greater than that of a purely electronic conducting electrode. Schematically, these two electrode function principles are displayed in

Figures 2.8a and 2.8b (even though the figures display the principles by the cathode reaction, the mechanism remains the same). However, in composite-electrodes, percolating transport pathways have to be guaranteed [47].

Various anode material compositions can be found in literature [48, 49, 50, 51]. Among these is Ytrium stabalized ZrO_2 (YSZ). It is the most commonly used and therefore most investigated anode material system. It is a composite of electronic conducting and catalytic active nickel (Ni) and ionic conducting Ytrium. Ni is a relatively cheap metal and can be processed with YSZ by standard ceramic fabrication methods (i.e. tape casting) at low costs [52, 53]. In state-of-the-art ASCs, a composition of 25 mol% Ni and 75 mol% 8YSZ is employed for both anode support and functional layer, which is also the anode design in the cells used in this work. Another advantage of Ni/YSZ anodes is the high catalytic activity enabling the conversion of hydrocarbon fuels (Section 2.4.3).

It has to be noted, that Ni/YSZ anodes also posses a disadvantage: at standard SOFC operating temperatures the Ni oxidation kinetics is relatively fast [54, 55]. Operating an SOFC at high fuel utilization increases the water vapor in the fuel gas with the danger of NiO formation. In addition to the deactivation of catalytic active surface area, the volume increase from Ni to NiO is substantial. This generates mechanical stress within the MEA which can result in complete cell failure.

Recently, research into new anode material systems has intensified. The focus now lies on metal-oxide based anodes [56, 57, 58] as well as on metal supported cells (MSCs) [17, 59, 60, 61, 62]. An increased re-oxidation reliability is promised by the first, but additional catalyst such as Ni or Pd have to be added to increase the electrochemical activity. Metal-supported cells potentially offer an increased redox stability with increased robustness against mechanical and thermal stress. In addition, metallic substrates can be produced at lower cost. All these features recommend the MSC for use in auxiliary power unit (APU) applications such as heavy trucks or ships.

Cathode

The main purpose of the cathode in SOFC operation is to reduce oxygen to oxide ions (O^{2-}) in the most efficient way (cathode reaction Equation 2.2). A widespread cathode material system in SOFCs with a high catalytic activity for the O_2-reduction reaction are ABO_3-type perovskite manganites, cobaltates or ferrates (A = La, Sr, Ca; B = Mn, Co, Fe). Their chemical stability in long-term operation at oxidizing conditions and high temperatures, thermal expansion characteristics as well as the electrochemical performance can individually adjusted by the perovskite configuration. Comprehensive discussions about the resulting properties can be found in Refs. [45, 63, 64, 65].

More importantly, the composition also determines, whether a perovskite exhibits pure electronic conductivity or additionally ionic conductivity. Hence, three different electrode working principles can be realized, which are displayed schematically by Figure 2.8. In the beginning of SOFC research, purely electronic conducting cathodes (Figure 2.8a) were implemented in the MEA. A common configuration was a strontium-doped lanthanum manganite (LSM) in the composition $La_{1-x}Sr_xMnO_3$ with $x = 0.15...0.25$ [21]. The electrochemical active sites (TPBs) are however reduced to the interface between electrode and electrolyte and the electrodes. An increase of TPBs can be achieved by creating a composite-electrode (Figure 2.8b), similar to the working principle used for anodes (previous paragraph). Adding an ionic conducting phase (i.e. YSZ) is spatially extends the electrochemical active area into the electrode volume and consequently increases the electrochemical performance [66, 67]. A downside to these LSM/YSZ composite-cathodes is a tremendously decreased electronic conductivity, creating additional ohmic losses due to in-plane conduction limitations [68, 37]. The issue is discussed further in Section 2.5. A way of countering the effect is to add an additional current collector layer of pure LSM on top of the composite-layer.

Nowadays, state-of-the-art SOFC-systems use mixed-ionic-electronic conductive (MIEC) cathodes [63]. In principle, the entire surface area of these cathodes are accessible from the pore phase can act as TPB. Figure 2.8c displays this increase of electrochemically active sites schematically. As a consequence, the electrochemical performance is increased in MIEC-cathodes compared to composite-cathodes or purely electronically conducting cathodes [54, 65]. A more detailed description of the MIEC-cathodes reaction mechanism is given in the modeling Section 4.3.3.2. Examples of MIEC-cathodes are lanthanum strontium cobalt ferrite (LSCF), lanthanum cobalt (LSC) and barium strontium cobalt ferrite (BSCF). The materials differ in their electrochemical performance as well as their charge carrier transport properties. Detailed information is given in Ref. [45]. It is important to note at this point that it is one of the main goals of this thesis to present in detail how these properties effect the overall LSCF-stack performance. The results are described and discussed in Section 7.2.

Furthermore, a disadvantage of MIEC-cathodes should be noted. The direct contact of these strontium containing materials with the electrolyte material YSZ leads to the formation of an insolating $SrZrO_3$ secondary phase at the interface [39, 69, 70, 71, 72]. The solution is to apply a gadolinium-doped ceria (GDC) interlayer [73, 74], which is, however, highly dependent on the applied fabrication parameters (e.g. temperature, grain size distribution) in order to function properly [72].

The cathodes applied to the ASCs used in this thesis have the stoichiometric composition $La_{0.58}Sr_{0.4}Co_{0.2}Fe_{0.8}O_{3-\delta}$, which has been extensively studied [27, 31, 63, 73, 74, 75, 76, 77].

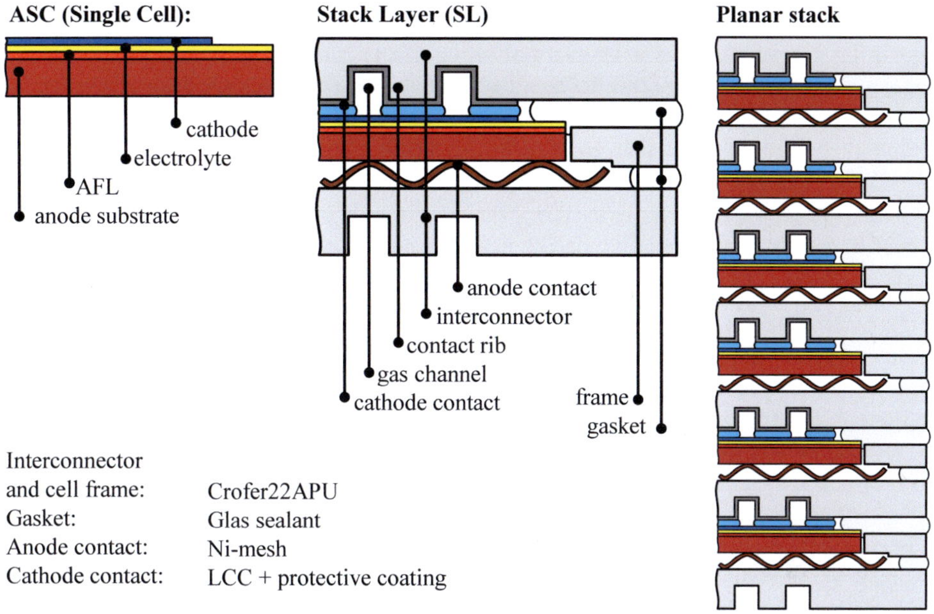

Figure 2.9.: Illustration of an ASC in a single stack layer (SL) and stack. The schematic shows in 2D the individual components and their serial connection in the planar stack design (adapted from [78]).

2.4.2. Stack Design

The cell voltage, U_{cell}, of an SOFC single cell in operation depends on the applied operating parameters (e.g. temperature, fuel/oxidant, fuel utilization, current load) and will always be lower than the corresponding OCV due to the inherent cell losses (Section 2.2). At $U_{cell} = 0.8 \ldots 0.7$ V, higher single cell current densities are achieved. In order to increase the overall power output to be technically relevant, single cells have to be connected in series, this is called a *stack*.

In general, two design concepts exist: The tubular design, where tube-shaped MEAs are connected in series and the planar design, where planar MEAs are sandwiched between ceramic or metallic interconnectors (IC). The planar stack design is the verified concept of choice especially for mobile applications (APUs) due to a higher volumetric power density generation [81]. However, research on the tubular concept continues because the volumetric power density output is less important in stationary applications [10, 82]. Still, the planar design is considered to be state-of-the-art for high power density stacks and is therefore the relevant stack design concept in this work.

Furthermore, planar stacks equipped with ASCs allow operation at lower temperature compared to ESCs due to decreased ohmic losses in the thinner electrolyte. This has the advantage of a lower degradation rate and the use of cheaper metal as interconnector material. A typical IC material is Crofer22APU from Thyssen Krupp [83, 37]. Unfortunately,

ICs fabricated purely from this material evaporate chromium (Cr) at the relevant elevated operating temperatures, leading to a poisoning of the cathodes electrochemical reaction kinetic [68, 37]. To counter this degradation mechanism, a protective coating can be applied to the IC surface to hinder Cr-molecules from leaving the metallic compound [84]. However, even though recently developed coatings show incredible performance in trapping the Cr [85], the downside is, that any of these material compositions exhibit a low electronic conductivity, which further decreases over time [85]. Hence, the purely electronic conducting layer has a major impact on the IC flowfield design in terms of the contact rib thickness. The matter is further discussed in Section 7.3 as it pertains to the major topics of this work.

Figure 2.9 schematically illustrates the planar SOFC-stack design, developed at Forschungszentrum Jülich [78]. Single ASCs are embedded into a cell frame and form (together with the metallic IC and anode/cathode contact) a stack layer (SL). The ICs surface at the cathode side is thereby designed as a repeated, symmetrical array of ribs and channels, called flowfield or repeat unit (RPU) (Section 4.2). In contrast to this elaborate and expensive construct, the surface at the anode side flat and the room between the anode and the next cell layer is filled with a rough, very porous Ni-net. The electrical interconnection between cells and ICs is secured via the flowfield ribs and Ni-net, while the flowfield gas channels supply the oxidant to the cathode. At the anode no gas channels are required, because the fuel gas distribution is sufficient, unhindered by the Ni-net [78]. The cell contacting and gas supply at the cathode in each SL is further explained in Section 2.5, this is because one of the major goals of this thesis. The gas-tightness between the fuel and oxidant gas atmospheres is secured by gasket rings, placed between the cell frame and each stack layer. In technical applications, this sealant is typically made of glass or welded [78]. Figure 2.10 shows (schematically) the technical components of a state-of-the-art planar SOFC-stack design from Forschungszentrum Jülich, designed for stationary combined heat and power (CHP) application [79].

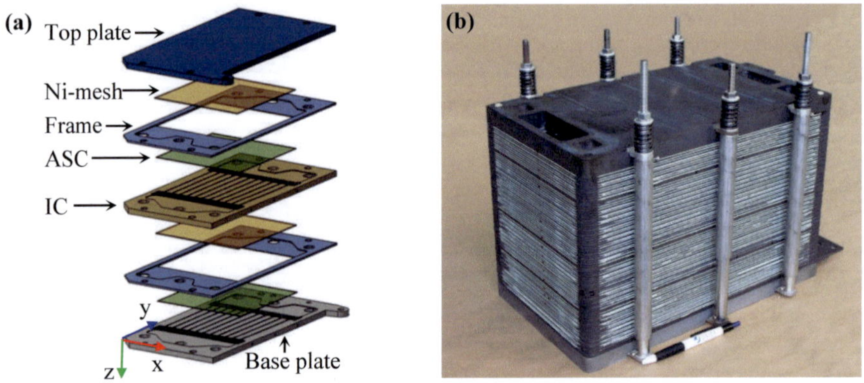

Figure 2.10.: Planar stack design for anode supported SOFCs: (a) Schematic view of individual stack layer repeat unit components from the F10-stack, designed by Forschungszentrum Jülich for stationary combined heat and power (CHP) application [79]. (b) Photograph of a completely assembled F10-stack unit with 40 layers [80].

2.4.3. SOFC-System Application

This section covers aspects relevant to SOFC applications. Briefly, the SOFC fuel flexibility and related mechanisms are described, as well as an overview of different SOFC-systems with respect to their area of application.

Fuel Flexibility

One of the major advantages of the SOFC offers over other fuel cell types is its ability use hydrocarbons (or derivatives thereof) as fuel. At high operating temperatures, Ni contained in the anode exhibits a high catalytic activity and natural gas [86], bio-derived fuels [87], evaporated liquid fuels (propane, diesel, kerosene, alcohol) [88, 89, 90, 91] and coal gas [92] can be fed to an SOFC. However, it should be noted that a stable operation has, so far, only been demonstrated on methane and natural gas. This superiority efficiency at using various fuels is demonstrated by the schematic in Figure 2.11, detailing the various fuel processing steps, required for the different fuel cell types.

The involved reaction mechanisms for hydrocarbon fuel conversion, SOFC operated with hydrocarbons as well as thermodynamic considerations and additionally created losses to be considered is given briefly in the following paragraphs. At the end there is a short description of occurring degradation mechanisms. Mainly, the information given is based on Ref. [17], further comprehensive information about hydrocarbon fuel conversion can be found in Refs. [53, 93, 94, 95, 96, 97]. Further information about experimental and modeling related works about the SOFC operation with hydrocarbons can be found in Refs. [53, 98, 99, 100, 101, 102, 103, 103, 26].

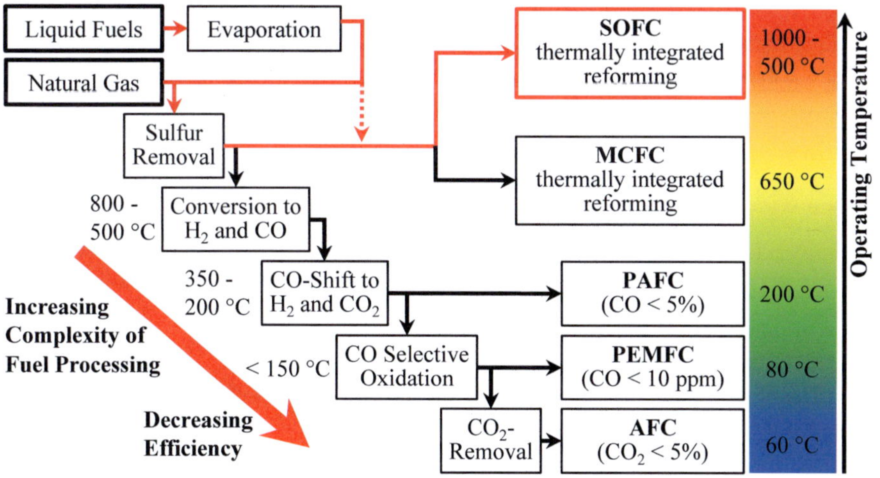

Figure 2.11.: Fuel processing steps for existing types of fuel cells from hydrocarbon fuels to hydrogen (adapted from [17]).

Direct internal reforming (DIR) in the SOFC proceeds by the *Steam Reforming* (SR) reaction mechanism, the most common industrial reforming method of producing hydrogen or syngas (mixture of hydrogen and carbon monoxide) from hydrocarbons. For methane, the reaction equation is given by [104, 105, 106]:

$$CH_4 + H_2O \rightleftharpoons CO + 3H_2 \qquad \Delta H^\circ = +206\text{kJ/mol}. \tag{2.16}$$

Associated with this exothermic main reaction of methane (CH_4) and steam (H_2O) to carbon monoxide (CO) and hydrogen (H_2) is the simultaneously occurring (slightly endothermic) *Watergas-Shift reaction* (WGS):

$$CO + H_2O \rightleftharpoons CO_2 + H_2 \qquad \Delta H^\circ = -41\text{kJ/mol}. \tag{2.17}$$

Here it is assumed, that both reactions 2.16 and 2.17 are reversible and proceed very quickly above 600 °C when in contact with a catalyst (Ni in the SOFC anode). They are therefore (for the most parts of the system) in chemical equilibrium [17]. It should be mentioned that dry steam reforming involving methane and carbon dioxide is also a possible reforming reaction mechanism [107], widely known as Fischer-Tropsch method, which is however not relevant for DIR in SOFCs.

Another well-established method of producing a syngas from hydrocarbons is the *Catalytic Partial Oxidation* (CPO). The reaction proceeds by the following equation:

$$CH_4 + \frac{1}{2}O_2 \rightleftharpoons CO + 2H_2 \qquad \Delta H^\circ = -247\text{kJ/mol}. \tag{2.18}$$

In the CPO reaction, it is important to provide only an under-stoichiometric amount of O_2 to avoid direct fuel combustion. A disadvantage of this mechanism is its unsuitability for DIR. Nevertheless, since in CPO no steam is required, it is the method of choice for an APU-system, by placing an external pre-reforming unit in the fuel gas upstream. The technique is hence called *indirect internal reforming* (IIR). The required process heat for the strongly endothermic reaction is provided by the fuel cell stack. Detailed reviews on this mechanism are in Refs. [108, 109].

There are several report in the literature about a *Direct Electrochemical Conversion* in Ni/YSZ SOFC anodes at higher temperatures (700 . . . 950°C) without the formation of solid carbon (coking) [110, 111, 112]. Stable operation has been demonstrated by Weber et al [110]. The reaction mechanism on methane proceeds according to [97] by the following equation:

$$CH_4 + O^{2-} \rightleftharpoons CO + 2H_2 + 2e^- \qquad \Delta H^\circ > 0. \tag{2.19}$$

Even though operation at lower temperatures (700°C) has been reported in literature, it has to be assumed that after an initial conversion to H_2 and CO the standard electrochemical reaction according to Equation 2.3 is preferred and H_2O and CO_2 are subsequently converted via Eqs. 2.16 and 2.17 [97].

A serious problem in hydrocarbon SOFC operation is *coking*, which describes the irreversible formation of solid carbon. Especially at lower operating temperatures ($<$ 700°C), the deposition of solid carbon from the gas phase on the anode surface leads to the deactivation of electrochemical sites and therefore a lower performance. Severe coking even blocks the pores and therefore the gas supply to the electrochemically active sites, which further decreases the performance. General information on carbon deposition is given in Refs. [105, 106] and more specific informations in Refs. [113, 114].

It was shown by A. Kromp [28, 17] by means of detailed impedance analysis that in SOFC operation with methane or syngas, only H_2 acts as electrochemical active species in the anode charge transport reaction (Equation 2.3. The equally electrochemical active CO is subsequently oxidized via the water-gas shift reaction (Equation 2.17). Furthermore, the experimental results revealed an additional arc in the low frequency range of the impedance spectra, denoted as P_{ref} (Section 2.3.2). It could be shown by a time-dependent FEM-model that this additional loss can be attributed to the interdependence of electrochemical charge transfer and WGS, linked via the gaseous species transport [28, 17, 29]. The model was validated by experimentally obtained data and gave further insights into the complex coupled network of electrochemical and reforming reactions kinetics within the multi-component gaseous species transport. This knowledge laid the foundation to the model development in this work.

It has to be mentioned, that fuel impurities have a negative impact on the performance and long term stability of the SOFC. Sulfur compounds have the strongest influence and are found in nearly every hydrocarbon related fuel. Sulfur acts as a strong catalytic poison by blocking the catalysts surface area and thereby slowing down or preventing the electrochemical anode reaction (Equation 2.3) and/or the reforming reactions (Equations 2.16 and 2.17). Comprehensive reviews on the issue can be found in Refs. [115, 116, 117].

		Power*	Requirements	
Mobile Application (APU¹)		<10 kW	**Operating life** **Temperature** **Start-up Time** **Fuel** **Efficiency** η_{el}	**4 000 h (160 000 km)** **(500) 600 – 800 °C** **< 30 min** **Gasoline, Diesel** **35 – 50 %**
Decentralized power supply (CHP²)		<10 kW	**Operating life** **Temperature** **Start-up Time** **Fuel** **Efficiency** η_{el}	**> 40 000 h (5 years)** **700° – 1000 °C** **< 3 000 min (48 h)** **Natural Gas** **45 – 70 %**
Cogeneration unit		>100 kW	**Operating life** **Temperature** **Start-up Time** **Fuel** **Efficiency** η_{el}	**> 40 000 h (5 years)** **650° – 1000 °C** **< 800 min (12 h)** **Natural Gas, Fuel Oil** **25 – 50 %**

¹ APU: Auxiliary Power Unit; ² CHP: combined heat and power (Communal Fuel Cell Heating Unit);
* Total Power: Electrical + Heat

Figure 2.12.: Application of SOFC-systems. Displayed are the requirements depending on the area of application [22, 37].

SOFC-Systems

The SOFC can be applied in different areas, which are displayed in Figure 2.12. The specific requirements for each application are listed. Especially in stationary operation, the internal reforming ability enables the use of (in a lot countries) well installed natural gas infrastructure. So, an elaborate fuel purification process with accompanied efficiency loss (Figure 2.11) is not necessary.

The high efficiencies stated in Figure 2.12 result from the use of waste heat, produced by the SOFC stack units irreversible losses as well as the subsequent combustion of exhaust gases in a hot-water boiler. Consequently, such systems are called combined heat and power system (CHP).

In mobile applications (APU), an external pre-reformer has to be included in the system as conventional fuels cannot be directly, internally reformed and require pre-treatment (Figure 2.11). The APU's system complexity thereby increases and the overall efficiency consequently decreases. Furthermore, an APU must meet higher standards in terms of system weight and volume as well as faster start-up times, compared to stationary systems.

Nevertheless, high power densities combined with low degradation rates are required in any of the described applications, necessitating a detailed modeling approach for a single stack layer. This promotes optimization and thereby increases performance for the whole system.

2.5. Contacting: Ideal and Stack

This section is solely dedicated to the contacting of SOFC single cells and its influence on performance in (i) ideal conditions and in (ii) technically relevant conditions. The information given in the following paragraphs is crucial for comprehending the results in Chapter 7. Substantial research on the topic of cell contact with respect to electric losses was carried out by M. Kornely [68, 118, 37]. Hence, the description here is based on the work of Kornely, but is kept in a more concise manner. The interested reader may find further information in Ref. [37].

Without cell contacting, no electronic transport, or gas supply and removal to and from the electrodes of a MEA is possible (i.e. no cell operation at all). The description of cell inherent losses occurring under load (Section 2.2) has so far been kept to the assumption of ideal operating conditions, which of course includes the contact design. These ideal conditions are not found in the technical application (or so to say, for a single cell implemented into a stack layer). A clear definition of ideal and stack contact design is now given the thereof deduced requirements on the contact components are described. There then follows an explanation of how the contacting influences the single cell losses and what the key features are for lessening the influences and sustainably improving the performance.

2.5.1. Requirements and Technical Realization

The general requirement on cell contacting is that it should not induce additional operation losses by inducing a new loss or by influencing the unavoidable, cell inherent losses (Section 2.2). However, the realization of such strict requirements depends on the intended purpose: if the goal is an electrochemical single cell characterization and inherent loss process analysis in a test bench (Section 3.2), then any influence due to the contacting is to be avoided, by all means. For an application relevant contacting however, different requirements apply.

2.5.2. Laboratory Design - *Ideal Contacting*

The laboratory contacting is designed for electrochemical measurements at the test bench, as described in Section 2.2. The aim thereby is to characterize the performance of single cells by means of current/voltage and impedance measurements and analyze the dependence of occurring losses to changes in relevant operating conditions. Furthermore, the gathered measurement data is used to determine certain material parameters required in a model development process, which is described in Sections 5.1.7 and 6.5. Last, the measurement data is used to validate individual modeling results, for example the anode gas species transport losses (Section 6.5), or measured and predicted current/voltage characteristics (Section 6.5).

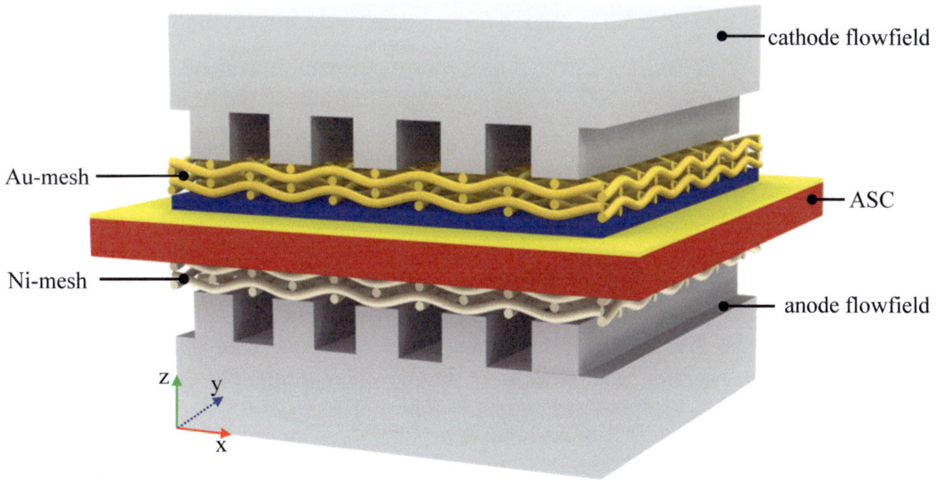

Figure 2.13.: *Ideal contacting*: Schematic illustration of how an ideal single cell contact design realized in the test bench by noble metal meshes, placed between each flowfield and corresponding electrode.

All these purposes require cell contact design to have no effect on the measured results or only in a negligible way, i.e., it has to guarantee *ideal* conditions. The following aspects are required from an ideal contact design in order to fulfill its purpose:

- low-resistance electronic contacting of the MEA electrodes,
- a sufficiently high gas permeability to enable a homogeneous gas distribution across the electrode surface,
- no chemical interaction with the electrodes,
- chemical stability in reducing and oxidizing atmospheres,
- reproducible performance.

All these requirements are fulfilled by the use of noble metal meshes, which are further described in Section 5.4. The use of noble metals implies that monetary aspects play no role. This does produce reliable measurement data for the above listed purposes.

Figure 2.13 illustrates schematically the technical realization of the laboratory contacting. A double-folded Au-mesh is placed between flowfield and electrode at the cathode side, while a Ni-mesh is used at the anode. More details about the mesh-structure parameters are given in the experimental Section 5.4. Both materials exhibit a very high electronic conductivity and mesh density (1024 meshes/cm^2) wherefore in-plane conduction losses can be excluded. The mesh structure also allows a nearly unhindered gas permeation, which combined with the relatively high gas flow rates (250 sccm) set for the mass flow controllers in the measurements, ensures an homogeneous gas distribution over the active electrode area. Furthermore, neither material interacts chemically with the electrode materials and both maintain their properties in oxidizing and reducing atmospheres.

As no further losses are induced or other occurring losses are influenced by the method of contact, it can be regarded as ideal and will be denoted as such in the following.

2.5.3. Technically Relevant Design - *Stack Contacting*

Stack unit contacting meant for practical application purposes is governed by different requirements compared to the ideal contact of a single cell in the lab. It is unquestionable that with increasing losses the efficiency of a stack, and thereby of the whole system, decreases. Consequently, for technical relevant contacting the basic requirement as stated in the beginning of this section, also hold: to minimize any additional loss. However, design guidelines for a technical relevant contact have to consider certain cost-benefit calculations, which are more or less irrelevant in the lab due to small quantities. Therefore, the contacting production process needs to be kept simple and the contacting must be easy to implement into the stack layer units. Any additional process step increases the costs and should be avoided unless beneficial to counteract long term degradation mechanisms (e.g. protective coating against Cr-poisoning [85] and corrosion). In summary, the following requirements are required of technical relevant stack contacting:

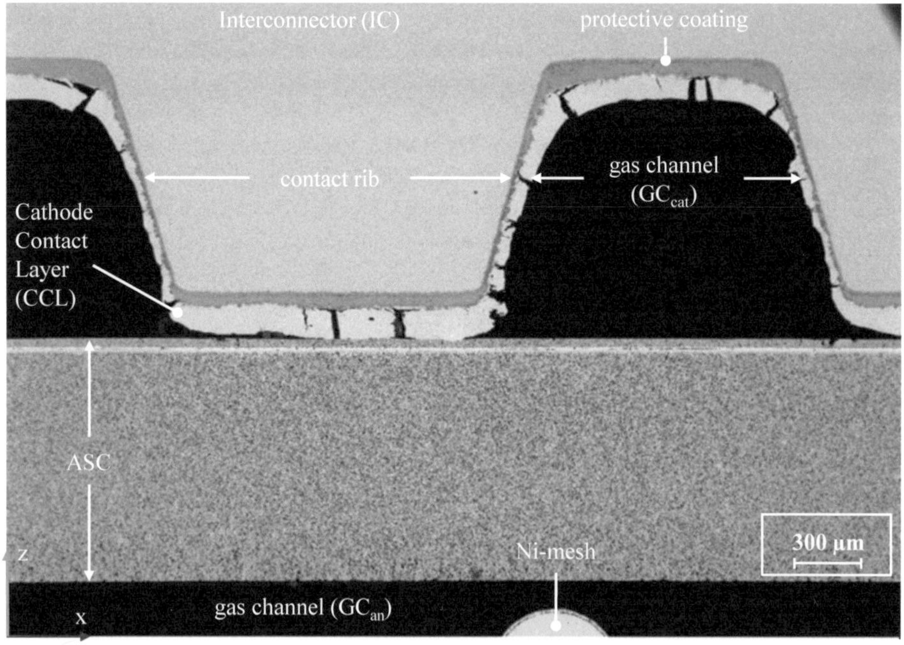

Figure 2.14.: *Stack contacting*: scanning electron microscopy (SEM) image of a F10 repeat unit (RPU) section in the x-z-plane (supplied by M. Menzler, Forschungszentrum Jülich). Displayed are at the top the IC with its cathode side flowfield (contact ribs + gas channels), the anode supported single cell SOFC (ASC) and parts of the anode gas channel (GC_{an}) and coarse Ni-mesh. The high SEM resolution enables the identification of the protective layer and the cathode contact layer (CCL) as well as the ASC details.

- low additional losses at the desired operating conditions,
- low material and production costs and complexity,
- high reproducibility,
- gas-tight,
- low, long-term oxidation-rate,
- high electronic and low ionic conductivities,
- comparable to MEAs thermo-mechanical properties,
- high thermal conductivity,
- mechanical stability.

Hence, the use of noble metals or other rare materials is undesirable as it increases costs.

The contacting design in the planar F10-design from Forschungszentrum Jülich (Figure 2.10) is regarded as state-of-the-art for stationary CHP applications (Section 2.4.3). Stacks built to this design have proven a reliable operation for several thousand hours of operation [79]. Displayed in Figure 2.14 is a scanning electron microscopy (SEM) image, taken from a repeat unit of such a F10-stack from Forschungszentrum Jülich. The SEM-image in Figure 2.14 displays the view on the x-z-plane, which is perpendicular to the gas flow direction. The interconnector (IC) is displayed on top with its repetitive array of gas channels and contact ribs. The IC itself is covered by a thin layer of manganese cobalt ferrite oxide ($MnCo_{1.9}Fe_{0.1}O_4$, MCF) or manganese oxide (MnO_2), a protective coating against corrosion and the evaporation of volatile chromium species. On top of the protection layer, a thicker layer of lanthanum manganese copper cobaltite (LCC10/12) or LSCF is applied, generally called cathode contact layer (CCL) or also current collector layer. Its purpose is to equalize certain, production related imperfections in the IC rib and cathode flatness and thus to guarantee a good electric interconnection between rip and electrode surface. Both layers are applied by atmospheric plasma spraying (APS) or wet powder spraying (WPS); experimental results from Jülich indicate, that the combination of MCF-LSCF applied by WPS promises the best combination for long term SOFC-operation [119]. What was not covered in the investigation carried out in Jülich [119] however, was the fact that the CCL should have a certain porosity as well in order to allow the oxidant to reach the electrochemical reaction zones beneath the rib from the gas channels. WPS however produces layers with a very dense microstructure, which might benefit the electronic conduction but is disadvantageous for gas diffusion. The matter will be discussed in detail in Section 7.3, as it is one of the key topics to be analyzed within this work.

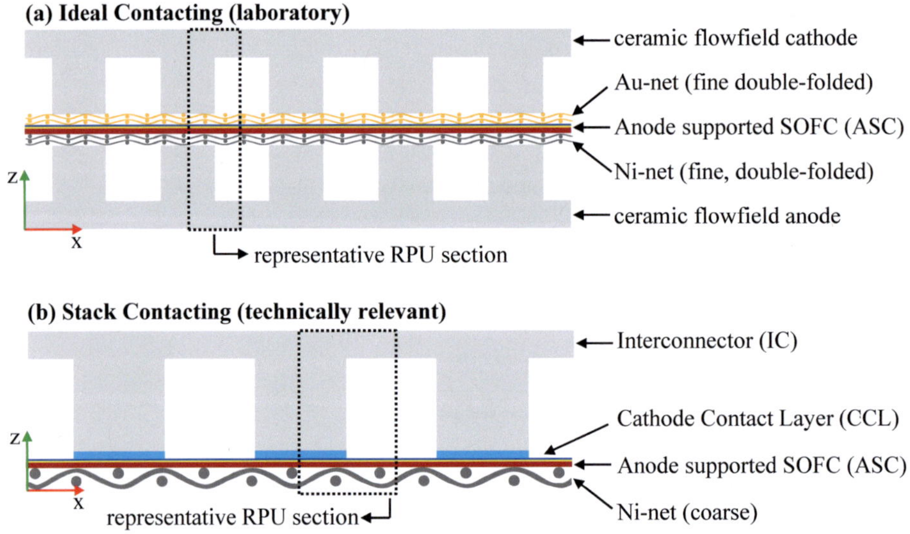

Figure 2.15.: Schematic cross-section display of a singe cell in (a) *ideal* and a single layer in (b) *stack* contacting.

At the anode side, the IC of the next stack layer has no flowfield. The electronic contact is realized by placing a coarse Ni-mesh in the resulting, wide single fuel gas channel, which can be identified in Figure 2.14 at the bottom. Coupled with the high electronic conductivity of Ni of the cermet anode, the coarse mesh ensures the electronic interconnection between the anode and the top of the IC in the next stack layer. Due to its high porosity the fuel gas stream can pass by unhindered.

Figure 2.15 displays a direct comparison between the ideal (Figure 2.15a) and stack contacting (Figure 2.15b) by a schematic cross-section illustration in the x-z-plane. As mentioned and displayed in Figure 2.15 the contact designs mainly differ in two respects:(i) fine meshes are used in the ideal design on both electrode sides and (ii) for the stack contacting a coarse mesh is applied on the anode side instead of a flowfield. Additionally, both illustrations have a black, dotted squared section which represents the whole repeat unit (RPU), shown due to symmetry features. This is further discussed in Section 4.2 by means of deriving the applied model geometry used in the numerical investigation of Sections 7.2, 7.3 and 7.4.

Hence, additional losses induced by the use of a stack contact-design as proposed by Forschungszentrum Jülich (Figure 2.14 and 2.15b) and the influence on the cell losses are discussed in the next section.

Losses Due to Stack Contacting

From the explanations given in the previous section it is clear that the F10 stack contacting on the anode side can be regarded as ideal, whereas the flowfield on the cathode side influences the processes occurring in the oxygen electrode. Hence, the following discussion will only focus on the latter.

In general, the stack contacting realized by an IC (as shown in the previous section in Figure 2.14) has to realize two competing functions: (i) conduct electrons to ensure the serial interconnection between the individual layers, and (ii) supply and remove the gaseous reactants for the electrochemical charge transfer reactions (Equations 2.2 and 2.3). However, perfect electronic contact would require a solid IC without flowfield, thus completely covering the electrode. In this way, a low-resistance contact with short conduction pathways from the IC to the TPBs would be realized, while gas transport supply to the electrode would be impossible. A perfect gas supply would require an IC, where the total electrode surface is accessible by a single wide gas channel, which of course would prevent any electrical interconnection between electrode and IC and therefore any normal SOFC operation.

An applicable realization must therefore be a compromise between both requirements, which is given by the flowfield design with its alternating array of ribs and channels. Contact ribs ensure electrical interconnection, while the gas channels supply the required gaseous reactants. The comparison of ideal and stack contacting is shown by the schematics in Figure 2.16.

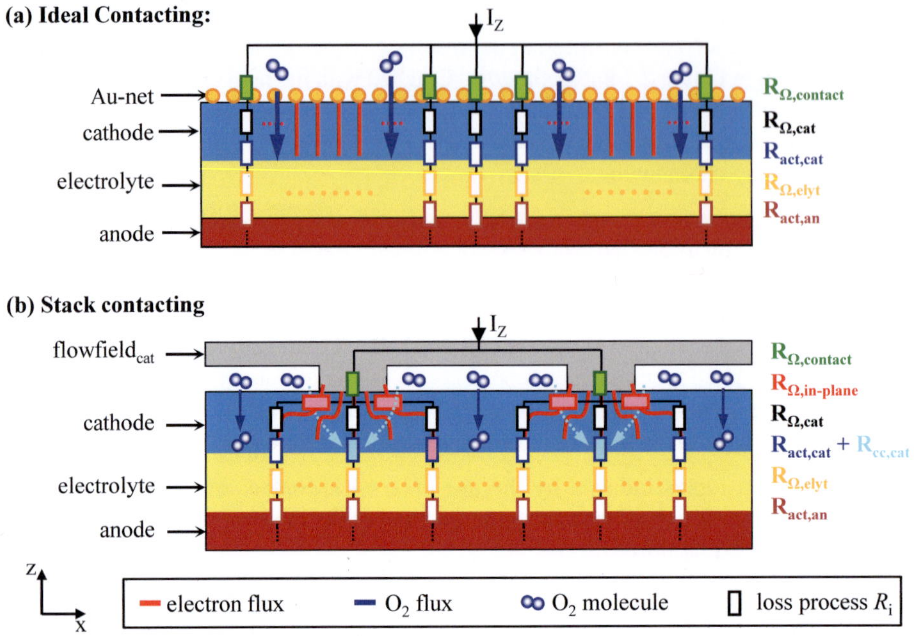

Figure 2.16.: Schematic illustration of (a) ideal and (b) stack contacting and corresponding loss processes.

Figure 2.16a illustrates the ideal contacting. The Au-mesh covers the cathodes surface entirely, thus ensuring a homogeneous electrical contact. The electrons can be conducted directly (red lines) in z-direction from the mesh to the electrochemically active zones near the electrolyte. Short conduction pathways lead to minimized ohmic losses in the cathode ($R_{\Omega,\mathrm{cat}}$). The high electronic conductivity of Au combined with the thin mesh distance and the high resistance against an oxide layer formation minimizes the contact resistance ($R_{\Omega,\mathrm{contact}}$), wherefore it can be neglected [37]. Therefore, only the electrolyte ($R_{\Omega,\mathrm{el}}$) contributes to the overall ohmic loss.

At the same time, the inherent permeability of the mesh leaves enough open space for gas transport, and oxygen molecules can be homogeneously distributed to the cathodes surface from the ceramic flowfield gas channels. Direct gas transport pathways (blue arrows) combined with the thin cathode thickness (≈ 45 µm) imply a low gas transport resistance ($R_{\mathrm{cc,cat}}$). The electrochemical losses in cathode ($R_{\mathrm{act,cat}}$) and anode ($R_{\mathrm{act,an}}$) are therefore not influenced by the IC flowfield and only depend on the inherent material kinetics of the corresponding electrode to execute the cathode half-cell reaction (Equations 2.2 and 2.3).

In the schematic (Figure 2.16a), the flowfields on anode and cathode side have been neglected for reasons of better clarity. It is however comprehensible, that the high porosity of the double-folded contact meshes offer enough open transport volume so that gas diffusion can occur unopposed.

Contacting the cathode with a flowfield (stack contacting, Figure 2.16b), only parts of the electrode surface is in contact and no direct electronic contact is given in the area beneath the gas channels. As a consequence, all electrons have to flow through the limited contact area given by the contact rib dimension, moreover, the local current density increases in that area. This has a direct impact on the created ohmic overpotential due to the increased R_{Ω},contact. Furthermore, the area specific contact resistance, ASR_{contact} (4.3.1.1), depends on the applied material used as protective coating applied against Cr-poisoning [119] and increases over time due to corrosion of the IC's metallic compound, caused by the high operating conditions in oxidizing conditions. The influence of an increasing ASR_{contact} on SOFC stack performance is further analyzed with the help of in this work developed model framework (Section 7.3.2).

The next important fact is shown by the red lines in Figure 2.16b. The electron transport is no longer solely directed in z-direction but also in x-direction as the electrons have to flow parallel to the electrolyte to reach TPBs beneath the gas channels. This induces additional ohmic, in-plane conduction losses ($R_{\Omega,\text{in-plane}}$) caused by the cathode's thin thickness and the limited electronic conductivity (σ_{cat}). This has a direct impact on the cathodic electrochemical reaction. It is displayed in Figure 2.16b by the red colored resistance symbol, symbolizing an increased electrochemical losses ($R_{\text{act,cat}}$). In addition, the ohmic losses in the electrolyte ($R_{\Omega,\text{el}}$) likewise increase, as the oxygen ion supply in the areas beneath the gas channels is slower, and faster in the areas beneath the ribs, causing (similar to the contact resistance) an increased ion flux in a limited area.

Lastly, additional gas diffusion losses ($R_{\text{cc,cat}}$) are created by the use of a flowfield. As can be seen in the illustration given in Figure 2.16b, the gas diffusion pathways (dotted blue arrows) from the limited area accessible from the gas channels are elongated to the reaction zones beneath the contact ribs in comparison to the direct pathway given in the ideal-contacting. This effect slows down the electrochemical activity in the areas beneath the ribs, which is marked by blue colored resistance symbols. Equally as for an under-supply of electrons correlates an under-supply of oxygen with increased electrochemical losses ($R_{\text{act,cat}}$) in the corresponding area and an increased ohmic loss in the electrolyte ($R_{\Omega,\text{el}}$) in the area of increased electrochemical activity where enough reactants arc available.

In summary, contacting a MEA using planar interconnector with a given flowfield design induces additional ohmic and gas diffusion losses. In reality, both loss mechanisms occur simultaneously and competitively, and are complicatedly coupled by the cathode's electrochemical kinetics . Furthermore, an interdependence exists between the cathode layer thickness and material (electrochemical kinetics, electronic conductivity), the flowfield design (ribs and channel width) and additional layers (CCL). To investigate the matter in detail, the model framework developed within this work is suited perfectly. The results are presented in Chapter 7.

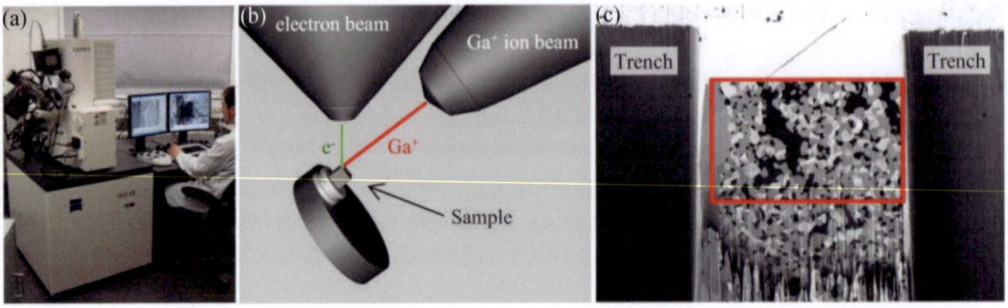

Figure 2.17.: (a) Photography of the SEM system used to acquire image data from cell samples and for the reconstruction procedure. (b) Schematic arrangement of the sample in the vacuum chamber of the SEM (taken from [121]). (c) An example SEM image of a prepared sample at the beginning of the sectioning. The red box indicates the maximum area which can be used for reconstruction (adapted from [120]).

2.6. 3D Reconstruction of Porous Electrodes

The electrodes of the cell samples used in this thesis (Section 3.1) are reconstructed into a numerical 3D structure using focused ion beam (FIB) tomography. Characteristic microstructural parameters are determined by applying appropriate numerical algorithms to the 3D electrode reconstruction. The whole procedure is too complicated to be explained in just one section, as development was based on several years of experience gathered at the Institute of Applied Materials (also the topic of various dissertation [120, 121]. However, essential results are obtained by the technique in collaboration with experienced colleges from the Institute and used in this work's model framework. Hence, the reconstruction and parameter determination procedure is here presented in a condensed manner, based on the work of Joos [120] and Ender [121].

The process of reconstructing a porous SOFC electrode can be subdivided into the following steps:

(1) Sample preparation,

(2) FIB/SEM procedure and data acquisition,

(3) Image processing,

(4) Segmentation,

(5) Parameter calculation.

Sample Preparation

First, the to be reconstructed cell is fractured and a small fragment ($\sim 4\times\ 7\ \mathrm{mm}^2$) is selected. Next, the sample is placed into a vacuum impregnation chamber to infiltrate the pore volume with a special epoxy resin. The resin functions later on as a material stabilization during the

FIB/SEM treatment and provides better contrast in the material phase separation process. After the resin is dried, the excess epoxy is removed and the sample is polished with SiC paper (grit < 1 μm), resulting in a $\sim 3{\times}6{\times}3\ \mathrm{mm}^3$ sample size. Thereby, it is important to have two perpendicular planes with clearly defined edges. In the final preparation step, the sample is glued to a sample holder with an Ag coating, and a thin Au layer (approx. 20–300 μm) is sputtered on top of the sample to ensure good electrical contacting between sample holder.

FIB/SEM Procedure and Data Acquisition

The prepared and mounted sample is placed into the vacuum chamber of the scanning electron microscope (a ZEISS 1540XB CrossBeam©, Figure 2.17a). The SEM is equipped with an electron beam and a Ga^+ ion beam, whereby the first is used to acquire the image data and the latter to remove or mill material from defined sections (Figure 2.17b). After milling two trenches to the left and right, and creating a smooth surface on the front and top of the volume of interest (Figure 2.17c), the actual procedure can start. An image is captured and in turn a small slice is milled away. The process is repeated until the required volume is achieved. The procedure has to be carefully monitored to avoid drifting, which leads to a misalignment of the captured images, and consequently, to a smaller reconstructed volume. Another problem is to avoid any charging of the electrical conduction material phase, which results in lower image quality. The resolution plays a very important part, given by pixel per length unit in 2D or by voxel (volumetric pixel) per volume in 3D. It is, on one hand, predefined by the particle and pore sizes in order to provide a high enough resolution to account for the specific sample features. On the other hand, the resolution should not be too low, because it limits the field of view (image size) and thus the reconstructed volume. However, a certain volume size is required so the subsequently determined parameters can be representative of the whole electrode (Section 5.1.6). Detailed information regarding resolution (minimum of 10 voxel per particle diameter), slice thickness and SEM settings (e.g. acceleration voltage, beam current, milling time, etc.) is given by Joos [120] for the different electrode types used in this work. In summary, the FIB/SEM procedure provides a consecutive stack of 2D images with a certain grey-scale value distribution, which are processed further in the next step.

Image Processing

Precise alignment of all captured 2D SEM images is essential for the 3D electrode reconstruction. This is carried out with the software package *ImageJ* [120] and defines the *region of interest* (ROI). To achieve a more pronounced phase separation in the three-phase anode reconstruction, image-information captured from both in-lens and Everhart-Thornley SEM-detectors are combined. For the cathode, images recorded with the Everhart-Thornley detector supply sufficient information depth. Depending on the image quality, certain filters (e.g. Anisotropic Diffusion, Non-Local Mean) are applied to reduce the noise or brightness gradients [121], which facilitates the segmentation process in the next step.

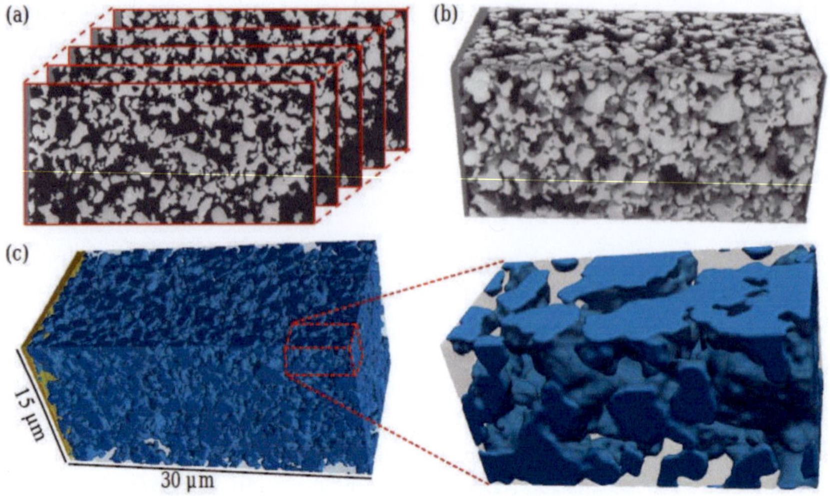

Figure 2.18.: Different steps of 3D reconstruction: (a) By stacking and aligning the 2D images in 3D space and (b) expanding the pixels in the slicing direction, a 3D reconstruction consisting of voxels is derived. (c) By assigning each voxel of the structure its corresponding phase, a 3D material distribution of the sample is obtained (taken from [120]).

At first however, the 2D image are stacked (in capturing order) into a 3D matrix using MATLAB, wherefore every matrix value corresponds to a voxel with a certain grey-scale value. Figure 2.18 shows an image-sequence captured from a porous cathode, which were then stacked and aligned to obtain a 3D reconstruction of the electrode structure.

Segmentation

The grey-scale information saved for every voxel in the 3D matrix must now be converted into the corresponding material or pore phase. This procedure is called segmentation. It is the most difficult step in the whole reconstruction process and is therefore the greatest potential error source. The most common method of carrying out this task is by using *threshold values* [122, 123]: a certain grey-value range is assigned (Figure 2.19a) to pore or material phase, divided by a distinct threshold value. In theory, all voxels with corresponding grey-values are assigned to the same phase. However, the method requires excellent image quality with distinct phase separation, marked in an image histogram by clearly separated peaks with at least a minimum space in between as threshold. In reality, the situation is not so simple, as can be deduced from Figure 2.19b. Even though the pore and LSCF phase can be identified clearly in the histogram, the peaks are too close, and the threshold is difficult to declare. Different algorithms exist to calculate a threshold: previously J. Joos found Otsu's method [124] the most robust algorithm for two-phase electrodes (LSCF cathode).

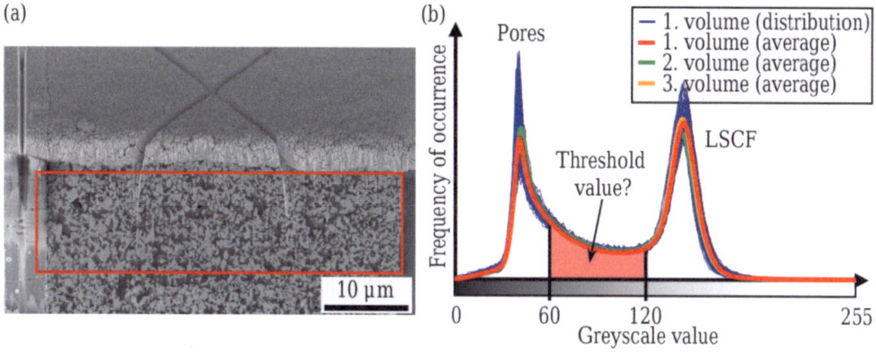

Figure 2.19.: (a) SEM image of a porous LSCF cathode. (b) Corresponding histogram of 213 images (thin blue lines) together with the averaged histograms of all these images (red line), as well as the average histogram of two other volumes (green and orange), acquired from the same sample (taken from [120]).

For three-phase electrodes (Ni/8YSZ anode) Joos [120] developed a more advanced method based on the *region growing method* [122], where not only a threshold is calculated, but also relations between neighboring voxels are taken into account. However, every electrode seems to have its own quality, so no clear universal segmentation guideline can be declared. Nonetheless, a manual inspection of the segmentation results is mandatory as the best segmentation algorithm (according to Joos) is still the human eye.

Completing all previously described steps with great care results in a realistic electrode microstructure reconstruction, as displayed in Figure 2.18c. The digitized structure can be analyzed by certain algorithms to determine the characteristic microstructural parameters (volume fractions, tortuosity, surface area, etc: described in Section 5.1).

2.7. Application of Numerical Simulation in Research and Development

The following section outlines how numerical simulations can be powerful tools in the research of complex multi-physical processes, and the subsequent development of interacting components. In principle, there are two pathways for analyzing a complex system: (i) The experimental (and original) pathway involves experiments designed to gather specific information about the system of interest with the purpose of drawing logical conclusions about the processes occurring within. On the example of the SOFC, this would be the measurement of C/V curves or impedance. Furthermore, experiments can be designed in a specific way to determine material parameters. An example for this would be to determine electronic conductivity using 4-point electronic conduction measurements. While experimental work has the advantage of dealing with actual physical properties, it also has several disadvantages. The measurement data accuracy is influenced by many environmental and test setup factors and therefore strongly depends on a well-designed setup with highly accurate sensors used in the recording of respective measurement data. Furthermore, and maybe even more importantly, the reproducibility of measured results determines the validity. Last but not least, time is a very important factor as well, since many experiments are very time-consuming to setup and conduct and it is simply impracticable to conduct greater parameter variations to evaluate impact.

This is where (ii) the numerical simulation exhibits maybe its greatest advantage. Once a model has been set up and validated, the inherent flexibility allows variations of any in the model containing parameter in very short time. However, the first two mentioned steps are of great importance to the numerical simulation pathway. By setting up a model is described, what processes are implemented and therefore regarded in the model. This however involves an *a priori*-knowledge about such. Furthermore, process implementation always involves inevitable compromises with respect to assumptions and simplifications, as either the regarding process is not completely understood or the effort to regard certain phenomena is simply too high. Usually, the consequence is an iterative model development as the impact of certain aspects may have or have not a greater influence on the model predictions. In the last step, the model validation is maybe the most crucial part. It is the nature of a numerical computation, that a solution is produced most of the time, but without the guarantee of any physical meaning. Therefore, it is the developer's responsibility to compare the model predictions with data gathered from experiments designed for this purpose. A thorough, multi-physic model validation should include a comparison of model and measurement results in distinct operating conditions, where the implemented physical processes individually exhibit an unambiguously identifiable impact on the result. Only in this way can a correct process implementation be verified and the model results later relied upon. It is therefore clear, that numerical simulations are always accompanied by experimental work, which may be more expensive in terms of cost and time but nevertheless inevitable. Furthermore, required material parameters often can only be reliably acquired by supervised experimental determination based on the necessary *a priori*-knowledge about occurring loss mechanisms.

Numerical Solution for Partial Differential Equations (PDEs)

Natural processes usually occur in a complex way, and so it is not always possible for a scientist to easily describe them in an adequate model. It is helpful to subdivide the problem into a finite number of elements, with well defined components. These problems are called *discrete*. A further subdivision leads to the mathematical function of infinitesimal elements and partial differential equations (PDEs), implying an infinite number of elements. These systems are called *continuous*. An analytical solution for both problems may be found for very simple geometries and boundary conditions. However, these do not apply to most real world problems. In real continuum systems, various physical processes usually occur simultaneously and interact with each other and an analytical solution may only be found by simplifying the problem. Hence, it may no longer reflect the real system with desired accuracy. Nowadays the world enjoys vast computational power, even on simple desktop PCs, so a numerical solution offers far more options to develop a suitable model and should therefore be the method of choice. The basic idea is to employ an algorithm to solve appropriate PDEs, which describe respective physical processes in a discrete space and/or time, whereby the solution is found stepwise in an iterative manner. The development of suitable approaches is an ongoing process in academic numerical mathematics, which has been recognized for some time now as an independent research branch. Commonly used algorithms are the Finite Volume Method (FVM), the Finite Element Method (FEM) and the Finite Difference Method (FDM). In this work, the software COMSOL Multiphysics v4.3b is employed, wherein the FEM is implemented [125]. The user has to define a model geometry, adequate equations for the desired physical process description, initial and boundary conditions. The subsequent solution is computed by the software and various desired post processing stages can be performed manually or even automatically via Matlab.

Finite Element Method (FEM)

The Finite Element Method (FEM) used nowadays in engineering is the result of combining various mathematical procedures, such as *Variational methods*, *Weighted residuals* and *Piecewise continuous trial functions*, which have partly been proposed in 1800 by Gauss [126]. The following short description is based on the introduction given in Ref. [127] and Refs. [65, 121].

The basis of the method is a weak or variation formulation of a boundary value problem. The derivatives of the differential operator will be replaced by the generalized Sobolev derivatives. For this purpose, the partial differential equation is multiplied by a function $v(x)$ from the Sobolev space and integrated over the area of consideration Ω. This results in the following weak expression

$$a(u, v) = f(v) \tag{2.20}$$

with $a(u, v)$ as elliptic bilinear form. The idea is that the function $u(x)$is an exact solution of the problem if it is also fulfilled for arbitrarily test functions $v(x)$) in the Sobolev space. As an example, the Poisson equation is used as

$$-\Delta u(x) = f(x) \tag{2.21}$$

with the weak form given by

$$\int_\Omega \nabla u \nabla v \cdot dV = \int_\Omega f \cdot v \cdot dV. \tag{2.22}$$

In the next step, $u(x)$ is written as sum of various functions $p_j(x)$, which are unequal to zero only in a small area:

$$u(x) = \sum_{j=0}^{n} u_j p_j(x). \tag{2.23}$$

The solution for $u(x)$ is consequently given by the determination of u_j. Hence, a linear dependency is given by

$$K \cdot \vec{u} = \vec{f} \tag{2.24}$$

with K as stiffness matrix and $\vec{f}$ as load vector.

For the herein presented FEM model framework, three different kinds of boundary conditions are used, described here by the example of electronic conduction:

- Dirichlet: Setting a fixed potential or current density.
- Newman: Electrical isolation between non-conducting domains or boundary flux due to charge transfer reactions.
- Robin: Coupling of different transport phenomena (gas diffusion, bulk diffusion, electronic conduction).

The spatial model geometry discretization is required to calculate the weak formulation Equation 2.22. For this, the computational area is divided into a finite number of elements by the numerical mesh. In this work, triangular elements are used for 2D models and tetrahedral elements in 3D geometries. COMSOL supplies prefabricated meshing algorithms, whereby the model geometry is approximated by the mesh. Thereby, a certain error is introduced, which depends on the elements form and the mesh resolution. At each mesh knot a solution for $u(x)$is calculated, hence with an increasing number of elements the solutions accuracy increases and, however, the required computational effort. In Section 6.5 these interdependencies are evaluated.

For time-dependent problems an additional time discretization is required, which can be performed before or after the spatial discretization. Common methods are the explicit Euler method or the implicit Crank-Nicolson-Method or Runge-Kutta-Approach. Detailed descriptions are given in Refs. [128, 129].

The numerical solution starts by setting up the stiffness matrix, K, and the load vector, $\vec{f}$, of Equation 2.24 and $\vec{u}$ can be calculated. For a simple problem with solution independent boundary conditions and source or sink term, the calculation of Equation 2.24 is quick. In the presented model framework three different transport phenomena (gas diffusion, electronic/ionic conduction and bulk diffusion) are regarded, which individually require separate equation systems (Equation 2.24). Furthermore, in the MIEC-cathode model domain all transport processes are coupled via Robin boundary conditions, wherefore the values in each K and $\vec{f}$ depend on the solution of the other equation systems. Hence, an iterative approach is required with the previous solutions as initial values for the next calculation. Well-chosen initial values are required to achieve convergence in an acceptable time range. Convergence is thereby a measure of accuracy, describing the difference between each iteration step's solution. Furthermore, spatially dependent transport coefficients calculated via matrix inversion (Section 4.3.2.3) coupled with highly non-linear transport equations enforce a linearization and thereby additionally increase the difficulty of reaching the required convergence criteria.

A comprehensive description of numerical solution methods and involved mathematics is given in Ref. [130].

3. Experimental Methods

This chapter addresses the experimental methods used in this work. The description of single-cell design and material composition is followed by the presentation of the different measurement setups used to record the complex cell impedance by electrochemical impedance spectroscopy (EIS), current/voltage characteristics (C/V) and gas conversion measurements. The next subsection describes the algorithm to characterize the single-cell, porous electrode microstructure. By combining FIB (focused ion beam) and SEM (scanning electrode microscopy) tomography techniques a digital electrode reconstruction in 3D can be generated, which allows the determination of characteristic electrode parameters with extremely high precision. The results obtained by an in-house code are further strengthened by additional results obtained from the commercial software-package GeoDict[1]. In the last section of this chapter the accuracy aspects for the described experimental methods are discussed.

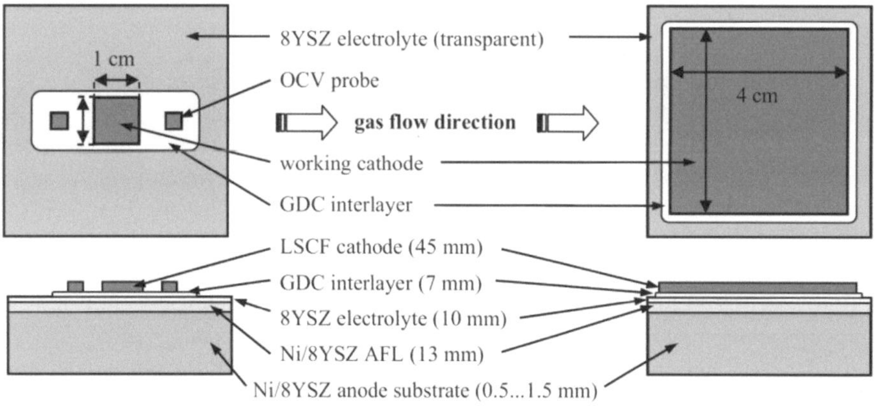

Figure 3.1.: Schematic of SOFC single-cell geometry (not to scale) used in this thesis: ASC with 1 cm^2 active electrode area (A_1, left) and with 16 cm^2 active electrode area (A_{16}, right).

[1]https://www.geodict.com/

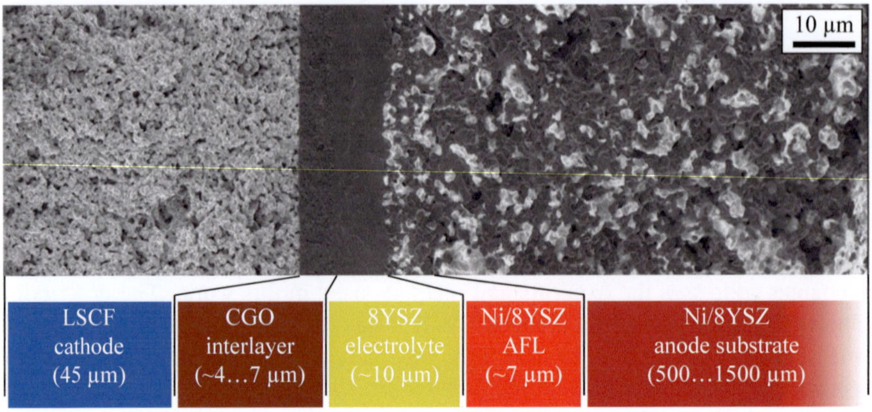

Figure 3.2.: Scanning electron micrograph of a cross-section from a fractured ASC, fabricated at Forschungszentrum Jülich. The image was captured with an Everhart-Thornley detector (ETD, or SE2 detector) at 1.3 kV and with a magnification of 5000.

3.1. Cell Samples

The solid oxide fuel cell (SOFC) single-cells analyzed in the scope of this thesis were fabricated by Forschungszentrum Jülich. The state of the art, planar anode supported cells (ASC) are among the best performing and (more importantly) most stable SOFC single cells currently available. Figure 3.1 displays the experimental layout of the membrane electrode assembly (MEA).

The single cells are supported on 50×50 mm^2 Ni/8YSZ anode substrate with a variable thickness of 500...1000 μm. On top of the tape casted substrate, the Ni/8YSZ anode functional layer (AFL) is deposited with an approximate thickness of $\sim$7 μm, followed by an approximately $\sim$10 μm thick 8YSZ-electrolyte. Both layers are applied via vacuum slip casting or screen printing with an subsequent co-firing step at 1400 °C. In the next step, the $Ce_{0.8}CG_{0.2}O_{2-\delta}$ interlayer (GDC) is screen printed on top of the electrolyte with an approximate thickness of $5-7$ μm [131] and sintered at 1300 °C.The GDC- interlayer functions as a barrier to prevent chemical reactions between the LSCF cathode and 8YSZ electrolyte, which (if allowed) would result in an insulating secondary phase formation (mainly $SrZrO_3$) [39, 70, 72]. A strong impact on the secondary formation and thus on the cell performance is linked to the GDC-interlayer sinter-temperature: a decrease of only 50 K lead to an increase in cell performance loss of up to two orders of magnitude [72]. The mixed-electronic-ionic conductive (MIEC) $La_{0.58}Sr_{0.4}Co_{0.2}Fe_{0.8}O_{3-\delta}$ (LSCF) cathode is deposited as the final layer on top of the GDC-interlayer via screen printing with a resulting thickness of approximately $\sim$45 μm. For the ASCs used in the EIS spectra recordings (A_1, Table 5.1), the cathode area is 10×10 mm^2, thus defining the active electrode area accordingly. This relatively small area combined with high gas flow rates ensures homogeneously distributed and well defined operating conditions in the EIS measurement setup and allows the application of equivalent circuit models (ECM) to recorded EIS. Cells with a larger cathode area of 16 cm^2 (A_{16}, Table 5.1) are regarded as more application oriented, although homogeneous operating conditions cannot be assumed. However, it will be shown in Chapter 6 that a FEM model parametrized with data obtained from small cells (A_1, where the operating conditions can be controlled precisely) is able to reproduced measured cell polarization of large cells (A_{16}) with high accuracy. . In order to monitor the open circuit voltage (OCV) during a measurement, two auxiliary electrodes are placed in front of and behind the cathode in gas flow direction.

Figure 3.2 displays a SEM cross-section of a fractured ASC (Type B, Table 5.1). The porous anode substrate is therein shown partly, whereas the porous AFL, dense electrolyte and GDC-interlayer and porous cathode are displayed in full scale. The image was captured with an Everhart-Thornley detector (ETD, or SE2 detector), which allows a phase separation of the electronic conductive Nickel (light colored phase) to the electrically non-conductive YSZ (gray colored phase) [120]. The pore phase is marked in the anode and cathode by the black phase. Comparing both electrodes qualitatively, it can be stated that in the cathode the pore and material phases are finer and more homogeneously distributed than in the anode. A quantitative comparison will confirm this, the results are given in Section 5.1. Further details about the cell preparation, microstructure and performance of this cell type can also be found in Refs. [132, 133, 134, 135].

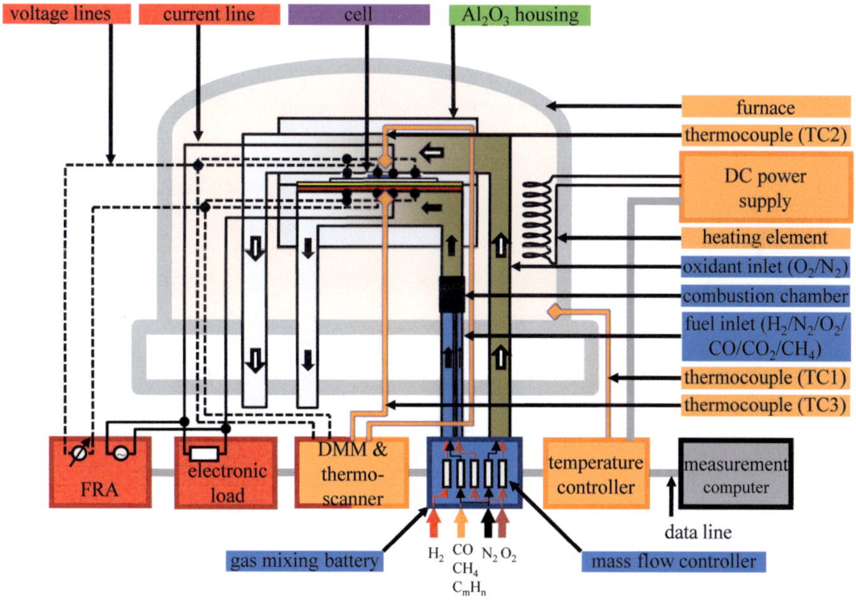

Figure 3.3.: Test bench schematic for electrolyte or anode supported SOFC cells with 1 cm^2 active electrode area (A_1) [136].

3.2. Setup for Electrochemical Testing of SOFCs

3.2.1. Test Bench A - Impedance and Performance Measurements

The single-cells (A_1) shown on the left in Figure 3.1 are mounted into an Al_2O_3 housing, which is schematically illustrated in Figure 3.4 and shown in Figure 3.5d. At the cathode side, two 10×10 mm gold meshes (1024 meshes/cm^2, 0.06 mm wire thickness) are placed above each other. Two nickel meshes are stacked at the anode with the same dimensions as the gold meshes, but with a slightly different texture (3487 meshes/cm^2, 0.065 mm wire thickness).This guarantees good electrical contact. Simultaneously, a homogeneous gas supply from the adjoining gas channels is secured due to the semi-porous and thus permeable texture. A repeating sequence of channels and ribs is carved from an Al_2O_3 block. Figure 3.5a/b detail the periodic flowfield design, called repeat unit (RPU, Section 4.2). A RPU section represents the total structure due to geometrical symmetry reasons, which enables a vast model geometry reduction. A gold seal separates the chambers' gaseous atmospheres and keeps O_2 leakage from cathode to anode side below $>1\%$.

The ceramic housing is constructed within a vertical tube furnace as displayed in Figure 3.5b. Defined and constant operating temperature are guaranteed, which is monitored by thermocouples built in both flowfields and outside the housing. Constant temperature distribution within the housing and thus throughout the entire cell during a measurement enables isothermal conditions in the modeling, even at higher current densities (Chapter 4).

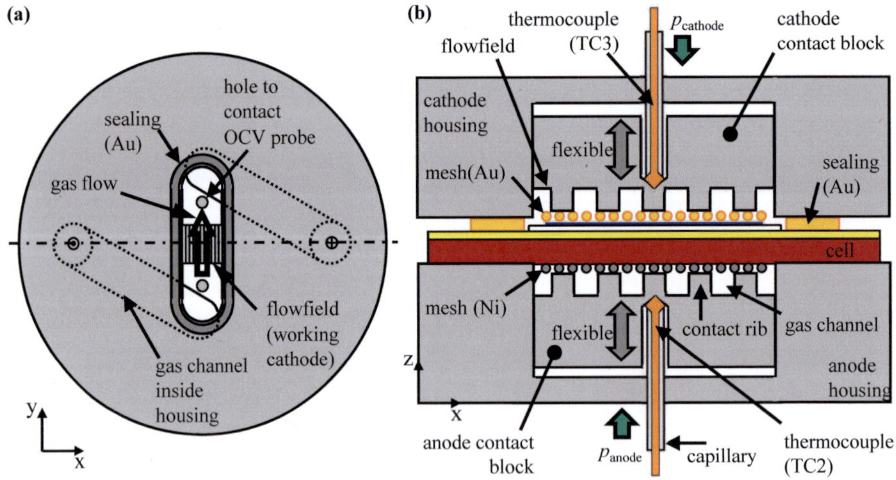

Figure 3.4.: Schematic of the housing: (a) top view of the anode housing, (b) cross-section of the cathode and anode housing with the cell in the center. Gas flow direction is normal to the paper plane [136].

Figure 3.3 schematically illustrates the test bench, which was developed at the Institute of Applied Materials (IAM-WET: Refs. [23, 37, 110, 137]). The test bench operates under ambient pressure, whereby pure O_2, synthetic air (O_2 + N_2) or ambient air can be fed from the gas mixing battery to the cathode side and various gas compositions (H_2, H_2O, CO, CO_2, CH_4 and N_2) to the anode side. Mass flow controllers from Bronkhorst deliver well defined gas composition mixtures with an error estimated $\sim 2\%$. The water vapor content in the fuel gas is realized by mixing H_2 and O_2in the fuel combustion chamber, where a Pt-mesh ensures (even at lower temperatures) a homogeneous and complete reaction to H_2O. The resulting gas mixtures are guided to the housing manifold where they pass through the gas channels of the flowfields in a co-flow manner.

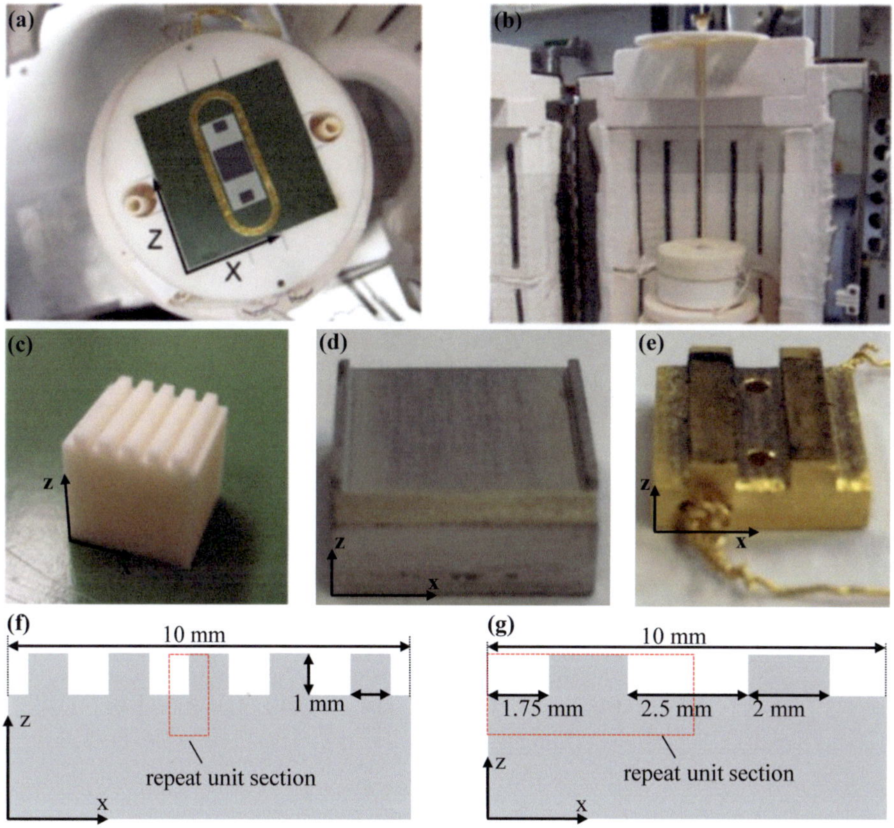

Figure 3.5.: (a) Top view at the bottom half of the ceramic housing with a prepared ASC and gold seal. (b) Side view on merged ceramic housing placed within the tube furnace with the cathode thermocouple already set in position. (c) Photo of a flowfield for 1 cm^2 active electrode area test bench (ideal contacting). (d) Photo of anode flowfield made out of Ni (stack contacting) and (e) associated cathode flowfield made out of Au (stack contacting, wide ribs). (f) Schematic cross section, ideal contacting. (g) Schematic drawing of Au cathode flowfield (Figure 3.5e) with geometric details.

3.2.2. Test Bench B - Performance and Gas Conversion Measurements

The test bench for large cells (A_{16}, Figure 3.1 on the right) basically resembles the test bench for small cells. The schematic in Figure 3.6 show similar system components. While both cell types have the same total cross-sectional area (Figure 3.1), they differ in the active electrode area. Accordingly, larger metallic contact meshes are installed. Furthermore, as illustrated schematically in Figure 3.6, 3 thermocouples are equally spaced in the middle of the anode flowfield along the gas channel length in order to monitor the temperature distribution in greater detail as (in contrast to the small cell test bench) larger temperature gradients manifest. Additionally, the local gas composition can be probed before and after the anode flowfield and at 3 equidistant places along the middle gas channel. The probed gas mixture can be subsequently analyzed with the help of a gas chromatograph.

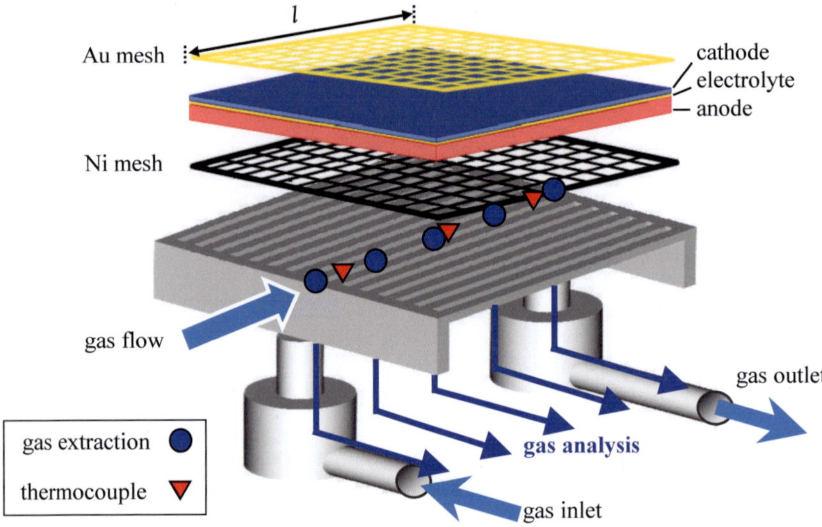

Figure 3.6.: Test bench schematic for SOFCs with 16 cm^2 active electrode area (A_{16}) used in this thesis to record C/V characteristics and to carry out gas conversion measurements. The fuel gas can be probed along the gas channel length at five different positions and its composition determined by a downstream gas chronograph.

3.3. Cell Measurement Details

The measurement protocol for cell characterization measurements with 1 cm^2 active electrode area (A_1) is designed especially for the purpose of determining the required model parameters from an application-oriented point of view. As was shown by Endler-Schuck *et al.* [69, 31], a strong degradation of the cathodic electrochemical reaction takes place in long term measurements (> 1000 h) under ambient air as oxidant, depending on the operation temperature. It was found that at $T = 750$ °C the strong initial degradation phase wore off after ~300 h of polarization and the cathode degradation continued at a near constant, low rate. The underlying mechanism has yet to be fully analyzed, it is assumed however that contaminants in the ambient air (e.g. sulfur and humidity) cause detrimental surface reactions. Nevertheless, it is the goal of this work to propose an application-oriented model framework, therefore such mechanisms have to be included in the model parameter determination in order to be representative for realistic SOFC stack operating conditions. Hence, the testing protocol starts after the heat up and anode reduction phase with an intermediate polarization of $j = 1$ A·cm^{-2} with 60 % humidified H_2 as fuel and ambient air as oxidant at $T = 800$ °C for ~300 h. Every 24h an EIS spectrum and C/V are recorded in order to monitor the cell degradation. Thereafter, the actual characterization protocol is started.

EIS spectra were recorded using a Solartron 1260 frequency response analyzer (FRA) in the frequency range from 10 mHz up to 1 MHz. A maximum voltage response of 12 mV is realized by a variable current stimulus, depending on the internal cell resistance at the set operating conditions (ranging from 75 m$\Omega$$cm^2$ at 900°C down to 10.4 $\Omega$$cm^2$ at 550°C for the cell types investigated). All impedance measurements were recorded at OCV with the exception of the spectra recorded to determine the charge transfer coefficients α_{an} and α_{cat} (Section 5.2.1, Ref. [9]), which were carried out under various current densities.

C/Vs were recorded in 200 mAcm−2 steps up to 2 Acm^{-2} with a pause of 20 sec at each current density step in order to give the system time to reach equilibrium. Furthermore, it was ensured that the cell voltage never dropped below $U_{cell} = 0.6$V to prevents irreversible cell damage due to NiO-formation. In this way, stable cell performance can be guaranteed at all throughout the whole experiment.

The oxidant and fuel gas composition are fed to the small cell test bench with a total gas flow rate of 250 sccm for each gas mixture. These relatively high rates eliminate the influence of gas conversion along the gas channels due to the electrochemical reaction, at least for EIS measurements under OCV. The results presented in Section 6.5 will show however, that at higher current densities, a certain gradient along the gas channel length and perpendicular to the gas flow has to be accounted for in the modeling.
In the test bench for the large cells, the gas flow is set to 1000 sccm at the cathode side and depending on the desired operating conditions to 300 – 1000 sccm at the anode side. In order to measure CV-curves under realistic conditions, the fuel utilization (f.u.) is controlled by lowering the fuel gas rate accordingly.

Two anode side fuel mixtures in various compositions were considered, which in this work are referred to as follows:

1. **Hydrogen operation:** Mixture of H_2, H_2O and N_2,
2. **Reformate operation:** Mixture of H_2, H_2O, CO, CO_2 and N_2,

To analyze the gas composition at the various gas extraction points along the fuel gas channel (Figure 3.6), a micro gas chromatograph (μGC) CP4900 from Varian was used. In order to have a representative mixture, the bypassing fuel gas is suctioned for 30 min from each extraction point and subsequently analyzed via a μGC [138].

The high operating temperatures and catalytic activity of Ni within the anode convert reformate fuels of arbitrary composition via the water-gas shift reaction (Equation 2.17) until equilibrium is reached. The same accounts for methane operation according the reaction mechanism of Equation 2.16. For the C/V measurements a realistic reformate composition (as produced by a CPOX-reformer) is used in order to achieve application oriented operating conditions on the anode side. The operating temperature was thereby kept above 700°C to avoid the formation of solid carbon (coking). The gas composition for C/V measurements under methane operation was set to a H/C-ratio of 2. In the gas conversion measurements to determine the WGS reaction kinetics, a gas mixture of CO and H_2O was fed to the cell with equal gas flow rates.

The fuel gas water vapor depends on the set O_2 flow delivered into the combustion chamber, the purity of the fed H_2, and the gas leakage from the cathode side.The latter value depends on the gas tightness of the electrolyte and gold seal, as well as on the manufacturing precision of the ceramic housing. The electrolyte gas leakage is tested in Jülich before applying the GDC-interlayer, and usually lies below 10^{-5} hPa dm^3 s^{-1} cm^2 (otherwise they are rejected). The influence of gold seal and ceramic housing cannot be separated, it is however wise to change the gold seal after some time. The real water vapor content in the anode gas atmosphere can be calculated by measuring the OCV for pure H_2-operation and using Equation 2.5. This procedure allows very precise anode gas atmosphere control at every operating point. Hence, highly accurate electrode characterization via impedance measurements are possible as each electrode exhibit a high sensitivity to applied gas composition.

4. Description of Finite Element Method Model Framework for SOFCs

It is common practice in academic research and in commercial research and development (R&D) to work with model concepts as a way of solving physically process related problems. The goal is to develop a deeper understanding of occurring processes and, if possible, to minimize thereby caused loss. Multiple physical processes occur simultaneously during SOFC/SOEC operation and therefore effect each other), so this provides an excellent example for a model-based approach. In this work, the finite element method (FEM, Section 2.7)) is employed in order to include not only material parameters and varying model equations but also geometric dependencies in the analysis of occurring loss processes.

This chapter is dedicated to describing the developed FEM model framework in detail. At first, a short overview about various model approaches provided by SOFC-literature is given with the focus set on level of detail and length scale. Based on that knowledge, the different applied model geometries applied in this work are presented with a description of included SOFC components. The main part of this chapter follows the implementation description of in the model regarded loss processes by providing used model equations and parameters. A special focus is set on gaseous species transport models used in literature. Inevitably required boundary conditions are subsequently given after each loss process description. The final section explains model framework function principles to calculate stationary performance predictions and a time-dependent solution to simulate a complex impedance response.

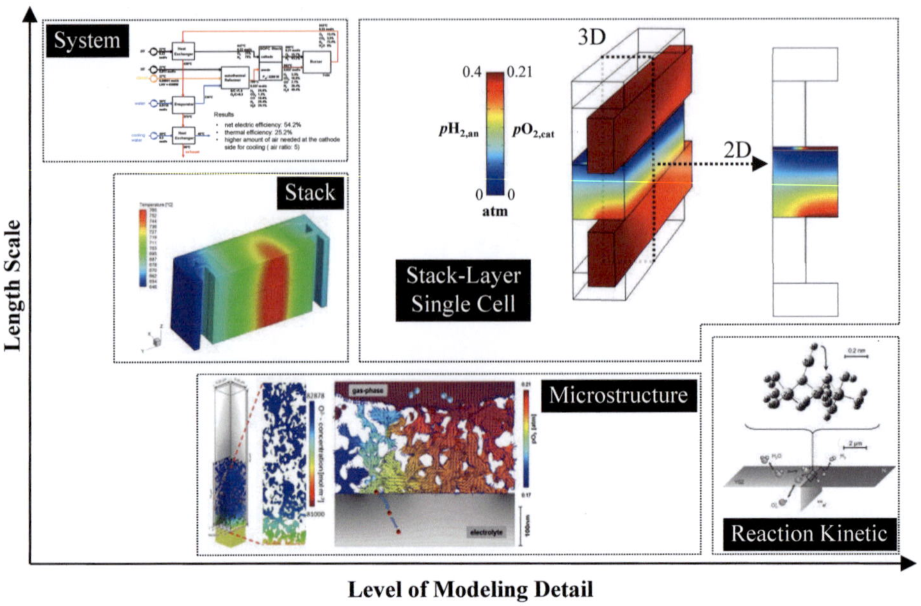

Figure 4.1.: Schematic illustration of different degrees of detail in fuel cell modeling in relation to the considered length scale.

4.1. Overview about Model Approaches in SOFC Research

Model approaches have become an integral part of modern research and engineering, especially with the vast computational power and memory supplied by state-of-the-art workstations. It is possible to simulate highly coupled multi-physic phenomena in great detail and gather information about their physical state at any desired spatial location under arbitrary chosen operating conditions, where comparable experimental work would exceed the necessary effort in terms of time, cost and realization. However, certain trade-offs with respect to modeling detail have to be made due to the accruing computational cost. Depending on the intended purpose, it is appropriate to simplify or disregard certain phenomena in a model and therefore decrease its complexity or to choose a different length scale in order to minimize the computational effort. Figure 4.1 displays a reasonable classification for SOFC modeling scales, ranging from complex system analysis down to models dealing with elementary reaction kinetics at atomic level. As modeling detail level increases with decreasing length scale, researchers choose an appropriate level depending on the designated purpose.

At *system level* (Section 2.4.3), the SOFC stack is one of many parts in an assembly of various components, forming a complex system to either supply heat and power (SOFC-mode) or to generate fuel (SOEC-mode). Such models are employed to analyze the dynamic interactions of regarded components with the goal of identifying system limitations caused by individual components and to find potential for further optimization. A literature example can be found under [139]. At this level of detail, the complete fuel cell stack is usually implemented as a rather simple performance model without deeper physical meaning regarding occurring losses within the stack.

A length scale smaller, at *stack level*, the modeling focuses on the whole SOFC stack unit, composed of its various components (Section 2.4.2). These models usually feature a 3D resolution of individual stack layers and gas supply. The model objective is to deliver temperature and mechanical stress distributions in sensitive areas such as at fuel inlets and sealing between the individual layers. A validation can be provided by comparing simulated and experimental temperature distributions within the stack, measured at various points in a specially prepared unit. Extensive and excellent work with extension to other SOFC-system components (reformer) and to system modeling has been carried out and published by Forschungszentrum Jülich [80, 140, 141, 142, 143]. As the scales of various stack components vary over several orders of magnitude, it is not expedient in terms of computational cost to resolve each component with equal detail. The solution is to homogenize various parts or the complete stack (Section 5.1.5). Hence, fitting parameters are required, especially in the description of electrochemical reactions, because the low spatial resolution at the charge transfer interfaces allows no realistic reaction kinetic description.

At *cell level*, numerous models exist in the literature, covering different aspects with regard to geometric dimensions and implemented physical processes. Models of this length scale are also referred to as *meso*-scale models, because they incorporate processes on macro-scale (convection of temperature, flow) and on micro-scale (gaseous diffusion, electrical/ionic conduction, reforming and electrochemical reactions). The latter are regarded by means of effective transport or reaction parameters, based on the homogenization of microstructure properties (Section 5.1.5). In the following only a few of the many approaches available in the literature are described, which have been selected as representative for their individual case.

By exploiting symmetry features, three-dimensional (3D) Repeat Unit (RPU) models cover a section of a complete stack layer (Section 4.2), and are thus able to spatially resolve species and temperature gradients in gas channels and cell layers. Electrochemical reactions are implemented by a 2D boundary-reaction approach. Literature examples for such a comprehensive model are presented by Andersson *et al.* [144] or Lin *et al.* [145]. Both works have a certain weakness in the gaseous species transport and fuel oxidation modeling, and in the fitting of sensitive modeling parameters, and therefore don't give a proper validation or further application-oriented results.

O. Deutschmann *et al.* [102] presented a pseudo three-dimensional model, by using a full 3D geometry, but restricting species transport flux description to 1D in channel flow direction and perpendicular in the homogenized porous electrode layers. Nevertheless, all relevant loss mechanisms are included and valuable results are obtained regarding fuel conversion via reforming reactions in the fuel electrode, based on a detailed elementary step kinetic reaction mechanism and the Dusty-Gas Model (DGM, Section 4.3.2.3) as gaseous species transport model. Due to the pseudo-3D approach, the model is limited in the prediction of species distribution under high fuel utilizations, where large species gradients along the gas channel direction govern electrode species fluxes. A profound model validation was also not presented. Another example of such a pseudo-3D model worth mentioning and is provided

by Bessler *et al.* [146].
Models restricted to a 2D gas channel geometry (2D GCh, Section 4.2) and with only hydrogen as fuel gas are provided among others by Bertei [147] and Tseronis [148]. In both works the DGM is used as gas transport model, although adopting several modeling parameters from the literature and lacking a validation. At least Bertei provides in Ref. [149] a numerical algorithm to predict porous microstructure parameters rather than just assuming the data validity (which is the case with Tseronis). 2D half-cell models, describing the anode fuel gas transport in gas channel and electrode are helpful for investigating reforming kinetic approaches [26] or further analyzing gas conversion effects under high fuel consumption [150, 151, 152].
Only very few works can be found in SOFC-literature regarding models with a 2D-RPU cross-section (Section 2.15), developed with the intent to analyze the impact of interconnector (IC) geometry and cell design on SOFC stack performance. Early work in this field was provided by Tanner and Virkar [153] and later on by Lin *et al.* [154, 155]. Both models lack a rigorous modeling approach or proper validation.

In between cell and microstructure models, 0/1-D models may be included. They function as segregated sub-models, incorporated into larger cell/stack approaches or as stand-alone models, but with corresponding restrictions due to their geometric limitations. Good examples are from A. Leonide [9] and in [17, 29]. Another example for 0-D modeling approaches is given by the equivalent circuit model (ECM, Section 2.3.2)), used to quantify individual loss processes incorporated in electrochemical impedance spectra [28, 36].

Real *microstructure models* on the next smaller length scale can only regard a cell fraction, usually restricted to a single electrode. To completely resolve the porous microstructure network an enormous amount of computational effort is required. Nevertheless, it is more than worthwhile, as with the help of such models it is possible to determine microstructure parameters with high precision (Section 5.1), thus supplying homogenized models with the required parameter input. Also, they give an interesting view on local species distribution at the electrochemically reaction zones. Very comprehensive work in this field was carried out by J. Joos and T. Carraro for both anode and cathode [120, 156, 157, 158, 159, 160]. At the smallest modeling scale, reaction kinetics are investigated by subdividing global reaction approaches into elementary steps with the goal of singling out rate determining steps [161, 162, 163, 164, 165].

In this context, the here presented FEM model framework is ranked in the *cell* length scale. Each step is presented in the following.

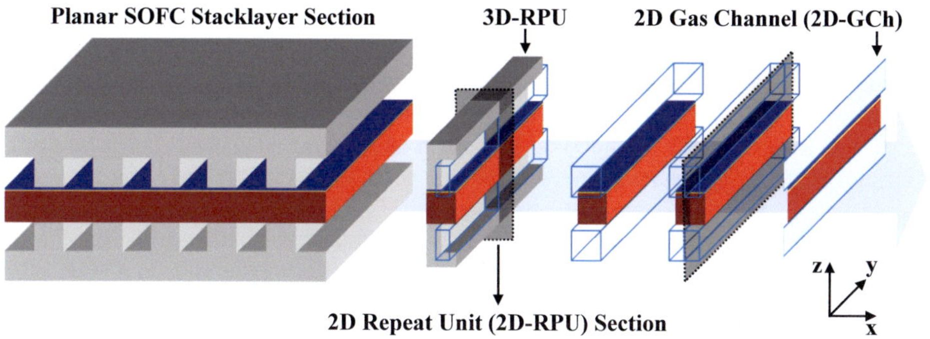

Figure 4.2.: Breakdown of FEM model geometry development.

4.2. Model Geometries

At first, a model geometry has to be defined on which the model framework can be applied. As described in the previous section, models are classified by different length scales. In the herein presented model, geometries rank in the *cell* length scale, built up of different sections from a single anode supported SOFC (ASC) with *ideal contacting* as in the laboratory (Section 2.5.2) and *stack contacting* as in a real, planar SOFC stack layer (Section 2.5.3). Figure 4.2 illustrates schematically from left to right a stepwise model geometry reduction from a complete stack layer down to various 2-/3D geometries. The underlying idea is to exploit symmetry features inherent in the planar stack design with its periodical array of gas channels and contact ribs and minimize thereby the required computational effort in the numerical calculation. Such a recurrent design is also called a *repeat unit* and a section consisting of half a gas channel and half of the IC contact rib with the corresponding ASC part in between is consequently called the RPU section (in the following called RPU). The RPU exactly realizes aforementioned symmetry features and enables a drastic geometry reduction. Basically, an RPU can be found anywhere within a single planar stack layer as part of the whole stack array (Figures 2.9 and 2.10). However, in order to assume a constant gas flow at the gas channel inlets and no external temperature gradients, the RPU is considered to be situated in the middle of a stack layer, which itself is in the middle of the whole stack.

Figure 4.2 further illustrates how the 2D geometries are derived from the *3D-RPU*. While the *2D-RPU* geometry represents a cross section of the 3D geometry and therefore only a slice anywhere along the gas channel length, describes the *2D Gas Channel* (GCh) geometry a longitudinal section along the channel. This results in certain advantages and disadvantages for each 2D model geometry:

The 2D-RPU cannot regard species transport in gas channel direction and is limited in the description of occurring chemical reforming reactions (Section 2.4 and 6.3) and gas conversion due to electrochemical fuel consumption (Section 6.5). On the contrary, it is possible to neglect temperature gradients and adopt an isothermal modeling approach, because larger thermal gradients are only expected along the fuel gas channel under high fuel utilizations or strong reforming reaction activity. Additionally, the lack of gas channels simplifies the required model equations and drastically reduces the computational effort (compared to models with gas flow). The 2D-GCh on the other hand can only regard species transport along the gas channel direction and not perpendicularly into the stack layer depth beneath the IC flowfields contact ribs, for example. As mentioned, modeling of gas flow necessitates higher computational demand, which is nonetheless still considerably lower than a full 3D-RPU model. At last, the 2D-GCh gas channel heights have to be divided by half, as illustrated in Figure 4.2 on the right. If the 2D model is supposed to represent the same cell area in depth as the 3D-RPU, the contact rib area has to be taken into account (Section 6.5).

Illustrated in Figure 4.3 is the detailed description of model geometries applied in this work . Relevant components are represented by computational domains (D), which are spatially denoted by width (b) and height (h), and by length (l) where needed. To supply a better overview and to minimize duplications, it is refrained from labeling all relevant domains and interfaces (IF) in each geometry. For example, only a detailed domain description with corresponding ASC interfaces is given in Figure 4.3c at the 2D-RPU geometry. Nevertheless, the notation accounts for all geometries. Figure 4.3a-c describe the various model geometry sections of a single ASC, built in the laboratory test bench with *ideal contact* (Section 2.5.2)and features contact meshes, accordingly, but no IC ribs. Figure 4.3d illustrates the *2D-RPU Stack Layer Model* geometry, without contact mesh and anode flowfield, but with additional current collector layer (CCL) and IC geometry. It represents the RPU section out of a planar stack layer (Section 2.5.3).

Geometric details for the various ASCs in ideal and stack contacting models are given in Table 5.1. Test bench flowfield details can be found in Figure 3.5, while stack flowfield dimensions are deduced from Figure 2.14. Contact mesh dimensions are described in Section 5.4.

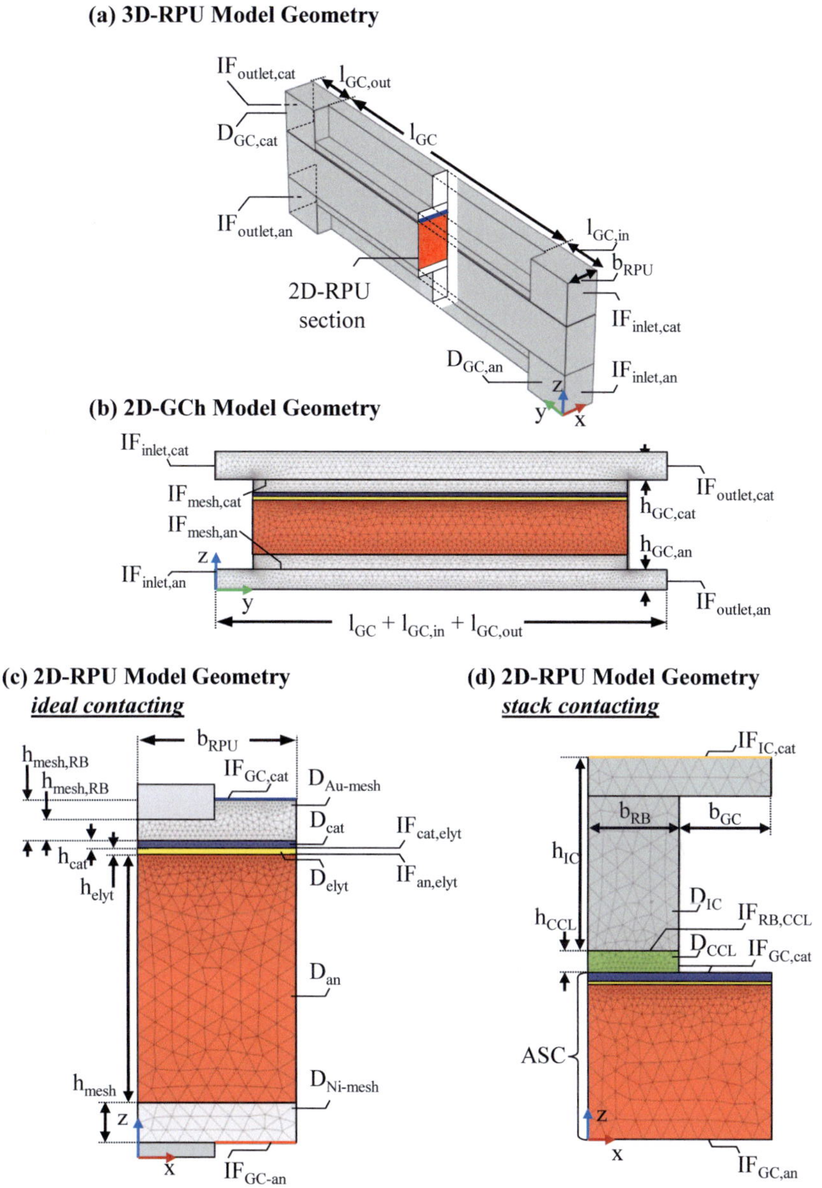

Figure 4.3.: Overview of FEM model geometries (and 2D numerical meshes) applied within this work. Individual domains (D), interfaces (IF), layer thicknesses (h) and layer widths (b) are labeled accordingly. Coloring: cathode (cat, blue), current collector layer (CCL, light blue), electrolyte (elyt, yellow), anode (an, red). (a) 3D Repeat Unit (RPU) geometry (*ideal contacting*, Section 2.5.2). (b) 2D Gas Channel (GCh) geometry: Longitudinal section (y-plane). (c) 2D Repeat Unit (RPU) geometry: Perpendicular cross section (z-plane). (d) 2D SOFC Stack Layer (SL) geometry (*stack contacting*): Perpendicular cross section (z-plane) of a planar SOFC stack layer (Section 2.5.3).

4.3. Loss Process Modeling

4.3.1. Electric and Ionic Charge Carrier Transport

This section describes how electric and ionic charge carrier transport is implemented in the model framework.

During operation, electrons (e^-) are transported in the anodes Ni-phase, in the cathodic electronically conducting material phase and in the cell-contacting (ideal contacting: mesh, stack contacting: interconnector). Within the electrolyte and interdiffusion-layer, oxygen-ions (O^{2-}) are the charge carrier.

When modeling a mixed-ionic-electronic conductive (MIEC) cathode, the cathodes material phase is equally capable of transporting electrons and ions (Section 2.4.1). Regardless, both transport mechanisms are non-ideal and driven by a charge carrier flux, $\vec{j}$, denoted as current density in A/m^2. It describes a relation derived from the fundamental Maxwell's Equations, also known as Ohm's Law:

$$\vec{j} = \left(\sigma_i + \epsilon_0 \epsilon_\mathrm{r} \frac{\delta}{\delta t} \right) \vec{E} \qquad \mathrm{A/m^2}. \tag{4.1}$$

Herein σ_i denotes the bulk conductivity in S/m set to material or computational domain i, ϵ_0 the permittivity in vacuum in F/m, ϵ_r the relative, material dependent permittivity in F/m and $\vec{E}$ the electric field vector in V/m. In the stationary case, the time dependent term vanishes. The electric field, $\vec{E}$, describes the force acting on the charge carrier and determines the direction of current transport. The electric/ionic potential $\phi_{ec/io}$ (ec: electronic, io: ionic) describes the potential energy of each charge carrier within an electric field, $\vec{E}$, given in its scalar form in V by $\phi_\mathrm{ec/io}$ and related to $\vec{E}$ by its gradient:

$$\vec{E} = -\nabla \phi_\mathrm{ec/io} \qquad \mathrm{V/m}. \tag{4.2}$$

The equation of continuity is the last required notation to describe the charge carrier transport problem fully:

$$\nabla \cdot \vec{j} = Q_i \qquad \mathrm{A/m^3} \tag{4.3}$$

wherein Q_i denotes a current source.

The required material parameter for a stationary simulation is σ_i assigned to the individual component represented by their computational domain and additionally for time-dependent solutions ϵ_r. While for the electrodes, $\sigma_\mathrm{an/cat}$ are determined experimentally from 4-point DC-measurements (Sections 5.3.1 and 5.3.2 and Ref. [68]) are values for σ_elyt determined from impedance measurements (Section 5.3.3). The conductivities of Ni and Au (contact-meshes, σ_mh) and for σ_IC (CroferAPU22) are adopted from literature [166]. An overview of set initial and boundary conditions is given in Tables 4.1 and 4.2.

4.3.1.1. Contact Resistance

For models regarding stack-contacting simulations (Section 2.5), non-ideal contacting between IC contact rib and cathode has to be assumed [167]Additionally, protective coatings on the IC designed against oxide scale growth and chromium evaporation also further lower the electronic conductivity [37, 85]. This is realized by implementing an area specific contact resistance ASR_{contact} between:

$$n \cdot \vec{j}_{\text{IC}} = \frac{1}{ASR_{\text{contact}}}(\phi_{\text{IC}} - \phi_{\text{cat}}) = -n \cdot \vec{j}_{\text{cat}} \qquad \text{A/cm}^2. \tag{4.4}$$

Values for ASR_{contact} are highly dependent on the applied coatings, operating temperature and time, and are very difficult to determine [168]. For LSCF as contact layer material, values have been experimentally determined at Forschungszentrum Jülich (Section 5.6, Tab. 5.7). Over time the oxide scale growth increases ASR_{contact} further, which can elevate the induced ohmic loss to become the predominant loss. The influence of ASR_{contact} on the remaining transport processes is further discussed in Section 7.3.2.

4.3.1.2. Domain and Boundary Condition Settings

The charge carrier transport modeling via Equation 4.1 requires parameter, initial value and boundary condition setting. In order to realize a continuous charge transport by regarding the electrochemical charge transfer reaction between electronic and ionic conducting phases, each phase is modeled in its own physical mode (electronic/ionic) with an extra set of charge transport equations (Equation 4.1). The charge transfer reactions and interconnection between both phases is realized by implementing global boundary conditions, which prevail in both physical modes simultaneously. Furthermore, isotropic distribution of material inherent conductivity is assumed.

The anodic electrochemical charge transfer reaction $\text{IF}_{\text{an-elyt}}$ is described by the Butler-Volmer approach (BVM, Section 4.3.3.1)), while at the cathode one can chose between BVM or a MIEC-cathode approach (Section 4.3.3.2). Hence, in Table 4.1 domain and initial value and in Table 4.2 boundary condition settings are listed according to the geometry illustration in Figure 4.3. Depending on the desired operation mode prediction (potentiostatic/galvanstatic), individual settings for U_{cell} or j_{cell} have to be chosen.

Table 4.1.: Domain parameter and initial condition settings for calculating electronic/ionic charge transport according to Equation 4.1.

Domain	Parameter	Variable / Setting	Initial Value	Comment
D_{IC}	$\sigma_{EC,IC}$	$\sigma_{Crofer22APU}$	$U_{occ}(T)$	Sec. 5.3
$D_{Au-mesh}$	$\sigma_{Au-mesh}$	$\psi_{mesh,mat} \cdot \sigma_{Au}$	$U_{occ}(T)$	Sec. 5.4, Tab. 5.6
D_{cat} / D_{CCL}	σ_{cat}	σ_{LSCF}^{eff}	$U_{occ}(T)$	Sec. 5.3.2, Eq. 5.27
		$\psi_{cat,mat} \cdot \sigma_{cat}$	$U_{occ}(T)$	Sec. 5.3.2, Tab. 5.4
	Q_{MIEC}	$a_{MIEC} \cdot k_{MIEC}^{\delta} \cdot (c_{eq} - c_O) \cdot 2F$		Sec. 5.2.2, Tab. 5.3
D_{elyt}	σ_{elyt}	$\sigma_{8YSZ/CGO}^{eff}$	0 V	Sec. 5.3.3, Eq. 5.28
D_{an}	σ_{an}	σ_{Ni8YSZ}^{eff}	0 V	Sec. 5.3.1, Eq. 5.26
$D_{Ni-mesh}$	$\sigma_{Au-mesh}$	$\psi_{mesh,mat} \cdot \sigma_{Ni}$	0 V	Sec. 5.4, Tab. 5.6

Table 4.2.: Boundary condition settings for charge carrier transport modeling using Equation 4.1. Separate physical modes are implemented (electronic/ionic), which are linked via the global boundary variable at $IF_{an/cat-elyt}$ respectively. Cathodic electrochemical charge transfer can be modeled via the Butler-Volmer (BVM) or MIEC-cathode approach.

Boundary		Physical Mode	Setting	Comment
ideal	stack			
$IF_{mesh-cat}$	IF_{IC-cat}	Electronic (ec)	U_{cell}	potentiostatic operation
			$-j_{cell}$	galvanostatic operation
	IF_{RP-cat}	Electronic (ec)	$ASR_{contact}$	Eq. 4.4
$IF_{cat-elyt}$		Electronic (ec)	$+j_{ct,cat}$	Equations 4.59 and 4.72
		Ionic (io)	$-j_{ct,cat}$	
		Boundary variable	$\eta_{act,cat}$	Equations 4.70 and 4.85
$IF_{an-elyt}$		Ionic (io)	$-j_{ct,an}$	Eqs. 4.59 and 4.60a
		Electronic (ec)	$+j_{ct,an}$	
		Boundary variable	$\eta_{act,an}$	Eq. 4.86
$IF_{mesh-an}$	IF_{GC-an}	Electronic (ec)	0 V	Ground

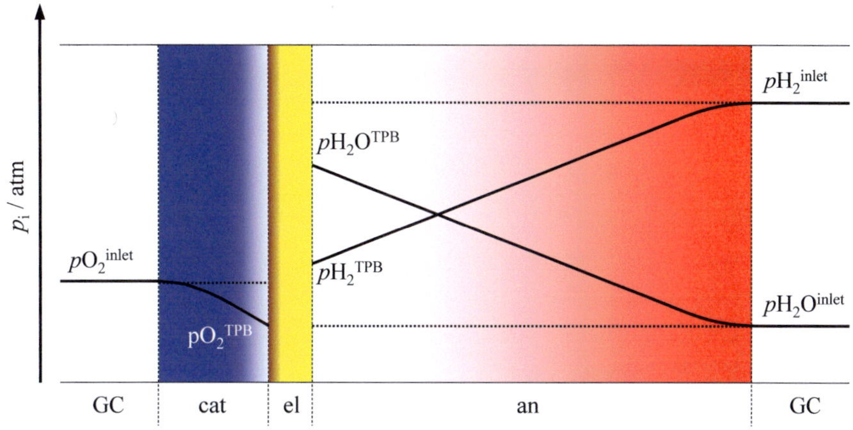

Figure 4.4.: Schematically illustrated partial pressure profiles within the porous electrode structures of an ASC under OCV (dashes lines) and under load in SOFC-mode (solid lines); GC: gas channel, Cat: cathode, El: electrolyte/interdiffusion-layer, An: Anode.

4.3.2. Gaseous Species Transport

This section will explain the modeling of gaseous species transport within the framework. Under *OCV*, the electrochemical reactions are in equilibrium and consequently no significant species concentration (or partial pressure) differences exist between gas channels and electrochemical interfaces (TPB) within the porous electrodes structure. Drawing (SOFC-mode) or applying (SOEC-mode) a current from or to a cell results in electrochemical activity at the TPBs. According to Equations 2.2 and 2.3, current direction reactants are consumed and produced at the TPB. Due to non-ideal gas transport properties within the electrode pore network, an instant supply or removal of these reactants is prevented, resulting in the formation of partial pressure (or concentration) gradients. This is illustrated schematically for steady state SOFC-operation with pure H_2 as fuel in Figure 4.4, where H_2 and O_2 are consumed and H_2O is produced.

In general, the relevant gaseous species transport processes are diffusion and convection, whereas the first process governs gas transport in the porous structure, while the second is dominant in the gas channels and realizes a constant supply and removal of reactants. Even though some approaches found in literature neglect convective transport in the electrodes [169], it is now common practice to include both processes in all regarding modeling domains. A significant error in the predicted losses is even possible, depending on the operating conditions [170]. Hence, gas transport is described in the model framework by a coupled convective and diffusive approach. Furthermore, when hydrocarbons are in the fuel, gas transport pathways in the anode are further complicated due an increased number of mixture components and additional complex coupling with reforming reactions occurring at the TPBs [17, 29, 171] and at the anode/gas channel surface. All this imposes challenging demands on the overall gas transport modeling.

In the following paragraphs, a general overview of relevant conservation laws with required definition is given. Thereafter, a closer look at diffusion and its distinctions is made, as it is the dominant transport mechanism in porous media and more complex to model. This is required to understand the subsequently presented sections, which present modeling equations. Implemented in the framework, these are required to describe gaseous species transport, depending on the various conditions prevalent in the regarding modeling domain:

(i) Gas channels: *convection* and *molecular diffusion* in an open volume,

(ii) Contact-meshes: effective *convection* and *molecular diffusion* in a semi-porous volume.

(iii) Porous Electrodes: effective *convection*, combined *molecular and Knudsen diffusion* in highly porous media.

Acquisition of required modeling parameters is described in Chapter 5 and an overview of initial and boundary condition settings are given in Tables 4.3 and 4.4.

4.3.2.1. Conservation Laws and Definitions

In general, three subjects are involved in gaseous species transport: Fluid dynamics, dealing with *momentum* transport, heat transfer to describe transport of *energy* and *mass* transport of various chemical species. Due to the isothermal approach applied within this work, heat transfer will not be further discussed in the following.

Conservation of Momentum The required balance equation are the *equation of motion* (for momentum balance) and the *equation of continuity* (for mass balance). In principle, both equations are derived by setting up a balance over a small volume element through which the medium (here gas) is flowing. Corresponding partial differential equation notations are then derived by decreasing the volume size until it becomes infinitesimal small [172].

Certain simplifications can be made to both equations, depending on the flow and geometry conditions. To begin with, a fluid or gas flow can be *laminar* or *turbulent*, depending on certain characteristics, which are described by the characteristic *Reynolds number* Re:

$$Re = \frac{\rho \vec{v} L}{\mu}. \tag{4.5}$$

Herein ρ denotes in kg/m^3 the fluids density, $\vec{v}$ in m/s its velocity, L in m a representative length and μ the fluids dynamic viscosity in Pa·s. Basically, it defines the ratio between inertial and viscous forces present in the fluid. At low Reynolds number, viscous forces dominate and dampen out all disturbances, while at high Reynolds numbers non-linear reactions within the flow can occur. For $Re < 2300$ pure laminar flow can be assumed [172], which is clearly the case for for this study, where $Re < 1$ is calculated for all operating conditions.

The dynamic viscosity μ describes a fluid's resistance towards applied shearing force [166]. The following expression is regarded as sufficiently precise to calculate μ for multi-component gas mixtures [166, 173]:

$$\mu_{\mathrm{el}} = \sum_{i=1}^{n} \frac{x_i \mu_i}{\sum_{j=1}^{n} x_j \sigma_{ij}} \qquad \mathrm{Pa} \cdot \mathrm{s} \tag{4.6}$$

with

$$\sigma_{ij} = \frac{\left[1 + \left(\mu_i/\mu_j\right)^{1/2} \left(M_j/M_i\right)^{1/4}\right]^2}{\left[8\left(1 + M_i/M_j\right)\right]^{1/2}} \tag{4.7}$$

and

$$\sigma_{ji} = \frac{\mu_j}{\mu_i} \frac{M_i}{M_j} \sigma_{ij}. \tag{4.8}$$

In Equation 4.6 $x_{i/j}$ denote the mole fractions and $\mu_{i/j}$ the pure viscosities of species i/j, which can be calculated by a polynomial function given in Section C or can be found in Ref. [173]. $M_{i/j}$ denote in mol/kg species i/j molar masses, also given in Section C or in Ref. [166].

Now, having clarified these initial characteristics, the *equation of motion* for laminar and incompressible fluids and without external forces is given by the following expression, which is also known as the (reduced) *Navier-Stoke Equation* [172]:

$$\rho_{el} \frac{\delta \vec{v}}{\delta t} + \rho_{el} (\vec{v} \cdot \nabla) \vec{v} = \nabla \cdot \left[-pI + \mu_{\mathrm{el}} (\nabla \vec{v} + (\nabla \vec{v})^T)\right] \qquad \mathrm{kg/m^3 \cdot s} \tag{4.9}$$

wherein $\vec{v}$ denotes the fluids velocity vector in m/s and ρ_{el} in kg/m^3 the corresponding gas mixtures density (subscript el denotes corresponding electrode; an: anode, cat: cathode).

If methane-steam reforming is regarded in the model (Section 4.3.4), ρ_{an} is not constant in the unit volume due to the increase in molecule number as a result of methane conversion (Equation 2.16). Consequently, Equation 4.9 is written for compressible fluids accordingly:

$$\rho_{\mathrm{an}} \frac{\delta \vec{v}}{\delta t} + \rho_{\mathrm{an}} (\vec{v} \cdot \nabla) \vec{v} = \nabla \cdot \left[-pI + \mu_{\mathrm{an}} (\nabla \vec{v} + (\nabla \vec{v})^T) - \frac{2}{3} \mu_{\mathrm{an}} (\nabla \vec{v})\right] \qquad \mathrm{kg/m^3 \cdot s}. \tag{4.10}$$

Equations 4.9 and 4.10 are applied in the open gas channel modeling domains of the 2D-GCh and 3D-RPU model (Figure 4.3). In the semi-porous contact-mesh domains, the Brinkmann Equation is applied in order to regard the momentum loss due to the semi-porous mesh structure. Derived from Equation 4.9 and neglecting Forchheimer drag and other external forces, it is written as [172]:

$$\frac{\rho_{el}}{\epsilon_{mh}} \left((\vec{v} \cdot \nabla) \frac{\vec{v}}{\epsilon_{mh}} \right) = \nabla \cdot \left[-pI + \frac{\mu_{el}}{\epsilon_{mh}} (\nabla \vec{v} + (\nabla \vec{v})^T) - \frac{2\mu_{el}}{3\epsilon_{mh}} (\nabla \cdot \vec{v}) I \right] - \left(\frac{\mu_{el}}{\kappa_{mh}}\right) \vec{v} \qquad \mathrm{kg/m \cdot s}. \tag{4.11}$$

Therein denoted variables ϵ_{mh} (pore volume fraction) and κ_{mh} (permeability) are calculated as described in Sections 5.1.8 and 5.4.

In the porous electrodes, it is common practice to use Darcy's law to model the mixture's mass averaged velocity $\vec{v}$:

$$\vec{v} = -\frac{\kappa_{el}}{\mu_{el}} \nabla p \qquad \mathrm{m/s}. \tag{4.12}$$

In a molar-based unit system the permeation flux of component i in a porous medium with a multi-component gas mixture can be expressed according to Arnost et al. by [174]:

$$\vec{N}_{\mathrm{i}}^{\mathrm{c}} = -x_{\mathrm{i}} \frac{\kappa_{el} p}{\mu_{el}} \nabla c \qquad \mathrm{mol/m^2 \cdot s}. \tag{4.13}$$

Therein denotes κ_{el} the effective permeability, which is calculated by Equation 4.48 in case of the mean pore transport model (MTPM) (Section 4.3.2.3) or the global permeability κ_{el} is used (Section 5.1.8). An overview of initial and boundary condition settings is given in Tables 4.3 and 4.4.

Mass and Molar Transport Fluxes At this point an important definition of unit convention has to be noted: A fluid's state can be described in *mass units* or *molar units*. Usually, when dealing with the equation of motion, mass units are preferred. However, when dealing with chemical reactions, the molar-unit system is preferred. Therefore, the following definitions apply:

$$\omega_i = \frac{\rho_i}{\rho} \qquad \mathrm{kg/m^3} \tag{4.14}$$

with

$$\rho = \sum_{i=1}^{N} \rho_i \qquad \mathrm{kg/m^3}. \tag{4.15}$$

Thereby ω_i denotes the mass fraction and ρ_i the mass density of species i.

Furthermore the molar concentration, c_i, of species i is defined as the number of moles present in the unit volume:

$$c_i = \frac{\rho_i}{M_i} \qquad \mathrm{mol/m^3}. \tag{4.16}$$

Similarly, a the molar fraction x_i is defined as

$$x_i = \frac{c_i}{c} \tag{4.17}$$

with

$$c = \sum_{i=1}^{N} c_i \qquad \mathrm{mol/m^3}. \tag{4.18}$$

With these definitions, the *mass-average* velocity $\vec{v}$ and *molar-average* velocity $\vec{V}$ are defined as

$$\vec{v} = \frac{\sum_{i=1}^{N} \rho_i \vec{v}_a}{\sum_{i=1}^{N} \rho_i} = \frac{\sum_{i=1}^{N} \rho_i \vec{v}_a}{\rho} = \sum_{i=1}^{N} \omega_i \vec{v}_a \qquad \mathrm{m/s} \tag{4.19}$$

and

$$\vec{V} = \frac{\sum_{i=1}^{N} c_i \vec{v}_a}{\sum_{i=1}^{N} c_i} = \frac{\sum_{i=1}^{N} c_i \vec{v}_a}{c} = \sum_{i=1}^{N} x_i \vec{v}_a \qquad \mathrm{m/s}. \tag{4.20}$$

Please note: $\vec{v}_a$ does not mean the velocity of an individual molecule of species a, rather an average velocity of all individual velocities of species i within the small unit volume. Consequently, $\rho \cdot \vec{v}$ means the local rate by which a unit mass passes through a unit cross section and $c \cdot \vec{V}$ means the local rate by which a unit mole a unit cross section.

The molecular mass flux $\vec{j}_i$ (not to be confused with the current density j) and molar flux, $\vec{J}_i$, of species i can therefore be described as the relative (or *diffusive*) flux per unit area with D_{ij} as diffusion coefficcient (Fick's 1st law of diffusion) [172]:

mass units:
$$\vec{j}_i = \rho_i \left(\vec{v}_a - \vec{v}\right) = -\rho D_{ij} \nabla \omega_i \qquad \mathrm{kg/m^2 \cdot s}, \tag{4.21}$$

molar units:
$$\vec{J}_i = c_i \left(\vec{v}_a - \vec{V}\right) = -c D_{ij} \nabla x_i \qquad \mathrm{mol/m^2 \cdot s}. \tag{4.22}$$

In 4.3.2.2 more details regarding diffusion are given. Using the expressions of Equations 4.21 and 4.22, the *total flux* of combined molecular and convective flux is given by

mass units:
$$\vec{n}_i = \vec{j}_i + \rho_i \vec{v} \qquad \mathrm{kg/m^2 \cdot s}, \tag{4.23}$$

molar units:
$$\vec{N}_i = \vec{J}_i + c_i \vec{V} \qquad \mathrm{mol/m^2 \cdot s}. \tag{4.24}$$

The question may now arise, as to why these conventions are necessary. In gas mixtures, where all contained species have a comparable molecular weight, M_i, the difference between $\vec{v}$ and $\vec{V}$ is negligible and in fact a differentiation is not required. This is holds true i.e. for the cathodic gas composition. At the anode however, where $M_{\mathrm{H_2}}$ is smaller by a factor of 8 than $M_{\mathrm{H_2O}}$, a significant error is made when the species flux equations of both unit systems are mixed. This is common practice in SOFC-literature, where the Dusty-Gas Model is applied [102, 148, 170]. The impact of mixing flux expressions from different unit systems on the species concentration distribution and therefore in gas diffusion losses and ultimately on performance predictions is further analyzed in Section 6.5.

Conservation of Mass / Species With the information given above in mind, and with Equations 4.23 and 4.24, the *equation of continuity* for species i is given for reacting, compressible multi-component gas mixtures of N species by [172]:

mass units:
$$\frac{\delta \rho_i}{\delta t} = -\left(\nabla \cdot \vec{n}_i\right) + s_i = -\left(\nabla \cdot \rho_i \vec{v}\right) - \left(\nabla \cdot \vec{j}_i\right) + s_i \qquad \mathrm{kg/m^2 \cdot s}, \tag{4.25}$$

molar units:
$$\frac{\delta c_i}{\delta t} = -\left(\nabla \cdot \vec{N}_i\right) + S_i = -\left(\nabla \cdot c_i \vec{V}\right) - \left(\nabla \cdot \vec{J}_i\right) + S_i \qquad \mathrm{mol/m^2 \cdot s}. \tag{4.26}$$

Herein s_i denotes the rate of mass production per unit volume due to chemical reactions in kg/m^3·s and S_i the rate of mol production in mol/m^3·s. For models with hydrocarbons in the fuel gas, reforming reactions occur, described by Equations 4.74 and 4.75. It is assumed that no chemical reactions occur in the gas channel volumes. While this is obvious for the air gas channel, it is a simplification for the fuel gas in the presence of hydrocarbons. Depending on the composition, steam reforming and/or water-gas shift reactions may occur every time the required molecules collide with the specific velocity needed to overcome the required activation energy [105]. However, in the standard operating temperature range $T = 600...900$ °C, the pure gas phase reaction rate is several orders of magnitude lower than the catalytic activated reaction rate at the Ni surface in the fuel electrode and can therefore be neglected [17, 28, 103]. Furthermore, it is assumed that the reforming reaction rates are much greater in the porous substrate than in the semi-porous Ni-mesh at the anode and are therefore only implemented in the porous anode domain.

Summing up both Eqs.4.25 and 4.26 for all N species, results in the *equation of continuity for the total gas mixture*:

$$\textit{mass units:} \qquad \frac{\delta \rho}{\delta t} = -\left(\nabla \cdot \rho \vec{v}\right) \qquad \mathrm{kg/m^2 \cdot s}, \tag{4.27}$$

$$\textit{molar units:} \qquad \frac{\delta c}{\delta t} = -\left(\nabla \cdot c\vec{V}\right) \qquad \mathrm{mol/m^2 \cdot s}. \tag{4.28}$$

4.3.2.2. Diffusion

Diffusion describes the random movement of molecules from a higher to lower level of molar or mass concentration, also denoted as relative velocity to the mixtures bulk velocity $\vec{v}$ (or $\vec{V}$, Section 4.3.2.1). Generally, it is described by Fick's 1st law, stated here again in its molar form:

$$\vec{J_i} = D_{ij} \cdot \nabla c_i \qquad \mathrm{mol/m^2 \cdot s}. \tag{4.29}$$

The diffusion coefficient D_{ij} governs the relationship between transport flux $\vec{J_i}$ and the molar concentration gradient ∇c_i. Indeed, D_{ij} is strongly dependent on the mixture's concentration and the transport medium.

In the beginning of this chapter, three types of relevant transport volumes are listed (open, semi-porous and porous), wherein the diffusion mechanisms *molecular diffusion* and *Knudsen diffusion* occur singularly or in combination. Figure 4.5 schematically illustrates how diffusion occurs in dependence to pore diameter, d_{por}.

Molecular Diffusion Pure molecular diffusion (Figure 4.5a) occurs in an open volume and describes how species A interacts with species B on its way from left to right. Each intermolecular collision signifies a momentum loss, whereby D_{ij} depends on the concentration of A and B and their thermodynamic characteristics. In a binary gas mixture, the Chapman-Enskog theory can be used to calculate D_{ij}, assuming spherical gas molecules and based on the kinetic gas theory [166, 172, 175]:

$$D_{ik} = 1.86 \cdot 10^{-3} \cdot T_{\mathrm{k}}^{3/2} \frac{\sqrt{M_i^{-1} + M_k^{-1}}}{p \cdot \sigma_{ik}^2 \cdot \Omega_D} \qquad \mathrm{m^2/s}. \tag{4.30}$$

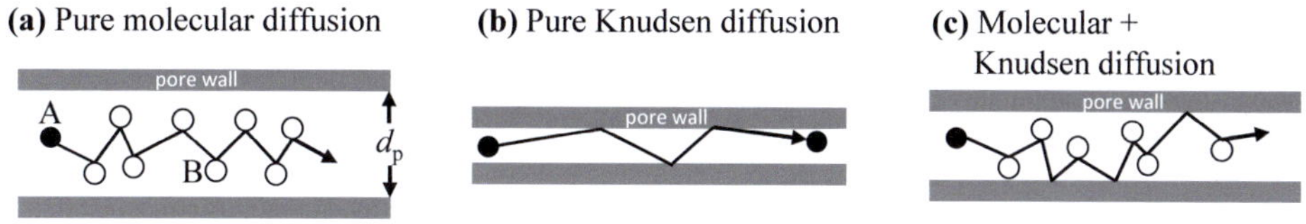

Figure 4.5.: Scematic drawing of gas species diffusion transport mechanism in a cylindrical pore with pore diameter d_{por}. (a) Pure molecular diffusion in a binary gas mixture, (b) pure Knudsen diffusion and (c) combined molecular and Knudsen diffusion in the transition regime.

Therein T_k denotes the absolute temperature in K, $M_{i/k}$ the molar weight of species i and k in kg/mol, p the absolute pressure in atm, σ_{ik}^2 the mean collision diameter (also known as Lennard-Jones-Length) in Å, which is the arithmetic mean of both gas molecules $(\sigma_{ik} = (\sigma_i + \sigma_k)/2)$. Ω_D denotes the temperature dependent collision integral. The values required to calculate D_{ik} are given in [175] and in Section B for the here relevant gas species.

Knudsen Diffusion Reducing the pore diameter down to a scale where the mean free path length λ_l (distance between the collision of two molecules) of a gas component exceeds d_{por}, additional wall collisions become noticeable. Beneath a certain pore wall distance limit, wall interactions become the dominating effect. In that region the resulting species transport is called Knudsen diffusion (Figure 4.5b).

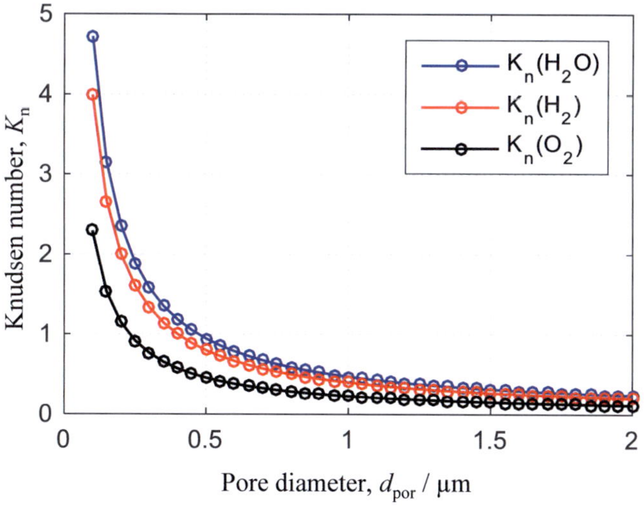

Figure 4.6.: Knudsen numbers K_n for various gas species, calculated using Equation 4.32 at $T = 800$ °C in the displayed pore diameter range d_{por}.

The Knudsen diffusion coefficient is calculated by the following equation [175, 172]:

$$D_{\mathrm{Kn,i}} = \frac{1}{3} d_{\mathrm{por}} \sqrt{\frac{8RT_{\mathrm{k}}}{\pi M_i}} \qquad \mathrm{m^2/s}. \tag{4.31}$$

The Knudsen number K_{n} denotes a measure of how strong the influence of Knudsen diffusion on the overall diffusion behaviour is, and can be calculated by the following relation:

$$K_{\mathrm{n}} = \frac{\lambda_{\mathrm{l}}}{d_{\mathrm{p}}} \tag{4.32}$$

with

$$\lambda_{\mathrm{l}} = \frac{RT_{\mathrm{k}}}{\sqrt{2}\pi d^2 N_{\mathrm{A}} p} \qquad \mathrm{m}. \tag{4.33}$$

where d is the spherical molecule diameter (for which the Lennard-Jones-Length can be adopted), N_{A} Avogadro's number and assuming ideal gas conditions. At $Kn << 1$ pure molecular diffusion prevails, while at $Kn >> 1$ pure Knudsen diffusion is dominant. Using Equation 4.32, the calculated Knudsen numbers for a typical SOFC electrode pore diameter range $d_{\mathrm{por}} = 100 \ldots 1500$ nm and the gas species H_2, H_2O and O_2 are depicted in Figure 4.6. It can be deduced from the figure that, for all gas species, the calculated Knudsen numbers lie in the transition regime where both molecular and Knudsen diffusion occur (Figure 4.5c). Hence, it is essential to choose a porous media gas transport model capable of regarding both diffusion mechanisms appropriately, which is discussed in the next section.

4.3.2.3. Multi-component Porous Media Gas Transport Models

In the following, four of the literature established approaches for multi-component porous media gas transport are presented and discussed. They are in chronological order, which corresponds somewhat naturally with their model implementation complexity. This refinement over time hints towards the effort to derive more precise solution methods. However, a greater model complexity implies increased implementation effort, with more parameters. Hence, the costs have to be weighed carefully against the gain in model precision and the effort to experimentally determine required parameters.

The models here regarded are (i) the *generalized Fick Model* (FM) in mass-based form [172] with a coupled Wilke/Bosanquet-approach to calculate the effective diffusion coefficients [9], (ii) the *Dusty-Gas Model* (DGM) in its originally derived form [176, 177] and in a purely molar-based form [174]. As last model (iv), the molar-based *Mean Transport Pore Model* (MTPM) [174] is described. The differentiation between mass and molar-based forms is a significant point in the validation process (Section 4.3.2.1 and Section 6.5). Other gas transport models frequently mentioned in the literature are the *Cylindrical Pore Interpolation Model* (CPIM) by Young and Todd *et al.* [178] and the *Binary Friction Model* (BFM) by Kerkhof [179]. However, the CPIM model is actually based on the same principles as the MPIM, only the flux density equations are presented in a slightly different way. Kerkhoff introduces within the BFM an additional term in the viscous flux density equation with extra parameters, which are hard to determine experimentally (if assumed, they seem rather to be additional fitting parameters). Moreover, Minkowycz *et al.* have already compared the CPIM, BFM and DGM model in their applicability to SOFC species mass transport modeling and found no advantages compared to the established DGM [180].

The well-known Stefan-Maxwell Model (SMM) [181] is not regarded in this work due to its inability to regard Knudsen diffusion and it is clear from Figure 4.6 that it should be considered for the pore size distribution given by SOFC electrodes. Although it is shown by A. Kromp *et al.* [171, 29], by introducing an experimentally determined factor ξ in the effective diffusion coefficient calculation equation (Equation 5.7), how the impact of Knudsen diffusion on anode diffusion polarization calculation can be regarded by the SMM. However, this work aims to present a more general approach capable to regard not only the anode but also the cathode side.

The porous anode and cathode with additional current contacting layer in the stack layer models (Figure 4.3) are described by their characteristic microstructural parameters (ϵ_{el}, $\tau_{\text{por},el}$) as listed in Table 5.1 for the individual electrodes of the here regarded cell types. With the help of Equations 4.30, 4.31 and 5.7, effective diffusion bulk (D_{ij}^{eff}) and Knudsen diffusion coefficients ($D_{\text{Kn},i}^{\text{eff}}$) are calculated, thus regarding the porous electrodes microstructure adequately (Section 5.1.5). After a short introduction of each transport model, relevant flux equations are described (and when required, certain calculation algorithms are explained).

Fick Model (FM) The most straight forward way to describe multi-component diffusive gas species transport in porous media is to use the well-known (generalized) Fick model [172]. To transform the required binary diffusion coefficient D_{ij} (Equation 4.30) into a multi-component form, the Wilke approach can be applied to calculate a spatially dependent diffusion coefficient [182, 9]:

$$D_{\text{bulk},i} = \frac{1 - x_i}{\sum_{j \neq i} x_i / D_{ik}} \qquad \text{m}^2/\text{s}. \tag{4.34}$$

With a further transformation, known as the Bosanquet approach, Knudsen diffusion is incorporated in the resulting coefficient [183, 184, 9]:

$$D_i = \left(\frac{1}{D_{\text{bulk},i}} + \frac{1}{D_{\text{Kn},i}} \right)^{-1} \qquad \text{m}^2/\text{s}. \tag{4.35}$$

To regard the porous electrodes microstructure, D_i is combined with Equations 4.31 and 5.7 and the following expression states the mass-based form as implemented in the model framework without thermal diffusion [172]:

$$\vec{j}_i = -\rho_{el} D_{i,el}^{\text{eff}} \left(\nabla \omega_i + \omega_i \frac{\nabla M_{el}}{M_{el}} \right) \qquad \text{kg}/\text{m}^2 \cdot \text{s}. \tag{4.36}$$

Herein M_{el} denotes the mean molar mass of the regarding electrodes gas mixture, which is calculated as:

$$M_{el} = x_i \cdot M_i \qquad \text{kg}/\text{mol}. \tag{4.37}$$

Hence, M_{el} is spatially dependent and consequently part of the derivative in Equation 4.36 (which is different to the molar FM). It is further assumed, that only $N-1$ species are independent and the Nth-species is calculated by:

$$\omega_N = 1 - \sum_{i=1}^{N-1} \omega_i \tag{4.38}$$

Inserting Equation 4.36 combined with 4.12 into Equation 4.25 delivers the mass-based transport equation of species i.

Dusty-Gas Diffusion Model (DGM) The Dusty-Gas Model (DGM) was derived by Jackson [185], Mason and Malinauskas [176] to describe species transport of multi-component gas mixtures in porous media with a certain pore diameter length scale, where both ordinary and Knudsen diffusion have to be accounted for (Section 4.3.2.2). Taylor and Krishna [186, 181] summarized the DGM derivation and developed it further by proposing a derivation for an explicit flux expression. As first, Suwanwarangkul *et al.* [169] adopted the DGM to SOFC modeling with an isobaric 1D fuel electrode diffusion model for binary and ternary gas mixtures. Zhu *et al.* [177] presented a 2D cell model with 1D-DGM electrode diffusion equations, incorporating multi-component hydrocarbon containing fuel gas mixtures and associated reforming chemistry reactions. Tseronis *et al.* [187] also presented a 2D cell model with 2D-DGM electrode gas diffusion equations for ternary fuel gas mixtures. Kong *et al.* [155] presented a modified DGM model for binary gas compositions and Bertei *et al.* [170] discussed the importance of regarding pressure gradients. Even though the DGM has been criticized by Kerkhoff [179] and Young & Todd [178], it has proven its applicability [169, 174, 177, 188, 189].

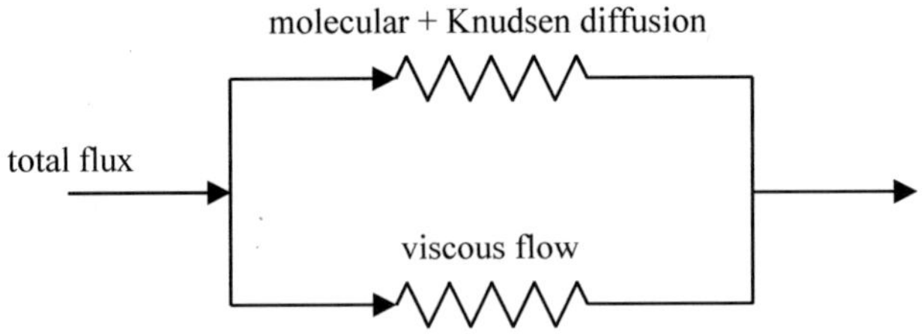

Figure 4.7.: Electric equivalent circuit, depicting the flux of diffusion species within porous medium (adapted from Ref. [186]).

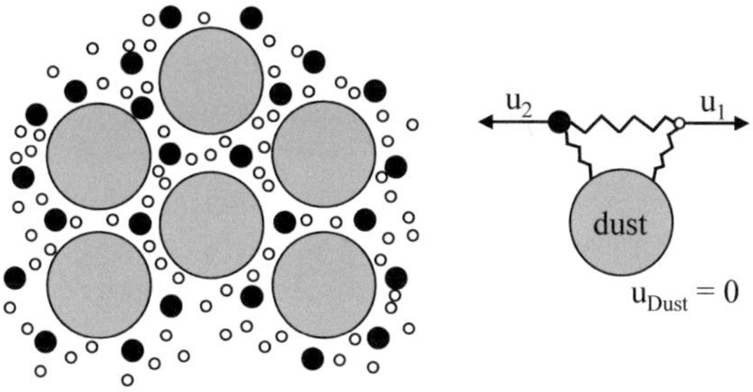

Figure 4.8.: Schematic picture of the dusty gas model in which the pore wall is modeled as giant dust molecules held motionless in space (adapted from Ref. [181]).

The DGM is essentially a further development of the generalized Maxwell-Stefan diffusion equations [181] based on the idea of modeling porous media species flux consisting of parallel occurring fluxes as depicted by the schematic in Figure 4.7. Namely, these are the combined molecular and Knudsen diffusion flux and viscous flow. The key idea of Mason and Malinauskas was to introduce into a N-component mixture a $N + 1^{th}$ pseudo species of giant ("dust") molecules as a surrogate for the porous medium pore walls (Figure 4.8) and assign them certain features: (i) Giant molecules with an infinite mass ($M_{N+1} \rightarrow \infty$), (ii) uniformly distributed in the relevant domain ($\nabla c_{N+1} = 0$) and (iii) held motionless by an unspecified external force ($c_{N+1}F_{N+1} = \nabla p$). Furthermore, in the present case, isothermal conditions, non-polarized species and no impact of external forces are assumed. On this basis (and adopting the generalized Maxwell-Stefan relations, see Jackson [185] and Jakobsen [190] for detailed derivation), the Dusty Gas Model is derived in its generalized form [191]:

$$\sum_{\substack{k=1 \\ k \neq i}}^{N} \frac{x_i \vec{J_k} - x_k \vec{J_i}}{D_{ik}^{\text{eff}}} - \frac{\vec{J_i}}{D_{\text{Kn},i}^{\text{eff}}} = \frac{\nabla p_i}{RT_{\text{k}}} \qquad \text{mol/m}^2 \cdot \text{s}. \tag{4.39}$$

Herein x_i and x_k denote the mole fraction of species i and k, respectively. Furthermore, $\vec{J_i}$ denotes the diffusive flux of component i, including bulk and Knudsen diffusion flux and depending on the individual driving force, expressed by the partial pressure gradient ∇p_i. D_{ik}^{eff} and $D_{\text{Kn},i}^{\text{eff}}$ denote the effective bulk and Knudsen diffusion coefficients, calculated by Equations 4.30 and 4.31 with Equation 5.7.

It is noteworthy that Equation 4.39 is described in molar units, but was derived by mass-based generalized SMM equations. It is further explained, that in Equation 4.39 only diffusive fluxes are considered and not the viscous contribution arising from the "mixture as a whole" [191]. Based on the assumption of parallel occurrence, the total species flux $\vec{N_i}$ is defined by [191]:

$$\vec{N_i} = \vec{J_i} + \vec{N_i^{\text{c}}} \qquad \text{mol/m}^2 \cdot \text{s} \tag{4.40}$$

wherein $\vec{N_i^{\text{c}}}$ denotes the viscous flow contribution (or permeation flux) of the relative total mixture movement adding to the total species flux. This assumption is criticized by

Kerkhoff [179], where it is argued that diffusive and viscous flow cannot be regarded as separate, otherwise one overestimates the Knudsen impact on the total flux calculation. Suwanwarangkul *et al.* [169] state that viscous flow is too small to have an effect and is therefore negligible. However, Bertei *et al.* [170] showed that this simplification can lead to erroneous model predictions and the viscous flux component should be regarded. Krishna [191] implemented the Darcy equation (Equation 4.12) for $\vec{N}_i^{\mathrm{c}}$, which was then adopted by Zhu *et al.* [177], Tseronis *et al.* [148], Kong *et al.* [155] and Bertei [170]:

$$\vec{N}_i^{\mathrm{c}} = -x_i \frac{\kappa_0 p}{\mu_{el} R T_{\mathrm{k}}} \nabla p \qquad \mathrm{mol/m^2 \cdot s} \tag{4.41}$$

wherein κ_0 denotes the characteristic permeability in m^2 (Section 5.1.8). Reshaping Equation 4.40 and inserting the resulting expression into Equation 4.39 delivers for the RHS an expression, where the viscous term $\vec{N}_i^{\mathrm{c}}$ is canceled out in the summation but remains in the second fraction with the Knudsen term. Sorting the total flux expression to one side and the rest to the other side and inserting Equation 4.41 delivers the "working form" of the DGM [191]:

$$\sum_{\substack{k=1 \\ k \neq i}}^{N} \frac{x_k \vec{N}_i - x_i \vec{N}_k}{D_{ik}^{\mathrm{eff}}} - \frac{\vec{N}_i}{D_{\mathrm{Kn},i}^{\mathrm{eff}}} = -\frac{p}{RT_{\mathrm{k}}} \left[\nabla x_i + x_i \left(1 + \frac{\kappa_0 p}{\mu_{el}} \frac{1}{D_{\mathrm{Kn},i}^{\mathrm{eff}}} \right) \frac{\nabla p}{p} \right] \qquad \mathrm{mol/m^2 \cdot s.} \tag{4.42a}$$

This equation is found in one or the other form in mentioned literature sources, whereby Zhu *et al.* [177] used a form with molar concentration gradients and Tseronis et al [148] partial pressure gradients. While this can be attributed to reshaping Equation 4.42a by using the ideal gas law ($p_i = c_i RT$), it is here noted that ∇p is calculated by the Darcy equation (Equation 4.12) and thus in the mass-based reference frame, while $\vec{J}_i$ is based on the molar-based reference frame.

In order to avoid a variable calculation of mixed inertial systems, $\vec{N}_i^{\mathrm{c}}$ (Equation 4.13) was proposed by Arnost and Schneider [174]. The resulting expression of the purely molar-based DGM form is then given by:

$$\sum_{\substack{k=1 \\ k \neq i}}^{N} \frac{x_k \vec{N}_i - x_i \vec{N}_k}{D_{ik}^{\mathrm{eff}}} - \frac{\vec{N}_i}{D_{\mathrm{Kn},i}^{\mathrm{eff}}} = -\left[\nabla c_i + x_i \frac{\kappa_0 p}{\mu_{el}} \frac{1}{D_{\mathrm{Kn},i}^{\mathrm{eff}}} \nabla c \right] \qquad \mathrm{mol/m^2 \cdot s.} \tag{4.42b}$$

A purely mass-based DGM form was proposed by Jacobsen [190], however it delivered unstable calculation results, and so was not further pursued.

The implicit flux formulations in Equation 4.42 complicate an easy numerical solution and cannot be used in commercial software packages, where an explicit flux formulation is often required. Analytical solutions are given for some special cases [185, 176], but cannot function as a general framework. Suwanwarangkul [169] neglects ∇p and by summing up

Equation 4.42a over all N species derives the following expression, also known as Graham's law [181]:

$$\sum_{i=1}^{\mathrm{N}} \vec{N}_i \sqrt{M_i} = 0. \tag{4.43}$$

Inserting Equation 4.43 back into Equation 4.42a delivers an explicit expression for $\vec{J}_i$. However, Equation 4.43 is only true, if all species have the same molecular weight, which is clearly untrue because $M_{\mathrm{H_2}} << M_{\mathrm{H_2O}}$. Hence, the proposed method delivers questionable numerical solutions (as was also shown by Bertei [170]). Fortunately, Krishna [191] proposes an alternative method, which was also described in detail by Jacobsen [190] and used by Zhu [177]. At first Equation 4.42 has to be written in N-dimensional matrix notation, shown here for Equation 4.42b:

$$\begin{pmatrix} \vec{N}_1 \\ \vec{N}_2 \\ \vdots \\ \vec{N}_N \end{pmatrix} = -\left[\vec{B}\right]^{-1} \begin{pmatrix} \nabla c_1 + x_1 \frac{\kappa_0 p}{\mu_{el} D^{eff}_{\mathrm{Kn},1}} \nabla c \\ \nabla c_2 + x_2 \frac{\kappa_0 p}{\mu_{el} D^{e}_{\mathrm{Kn},2}} \nabla c \\ \vdots \\ \nabla c_N + x_N \frac{\kappa_0 p}{\mu_{el} D^{eff}_{\mathrm{Kn},N}} \nabla c \end{pmatrix} \quad \mathrm{mol/m^2 s} \tag{4.44a}$$

whereby the elements of $[\vec{B}]$ are

$$B_{ii} = \frac{1}{D^{eff}_{\mathrm{Kn},i}} + \sum_{\substack{k=1 \\ k \neq i}}^{N} \frac{x_k}{D^{eff}_{ik}} \qquad i = 1, 2, ..., N, \tag{4.44b}$$

$$B_{ik} = -\frac{x_k}{D^{eff}_{ik}} \qquad i, k = 1, 2, ..., N \quad (i \neq k). \tag{4.44c}$$

(4.44d)

The method requires the following assumptions to be taken into account [191]:

(i) The elements of matrix $[\vec{B}]$ are constant along the diffusion pathway and an averaged value $\overline{x}_i$ for the mole fractions must be used in Equations 4.44

(ii) *Linear* composition (and pressure) profile along the length of the diffusion pathway δ:

$$-\delta_i \cong \frac{\Delta c_i}{\delta} + \overline{x}_i \frac{\kappa p}{\mu D^{\mathrm{eff}}_{\mathrm{Kn},i}} \frac{\Delta c}{\delta}. \tag{4.44e}$$

The linearization procedure has been validated by Krishna [191] against published experimental data. Considering a fine numerical mesh where the node distance is analogous to the diffusion pathway length, the assumption is regarded as legitimate. Hence, with this approach it is possible to come up with explicit flux expressions for each species, i, in a N-component mixture, given that adequate computing power is provided due to the required matrix notation in Equation 4.44.

Contrarily to the previous flux models with $N-1$ independent species, Equation 4.44 must be solved for all n species. Therefore, additional constraints must be imposed to ensure $\sum_{i=1}^{N} x_i = 1$, $\sum_{i=1}^{N} c_i = c$ or $\sum_{i=1}^{N} p_i = p$, respectively.

Concluding this section, it is possible to derive an explicit flux expression for multi-component species gas transport in porous media based on the DGM equations. The model complexity is somewhat higher compared to the FM, especially due to the necessary matrix inversion, which is not state-of-the-art in commercial FEM software packages. Another important point is that molar- and mass-based variables are mixed in the literature for viscous flux consideration in the DGM equations. As described in this section, a strict derivation is possible, a physically correct solution depends on the compliance of staying in one inertial system (see Section 6.5).

Mean Transport Pore Diffusion Model (MTPM) This was first suggested by Rothfeld [192] as a transport model for multi-component gas mixtures in porous solids and validated for steady-state with experimental data. Later, Schneider et al. [193, 194, 174] showed in various publications how the MTPM can be applied to determine the microstructural parameters of porous catalysts. Lehnert *et al.* [99] showed with the MTPM that numerically determined microstructure parameters of SOFC anodes correlate well with experimental data (similar to the samples used in this work).

Comparable to the DGM, the MTPM is based on modified SMM equations, whereby the diffusive flux expression is equal, but the viscous flux expression differ: In the DGM equation, viscous flow is described by the standard Darcy expression (4.12) and in the MTPM by a modified darcy expression [174], which includes a term that regards an additional pore wall slip. Unlike in the original DGM derivation by Jackson [185], Mason and Malinauskas [176] and later Krishna [191], the MTPM is kept strictly in the molar unit system.

The total species flux is described as:

$$\vec{N}_i = \vec{N}_i^{\mathrm{c}} + \vec{J}_i \qquad \mathrm{mol/m^2 \cdot s} \tag{4.45}$$

summing up diffusive and permeation flux for all species. The diffusion flux is as mentioned, based on the Maxwell-Stefan equations and is equal to Equation 4.39:

$$\sum_{\substack{k=1 \\ k \neq i}}^{N} \frac{x_k \vec{J}_i - x_i \vec{J}_k}{D_{ik}^{\mathrm{eff}}} + \frac{\vec{N}_i}{D_{\mathrm{Kn},i}^{\mathrm{eff}}} = -\nabla c_i \qquad \mathrm{mol/m^2 \cdot s.} \tag{4.46}$$

The permeation or viscous flux is described with the modified Darcy expression (Equation 4.47), whereby its notation is slightly different:

$$\vec{N}_i^{\mathrm{c}} = -x_i \kappa_i \nabla c \qquad \mathrm{mol/m^2 \cdot s.} \tag{4.47}$$

Here κ_i denotes the effective permeability coefficient of component i [174]:

$$\kappa_i = D_{\mathrm{Kn},i}^{\mathrm{eff}} \frac{\omega \nu_i + K_i}{1 + K_i} + \frac{\Psi_{el} \left\langle r_{\mathrm{por},el}^2 \right\rangle p}{8 \mu_{el}} \qquad \mathrm{m^2} \tag{4.48}$$

with

$$\nu_i = \sqrt{M_i / \sum_{j=1}^{N} x_j M_j} \tag{4.49}$$

and

$$K_i = \frac{\lambda_i}{2 \langle r_{\text{por},el} \rangle}. \tag{4.50}$$

Compared to the DGM (Equation 4.42b), the only difference is the first term on the LHS, which represents a combined contribution to Knudsen and slip flow, while the second term is equal to the Poiseuille-like viscous flow term in the DGM. Thereby ω denotes a numerical coefficient, which depends on the wall-slip description ($\omega = 0.9, \omega = \pi/4, \omega = 3\pi/16$, etc., see [174]). Varying ω in the suggested parameter range did not show a significant impact, wherefore in this work $\omega = 1$ is adopted. λ_i denotes the mean free path length (Equation 4.33). The net molar flux density $\vec{N}$ can be expressed in $n \times n$ matrix notation as well:

$$\vec{N} = [\vec{F}]^{-1} \nabla c \qquad \text{mol/m}^2 \cdot \text{s} \tag{4.51}$$

whereby the $n \times n$ matrix elements f_{ij} are defined as:

$$f_{ii} = \frac{1}{D^{\text{eff}}_{\text{Kn},i}} + \frac{1}{c} \sum_{\substack{k=1 \\ k \neq i}}^{N} \frac{c_k}{D^{\text{eff}}_{ik}} + \frac{c_i}{D^{\text{eff}}_{\text{Kn},i}} \alpha_i, \tag{4.52}$$

$$f_{ik} = -\frac{1}{c} \frac{c_i}{D^{\text{eff}}_{ik}} + \frac{c_i}{D^{\text{eff}}_{\text{Kn},i}} \alpha_i \tag{4.53}$$

and the parameter α_i is

$$\alpha_i = \frac{\left[1 - \frac{\kappa_i}{D^{\text{eff}}_{\text{Kn},i}} + \frac{1}{c} \sum_{\substack{k=1 \\ k \neq i}}^{N} \frac{c_k (\kappa_k - \kappa_i)}{D^{\text{eff}}_{ik}} \right]}{\sum_{k=1}^{N} \frac{c_k \kappa_k)}{D^{\text{eff}}_{\text{Kn},k}}}. \tag{4.54}$$

As mentioned, the MTPM and DGM differ only in their permeation flux coefficient κ_i. For the sake of completeness, α_i to describe the DGM with the above-mentioned model equations is:

$$\alpha_i = -\frac{\kappa_{el} / D^{\text{eff}}_{\text{Kn},i}}{1 + \sum_{k=1}^{N} c_k / D^{\text{e}}_{\text{Kn},k}} \tag{4.55}$$

with

$$\kappa_{el} = \frac{\Psi \left\langle r^2_{\text{por},el} \right\rangle p}{8 \mu_{el}}. \tag{4.56}$$

Combining Eqs. 4.51 with a molar continuity type equation describes combined diffusive and convective multi-component species transport by the MTPM model (or DGM) in porous media (i.e. SOFC anode and cathode):

$$\epsilon_{el}\frac{\delta c}{\delta t} = \nabla\left[F^{-1}\nabla c\right] \qquad \mathrm{mol/m^3\cdot s.} \tag{4.57}$$

4.3.2.4. Domain and Boundary Condition Settings

This section details the domain and boundary condition settings required to describe gaseous species transport based on equations presented in the previous section. Depending on the model geometry, different settings of physical phenomena are required. Models with gas channels (3D-RPU, 2D-GCh, 4.3) require free flow based on Equation 4.9, semi-porous flow based on Equation 4.11 and porous media flow based on Section 4.3.2.3. The 2D-RPU Model does not require free flow model equations, while semi-porous or viscous flow in the meshes is described by Equation 4.12 or Equation 4.13 in combination with a selected porous media diffusion model (Section 4.3.2.3. Regardless of the model geometry, mass or species continuity is secured via Equations 4.25 and 4.27 in the mass-based or via Equations 4.26 and 4.28 in the molar-based description (Section 4.3.2.1).

Table 4.3.: Domain parameter and initial condition setting to calculate electronic/ionic charge transport according to Equation 4.1.

Domain	Parameter	Variable	Initial Value	Reference Comment
$D_{GC/mesh/an/cat}$	ρ	$c_{an/cat}\cdot M_{an/cat}$	-	Eq. 4.18 and 4.37
	μ	$\mu_{an/cat}$	-	Eq. 4.6
	p	-	p_0	-
$D_{GC,an/cat}$	$\vec{v}$	$\vec{v}_{y,an/cat}$	$v_{0,an/cat}$	Eq. 4.58
$D_{mesh,an/cat}$	ϵ	$\epsilon_{mesh,por}$		Sec. 5.4, Tab. 5.6
	$\vec{v}$	$\vec{v}_{y,an/cat}$	0	-
$D_{an/cat/CCL}$	ρ_{Darcy}	$\frac{pM_{an/cat}}{RT}$	-	Eq. 4.37
	ϵ	$\epsilon_{an/cat,por}$	-	Sec. 5.1, Tab. 5.1
	κ	$\kappa_{an/cat}$	-	Sec. 5.1.8, Eq. 5.19
$D_{cat/CCL}$	R_i^s	$-a_{MIEC}\cdot k^{\delta}\cdot(c_{eq}-c_O)\cdot\frac{1}{2}$		Sec. 4.3.3.2 Sec. 5.2.2

Table 4.4.: Boundary condition setting for charge carrier transport modeling using Equation 4.1. Separate physic modes are implemented (electronic/ionic), which are linked via the global boundary variable at $\mathrm{IF}_{\mathrm{cat/an-elyt}}$ respectively. Cathodic electrochemical charge transfer can be modeled via the Butler-Volmer (BVM) or MIEC-cathode approach.

Boundary	Variable / Setting	Comment / Reference
$\mathrm{IF}_{\mathrm{GC,an/cat}}$	$c_i = \frac{x_{i0/\mathrm{GC,an/cat}} \cdot p_0}{RT}$	2D-RPU Model, GC: Eq.6.6
$\mathrm{IF}_{\mathrm{an,elyt}}$	$\vec{n} \cdot \vec{v}_{\mathrm{H_2}} = -\frac{j_{\mathrm{ct,an}}}{2F} \cdot \frac{M_{\mathrm{H_2}}}{\rho_{\mathrm{an}}}$ $\vec{n} \cdot \vec{v}_{\mathrm{H_2O}} = \frac{j_{\mathrm{ct,an}}}{2F} \cdot \frac{M_{\mathrm{H_2O}}}{\rho_{\mathrm{an}}}$	$\mathrm{FM}_{\mathrm{mass-based}}$, Eq. 4.60a
	$S_{\mathrm{H_2}} = -\frac{j_{\mathrm{ct,an}}}{2F}$ $S_{\mathrm{H_2O}} = \frac{j_{\mathrm{ct,an}}}{2F}$	Eq. 4.60a
$\mathrm{IF}_{\mathrm{cat,elyt}}$	$\vec{n} \cdot \vec{v}_{\mathrm{O_2}} = -\frac{j_{\mathrm{ct,cat}}}{4F} \cdot \frac{M_{\mathrm{O_2}}}{\rho_{\mathrm{cat}}}$ $S_{\mathrm{O_2}} = -\frac{j_{\mathrm{ct,cat}}}{4F}$	Eq. 4.60b
Walls	$-\vec{v} \cdot \rho_{\mathrm{an/cat}} = 0$	remaining boundaries

Hence, the domain and initial value in Table 4.1 and the boundary conditions settings in Table 4.2 are all listed according to the geometry illustration in Figure 4.3. Individual settings are chosen depending on the desired operation mode.

The gas channel inlet entrance velocity is calculated by the following relation:

$$\vec{v} = -\left(\dot{v}_0 \cdot \frac{T}{T_{\mathrm{STP}}} \cdot A_{\mathrm{ch}} \cdot \frac{1}{n_{\mathrm{ch}}}\right) \cdot \vec{n} \qquad \mathrm{m/s}. \tag{4.58}$$

Therein, denotes $\dot{v}_0$ the total mass flow rate supplied by the mass flow controllers (Figure 3.4), $T_{\mathrm{STP}} = 273.15\,\mathrm{K}$ is the standard temperature, A_{ch} the gas channel cross sectional area in the applied flowfield with n_{ch} denoting the number of gas channels (Table 4.5).

Table 4.5.: List of geometric details according to the regarded flowfield design used in A_1 or A_{16} testbench setup (Section 3.2) to calculate the entrance velocity in a single gas channel via Equation 4.58.

Flowfield type	A_1-cell	A_{16}-cell anode	A_{16}-cell cathode
Number of gas channels, n_{ch}	10	19	16
Cross sectional area (single channel), A_{ch} / mm^2	1	1	2.25

4.3.3. Electrochemical Charge Transfer Reactions

This presents, how the implementation of the electrochemical charge transfer reactions at fuel and air electrode is realized in the model framework. As described in Section 2.2, all regarded physical processes are involved in order to transport the required reactants. Hence, the electrochemical reaction at each electrode is the core process, which drives all other occurring processes and therefore plays the key role in modeling SOFC performance.

Two separate models are implemented into the model framework to describe the charge transfer reactions of Equations 2.2 and 2.3:

(a) The Butler-Volmer Model (BVM), a generally valid approach from literature [9], implemented as fixed boundary condition at the electrode/electrolyte interfaces of anode and cathode.

(b) The MIEC-Model, a homogenized approach for mixed ionic-electronic conducting (MIEC) electrodes, based on the works of Adler [30] and Rüger [65]. It offers an alternative approach of modeling the cathodic charge transfer reaction with a variable reaction penetration depth for various MIEC electrode material.

In the following, the required model equations for both approaches are presented together with made assumptions and required boundary condition settings.

4.3.3.1. Butler-Volmer Model (BVM)

It has been shown in previous studies [27, 9] by EIS measurements, recorded on SOFCs with Ni/8YSZ fuel electrode but various anode functional layer (AFL) thicknesses, that the penetration depth of the electrochemical charge transfer reaction into the anode is approximately $h_{\mathrm{an}} = 8 - 12\ \mu\mathrm{m}$. A similarly low value for the cathode with $h_{\mathrm{cat}} \approx 6\ \mu\mathrm{m}$ was predicted by detailed 3D-FEM electrode microstructure simulations [65, 157, 120]. Hence, the electrode volume in non-equilibrium is relatively small compared to the full electrode thicknesses (Table 5.1, $h_{\mathrm{an}} = 500 - 1500\ \mu\mathrm{m}$ and $h_{\mathrm{cat}} = 40 - 45\ \mu\mathrm{m}$). It is therefore justifiable to neglect the AFL and penetration depth, thus modeling the charge transfer reactions as interface boundary conditions between the corresponding electrode and electrolyte interfaces ($\mathrm{IF}_{\mathrm{an/cat,elyt}}$, Figure 4.3). This leaves the anode substrate as single fuel electrode in the framework, assuming no distinct gaseous species concentration gradients within the AFL. The validity of the BVM has been shown by Leonide [9] and Klotz [23] by comparing measured current/voltage (C/V) characteristics and EIS spectra using a 0D electrochemical model based on the BVM here presented.

Hence, the BVM regards the charge transfer current density $j_{\mathrm{ct,el}}$ at anode and cathode using an exponential expression, known as the Butler-Volmer equation [9]:

$$j_{\mathrm{ct},el} = j_{0,el} \left[\exp\left(\alpha_{el} \frac{n_{el} F \eta_{\mathrm{act},el}}{RT} \right) - \exp\left(-(1-\alpha_{el}) \frac{n_{el} F \eta_{\mathrm{act},el}}{RT} \right) \right]. \qquad (4.59)$$

In this fundamental equation $j_{0,el}$ denotes the (partial pressure and temperature dependent) exchange current density, n_{el} the number of electrons participating in the reaction ($n_{\mathrm{an/cat}} = 2$ with H_2/CO as electrochemical active species), α_{el} the charge transfer coefficient, which accounts for a asymmetric reaction mechanism [195] and $\eta_{\mathrm{act},el}$ the activation overpotential. The BVM describes the multiple occurring elementary reaction steps of the electrochemical charge transfer reaction [146] in a global one-step reaction. It has been widely used in literature for both anode and cathode electrode reactions [9, 100, 184, 196, 197] and has proven its excellent applicability to reproduce non-linear experimental data with high accuracy [9, 198].

The parameters $j_{0,el}$ and α_{el} have to be determined experimentally as they are highly material dependent. It has been shown by Leonide [9] that determination is a complex process, involving numerous measurements on a highly complicated test setup and with a stable cell and subsequent detailed parameter quantification with suitable software tools. The lack of any of these components makes it rather difficult, or even impossible, to come up with reasonable data by oneself. As a consequence, models are commonly parametrized with data taken from different literature sources, as a complete parameter set for one SOFC type (besides ASC Type-B, determined by Leonide [9]) is not available in literature. Hence, the mixing of modeling parameters is unavoidable, however with the consequence of mixing different electrode kinetics. Furthermore, measurement procedure and/or cell production and history are often only partially documented, adding to corresponding model prediction uncertainty. Another problem of using literature data for model parametrization is the validation with experimental data from other literature sources. As will be shown in Chapter 6, $j_{0,\mathrm{el}}$ is especially sensitive and affects the models performance prediction, entitling it to be a fitting parameter to match regarding validation measurements. The same accounts for α_{el}: when not adopted from the literature, it is often simply assumed to be $\alpha_{el} = 0.5$ [161, 162, 199, 200].

The exchange current density, $j_{0,el}$, can be described with a semi-empirical approach, incorporating the partial pressure dependency with a power-law ansatz and the temperature dependency by an Arrhenius-type expression [9, 201]:

$$j_{0,\mathrm{an}} = \gamma_{\mathrm{an}} \cdot \left(\frac{pH_{2,\mathrm{an}}}{p}\right)^{a} \cdot \left(\frac{pH_2O_{\mathrm{an}}}{p}\right)^{b} \cdot \exp\left(\frac{-E_{\mathrm{act,an}}}{RT}\right) \qquad \mathrm{A/m^2}, \tag{4.60a}$$

$$j_{0,\mathrm{cat}} = \gamma_{\mathrm{cat}} \cdot \left(\frac{pO_{2,\mathrm{cat}}}{p}\right)^{m} \cdot \exp\left(\frac{-E_{\mathrm{act,cat}}}{RT}\right) \qquad \mathrm{A/m^2}. \tag{4.60b}$$

The here listed pre-factors (γ_{an}, γ_{cat}), the exponents (a, b, m) as well as the activation energies ($E_{\mathrm{act,an}}$, $E_{\mathrm{act,cat}}$) have to be experimentally determined. The procedure proposed by Leonide [9], where individual operating parameters such as gas flow rates or temperature are carefully varied, is described in Section 3.3.

Explicit boundary conditions are not required as $j_{\mathrm{ct},cat}$ is regarded in the description of electronic/ionic charge and gaseous species transport BC setting (Tables 4.2 and 4.4).

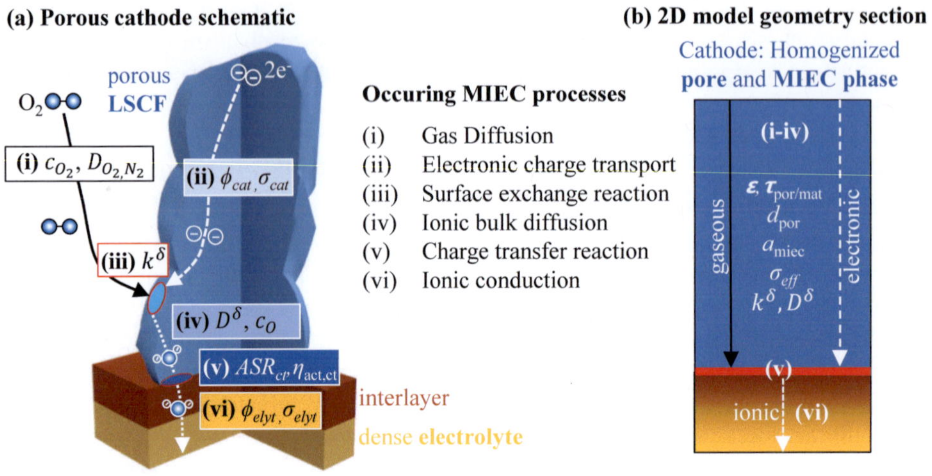

Figure 4.9.: (a) Schematic display of MIEC-cathode model function principle and (b) exemplary 2D model geometry section with homogenized microstructure (Section 5.1.5).

4.3.3.2. Mixed Electronic Ionic Conductive (MIEC) Cathode Model

The reactions penetration depth $\delta_{\mathrm{ct,el}}$ into the corresponding electrode material is neglected in the Butler-Volmer Model (BVM), thus considering only electrical conduction and porous media gas transport as relevant loss processes within the electrodes. While this is a reasonable modeling approach (where $\delta_{\mathrm{ct,el}} << h_{\mathrm{el}}$) this assumption only holds for ideally contacted single cells in the laboratory with an additionally Au contact mesh placed in between cathode and flowfield (Figure 2.13), thus ensuring an ideal electrical contact and gas supply [9].

Nevertheless, it is one of this work's goals to analyze how non-ideal current and gas supply affects SOFC single cell and stack performance. Hence, it is desirable to have a more detailed modeling approach, capable of regarding all physical loss processes occurring in a mixed ionic-electronic conduction (MIEC) electrode, which is state of the art in advanced SOFC stacks [79]. The working principle of the MIEC electrode is schematically shown in Figure 4.9 and described in Section 2.4.1). The occurring loss processes are: (i) porous media gaseous species transport of oxygen (O_2) in the open pores, (ii) electrical conduction of electrons (e^-) in the electronic conducting material phase, (iii) surface exchange of O_2 into the bulk material, (iv) bulk diffusion of oxygen ions (O^{2-}) in the bulk material phase and (v) the charge transfer at the electrode/electrolyte interface.

The modeling of gas diffusion and electrical conduction are described in the previous sections. An innovation in this approach is to regard the MIEC-feature in a homogeneous model, requiring a coupling of gaseous and electronic species transport with ion bulk diffusion via an area dependent surface exchange rate. This is realized by implementing adequate model equations of an established 3D model for MIEC-cathodes with a simplified microstructure implemented in MATLAB/COMSOL Multiphysics, proposed by Rüger et al. [65]. Further use of the approach was made in a real microstructure 3D FEM model for SOFC cathodes by Joos et al. [120], implemented in ParCell3D. This highly sophisticated model delivers results on a very detailed microstructural level (Section 4.1), but also at high computational costs and therefore limiting the size of investigated electrode volume. To investigate MIEC-cathodes at full cell and SOFC stack layer level, the required computational cost is too high to use the model in any practical way. Hence, homogenizing the microstructure realizes MIEC-model functionality at a larger scale, but preserving precise microstructural data regarding the active surface area a_{MIEC}, determined by 3D-reconstruction [120], which is essential for the model accuracy.

In the following, the required model equations are explained and it is shown how their implementation together with set boundary conditions is realized in this work's model framework. The acquisition of required modeling parameters from electrochemical impedance measurements is described in Section 5.2.2.

Surface Exchange The surface exchange of oxygen is a general term for a multiple-process reaction, which can be classified into the following steps [45]: (i) assuming a sufficient gas supply in the pores, molecular O_2 adsorbs at the MIEC-surface, (ii) dissociates into mono-atomic oxygen, (iii) which is then ionized by accepting two electrons and becoming O^{2-}. In a final step (iv) the O^{2-} ions are incorporated into the perovskite lattice of the corresponding MIEC-material, thus occupying a free oxygen vacancy, $\ddot{V}_{\mathrm{O}}$. The sum of individual reaction steps is described by the surface exchange coefficient, k^{δ}, in order to supply a quantifiable measure for the reaction rate. So far, a complete understanding of the entire process remains elusive, especially as the rate determining step changes depending on the reaction conditions [202].

However, it is well-established practice to assume a linear behavior for the surface exchange rate $R_{\mathrm{i}}^{\mathrm{s}}$ [45, 65, 120], here given as source terms for the corresponding model equations:

MIEC phase:
$$R_{\mathrm{O}}^{\mathrm{s}} = a_{\mathrm{MIEC}} \cdot k^{\delta} \cdot (c_{\mathrm{eq}} - c_{\mathrm{O}}) \qquad \mathrm{mol/m^3 \cdot s.} \tag{4.61a}$$

Gas phase:

(species)
$$R_{\mathrm{O_2}}^{\mathrm{s}} = -a_{\mathrm{MIEC}} \cdot k^{\delta} \cdot (c_{\mathrm{eq}} - c_{\mathrm{O}}) \cdot \frac{1}{2} \qquad \mathrm{mol/m^3 \cdot s,} \tag{4.61b}$$

(momentum)
$$R_{\mathrm{O_2}}^{\mathrm{s}} = -a_{\mathrm{MIEC}} \cdot k^{\delta} \cdot (c_{\mathrm{eq}} - c_{\mathrm{O}}) \cdot \frac{M_{\mathrm{O_2}}}{2} \qquad \mathrm{kg/m^3 \cdot s.} \tag{4.61c}$$

Electr. phase:
$$R_{\mathrm{ec}}^{\mathrm{s}} = a_{\mathrm{MIEC}} \cdot k^{\delta} \cdot (c_{\mathrm{eq}} - c_{\mathrm{O}}) \cdot 2F \qquad \mathrm{A/m^3 \cdot s.} \tag{4.61d}$$

Herein, a_{MIEC} denotes the surface area available for the surface exchange reaction between MIEC-material and pore gas phase in $\mu m^2/\mu m^3$. This parameter is essentially responsible for the homogenization approach and can only be determined by 3D-reconstruction [31, 120]. The parameter bears a strong impact on the cathodic polarization resistance $R_{\mathrm{act,cat}}$ and therefore on the overall loss (Section 6.1), coupled with a much higher demand on the numeric grid resolution (Section 6.5). k^{δ} is the surface exchange coefficient in m/s, using the notation used when determined by chemical potential experiments (further details are given below). c_{O} denotes in $\mathrm{mol/m^3}$ the current O^{2-}-ion concentration in the MIEC-material phase, while $c_{\mathrm{eq}}(T, pO_2)$, denotes the equilibrium concentration, which depends on the O_2 partial pressure in the gas phase and operating temperature T [65, 120]:

$$c_{\mathrm{eq}}(T, pO_2) = C_1(T) \cdot \log_{10}(pO_{2,\mathrm{gas}}/p) + C_2(T) \qquad \mathrm{mol/m^3}. \tag{4.62}$$

Herein $pO_{2,\mathrm{gas}}$ denotes the oxygen partial pressure in the gas phase and C_1 and C_2 are MIEC-material dependent constants, given in Table 4.6.

pO_2 dependency for k^δ is regarded by [120]:

$$k^\delta(T, pO_2) = k^\delta_{\mathrm{ref}}(T) \cdot (pO_{2,\mathrm{gas}}/pO_{2,ref})^{\alpha_\mathrm{k}} \qquad \mathrm{m/s} \tag{4.63}$$

where $k^\delta_{\mathrm{ref}}(T)$ was determined at $pO_{2,ref} = 0.21$ atm and α_k is the slope of $k^\delta(T)$ at various pO_2. $k^\delta_{\mathrm{ref}}(T)$ follows an Arrhenius-type behavior and is therefore given by:

$$k^\delta_{\mathrm{ref}}(T) = k_0 \cdot \exp\left(-\frac{E_{\mathrm{act,k}}}{k_\mathrm{B}T}\right) \qquad \mathrm{m/s} \tag{4.64}$$

where k_0 is a pre-exponential factor, k_B the Boltzmann constant and $E_{\mathrm{a,k}}$ the activation energy. However, Ref. [202] describes how the surface exchange is also dependent on the oxygen vacancy concentration, $\ddot{V}_O$, and that $E_{\mathrm{act,k}}$ can therefore not be solely ascribed to the surface exchange but also partly to the thermally activated vacancy formation.

In summary, Equation 4.61 describes the surface exchange reaction in the following way: At equilibrium (i.e. no current is drawn or under OCV), no concentration gradient exists between bulk and gas phase and consequently no surface exchange takes place. Drawing a current from the cell leads to lower or higher c_O because O^{2-} ions in the MIEC material are transported to or away from the electrolyte, depending on the current direction (SOFC/SOEC). Consequently, a surface exchange flux, R^s_i, is generated.

Bulk diffusion describes, in the case regarded here, the movement of O^{2-} ions only in the MIEC-material. O^{2-} transport functions by interchanging places in the MIEC-perovskite lattice with $\dot{V}_\mathrm{O}$ and is generally described by a bulk diffusion coefficient D^δ in m^2/s and Ficks 1st law [65, 120]:

$$J_{O^{2-}} = -\left(\frac{1-\epsilon_{\mathrm{cat}}}{\tau_{\mathrm{mat,cat}}} \cdot D^\delta(T, p_{O_2})\right) \nabla c_{O^{2-}} \qquad \mathrm{mol/m^2 \cdot s}, \tag{4.65}$$

where J_{bulk} denotes the O^{2-} ion flux within the MIEC-bulk and D^δ is the chemical bulk diffusion coefficient. The temperature dependency of $D^\delta_{\mathrm{ref}}(T)$ is described by an Arrhenius-type expression [120]:

$$D^\delta_{\mathrm{ref}}(T) = D_0 \cdot \exp\left(-\frac{\Delta H_\mathrm{m}}{k_\mathrm{B}T}\right) \qquad \mathrm{mol/m^3 \cdot s} \tag{4.66}$$

where ΔH_m denotes the enthalpy for the place interchange process. Strictly speaking, further considerations regarding the formation enthalpy of vacancy clusters at lower temperatures need to be considered [45]. However, at this point the more simplified approach given by Equation 4.66 is applied. Similar to Equation 4.63, a pO_2 dependency of D^δ is regarded by [120]:

$$D^\delta(T, pO_2) = D^\delta_{\text{ref}}(T) \cdot (pO_{2,\text{gas}}/pO_{2,ref})^{\alpha_\text{D}} \qquad \text{m/s}^2. \tag{4.67}$$

The diffusion equation in the bulk material is given together with Equation 4.65 by

$$\nabla J_{\text{O}^{2-}} = \nabla \cdot \left(-D^\delta(T, p_{\text{O}_2}) \cdot \nabla c_\text{O}\right) = R^\text{s}_{\text{O}^{2-}}, \qquad \text{mol/m}^3 \cdot \text{s} \tag{4.68}$$

$$J_{\text{O}^{2-}} = k^\delta \cdot (c_\text{eq} - c_\text{O}) \qquad \text{mol/m}^2 \cdot \text{s}, \tag{4.69}$$

while yet another boundary conditions is set by the charge transfer reaction at the electrode/electrolyte interface given Equation 4.73a. The source term $R^\text{s}_{\text{O}^{2-}}$ in Equation 4.68 is given by Equation 4.61a, which couples bulk diffusion, gas diffusion and electron transport. In this way, a spatially dependent surface exchange reaction is realized in this works model framework.

Charge Transfer The exchange of O^{2-} between the MIEC-material and electrolyte layer involves a charge transfer reaction. The local overpotential is calculated by Equation 4.85, which is now given again for the sake of completeness:

$$\eta_{\text{ct,cat}} = \Phi_{\text{Nernst,cat}}(p_{\text{O}_{2,MIEC-IF}}) - \Phi_{\text{ec,cat}} + \Phi_{\text{io,elyt}} \qquad \text{V} \tag{4.70}$$

where $\phi_{\text{ec,cat}}$ denotes the electronic potential in the MIEC material, $\phi_{\text{io,elyt}}$ the ionic potential in the electrolyte and $\phi_{\text{Nernst,cat}}$ the local Nernst-potential. $\phi_{\text{Nernst,cat}}$ depends on the local oxygen partial pressure in the MIEC-material and is calculated by:

$$\Phi_{\text{Nernst,cat}} = \frac{RT}{4F} \cdot \log\left(\frac{pO_{2,MIEC-IF}}{p}\right) \qquad \text{V} \tag{4.71a}$$

with

$$pO_{2,\text{MIEC-IF}} = 10^{\frac{c_{\text{O,IF}} - C_2(T)}{C_1(T)}} \cdot p \qquad Pa. \tag{4.71b}$$

The local oxygen partial pressure ($pO_{2,\text{MIEC-IF}}$) in the MIEC material at the electrolyte interface corresponds to the MIEC-material equilibrium partial pressure at the regarding local oxygen ion concentration, $c_{\text{O,IF}}$ [65, 203]. Hence. the charge transfer overpotential $\eta_{\text{ct,cat}}$ in Equation 4.70 is dependent on the local oxygen ion concentration in the MIEC-material ($\phi_{\text{Nernst,cat}}$, the electrical potential in the MIEC ($\phi_{\text{ec,cat}}$) and the ionic potential in the electrolyte ($\phi_{\text{io,elyt}}$) and is calculated at each numerical grid point along the interface. $\eta_{\text{ct,cat}}$ now serves within the following equation to calculate the local oxygen charge transfer current density:

$$j_{\text{ct,cat}} = \frac{\eta_{\text{ct,cat}}}{ASR_{\text{ct,cat}}} \qquad \text{A/cm}^2 \tag{4.72}$$

which is then used to model the boundary condition for the O^{2-} bulk diffusion (Equation 4.73a) and ionic potential calculation (Equation 4.2):

$$\text{Bulk diffusion:} \qquad J_{O^{2-}} \cdot \vec{n} = \frac{j_{\mathrm{ct,cat}}}{2F} \qquad \mathrm{mol/m^2 \cdot s}, \tag{4.73a}$$

$$\text{Ionic current density:} \qquad j_{\mathrm{io}} \cdot \vec{n} = j_{\mathrm{ct,cat}} \qquad \mathrm{A/cm^2}. \tag{4.73b}$$

The area specific charge transfer resistance, $ASR_{\mathrm{ct,cat}}$, in Equation 4.72 is a very difficult parameter to determine, and no data is to be found in literature. It depends strongly on the interface composition and fabrication method. For one thing, in the case of a LSCF/CGO interface, where the GCO layer is dense (e.g. applied with PVD-sputtering), it is negligible [203]. However, in the case of a not completely dense CGO layer (due to $T_{\mathrm{sinter}} < 1250$ °C) or in the case of a direct LSCF/YSZ interface, $ASR_{\mathrm{ct,cat}}$ can be significantly higher or even become the dominating loss due to the formation of insulating secondary phases (lanthanides and zirconates) [72]. The CGO layers in the ASCs used in this work were all sintered at $T_{\mathrm{sinter}} = 1300$ °C and the impact of secondary phase formation should be very low [72]. Furthermore, any small contribution is already regarded in the effective conductivity $\sigma_{\mathrm{elyt}}^{\mathrm{eff}}$, which is determined of recorded impedance spectra of in this work's validation measurements (Section 5.3.3). For other material combinations without secondary phase formation (e.g. LSC/YSZ), a loss-less charge transfer can be assumed and $ASR_{\mathrm{ct,cat}}$ neglected [45]. Hence, a very low value for the charge transfer resistance with $ASR_{\mathrm{ct,cat}} = 10^{-4}$ mΩ·cm² is used here in the model framework.

Required values for the calculation of k^{δ} and D^{δ} with Equations 4.63, 4.64, 4.66 and 4.67 for LSCF and alternative MIEC-materials are listed in Table 5.3. Furthermore, it is described in Refs. [31, 120] how $k_{\mathrm{ref}}^{\delta}$ and $D_{\mathrm{ref}}^{\delta}$ can be determined experimentally from impedance measurements with values for C_1 and C_2 given in Table 4.6. This completes the MIEC-cathode model description.

Table 4.6.: List of parameters given by Ref. [120] to calculate c_{eq} (4.62) and $pO_{2,MIEC-IF}$. (4.82) at various temperatures, T.

T / °C	C_1(T)	C_2(T)
600	85317	267
650	84891	396
700	84465	525
750	84039	653
800	83613	782
850	83187	910
900	82761	1039

4.3.4. Chemical Reforming Reactions

The SOFC features a unique capability: By not relying only on pure hydrogen as fuel but by also accepting hydrocarbons, it has a higher fuel flexibility, and this recommends the SOFC to a greater variety of applications. At elevated SOFC operating temperatures the Ni, as part of the cermet-anode (Section 2.4.1), becomes catalytically active and enables a direct (or internal) conversion of hydrocarbons to hydrogen via reforming reactions. In the presence of methane (CH_4), a combined reaction mechanism of methane-steam reforming (SR) (Equation 2.16) and water-gas shift reaction (WGS) (Equation 2.17) occurs, while during operation under pre-reformed or artificial syngas only the WGS takes place. In order to model these chemical reaction mechanisms in a realistic manner, adequate kinetic expressions are required to be determined and implemented in the model framework.

Reforming reaction-induced species conversion is implemented into the model framework via a global rate expression approach as source terms in the individual species mass balance equation (Equation 4.26 or 4.25). Based on the work of H. Timmermann [204], A. Kromp [17] and Lehnert et al [99] the reaction rate expression for the SR, $S_{\mathrm{sr,i}}$, is given by [204]:

$$S_{\mathrm{sr,i}} = \nu_{\mathrm{sr,i}} \cdot k_{\mathrm{sr}}^{+} \cdot \left(p\mathrm{CH_4} - k_{\mathrm{sr}}^{-} \frac{p\mathrm{CO} \cdot p\mathrm{H_2}^3}{p\mathrm{H_2O}} \right) \qquad \mathrm{mol/m^3 \cdot s} \tag{4.74}$$

and $S_{\mathrm{sh,i}}$ as rate expression for the WGS [99, 17]:

$$S_{\mathrm{sh,i}} = \nu_{\mathrm{sh,i}} \cdot k_{\mathrm{sh}}^{+} \cdot \left(p\mathrm{CO} \cdot p\mathrm{H_2O} - k_{\mathrm{sh}}^{-} \cdot p\mathrm{H_2} \cdot p\mathrm{CO_2} \right) \qquad \mathrm{mol/m^3 \cdot s}. \tag{4.75}$$

In both rate expressions based on molar units, species concentrations are expressed by partial pressures $p_i = p \cdot x_i$. Multiplying each equation with the corresponding species molar mass, M_i, delivers species, i, related mass-based expression. The term $\nu_{\mathrm{sr/sh,i}}$ denotes in Equation 4.74 and 4.75 the stoichiometric coefficients according to the chemical reaction mechanism of SR (subscript: sr) and WGS (subscript: sh) as listed in Table 4.7. Depending on the local thermodynamic conditions at the catalyst surface area, a *forward* (+) or *backward* (−) directed reaction pathway of SR and WGS occurs (Equations 2.16 and 2.17 as written from left to right). Hence, $k_{\mathrm{sr/sh}}^{+}$ and $k_{\mathrm{sr/sh}}^{-}$ denote in mol/m^3/s/Pa^2 the according reaction velocity constants, which are related by the equilibrium constant $K_{\mathrm{eq,sr/sh}}$ with the following expression [99]:

$$K_{\mathrm{eq,sr/sh}} = \frac{k_{\mathrm{sr/sh}}^{+}}{k_{\mathrm{sr/sh}}^{-}}. \tag{4.76}$$

Combining Eqs.4.74 and 4.75 with the corresponding values of Equation 4.76 delivers the final expressions for $R_{\mathrm{i,sr/sh}}$ as implemented into the model framework:

$$S_{\mathrm{sr,i}} = \nu_{\mathrm{sr,i}} \cdot k_{\mathrm{sr}}^{+} \cdot \left(p\mathrm{CH_4} - \frac{1}{K_{\mathrm{eq,sh}}} \frac{p\mathrm{CO} \cdot p\mathrm{H_2}^3}{p\mathrm{H_2O}} \right) \qquad \mathrm{mol/m^3 \cdot s} \tag{4.77}$$

and $S_{\mathrm{sh,i}}$ as rate expression for the WGR [17, 99]:

$$S_{\mathrm{sh,i}} = \nu_{\mathrm{sh,i}} \cdot k_{\mathrm{sh}}^{+} \cdot \left(p\mathrm{CO} \cdot p\mathrm{H_2O} - \frac{1}{K_{\mathrm{eq,sh}}} \cdot p\mathrm{H_2} \cdot p\mathrm{CO_2} \right) \qquad \mathrm{mol/m^3 \cdot s}. \tag{4.78}$$

This leaves $K_{eq,sr/sh}$ and $k^+_{sr,sh}$ as unknown parameters. Values for $K_{eq,sr/sh}$ can be calculated from thermodynamic databases (for example MALT [205]). Temperature dependent expressions regarding $K_{eq,sr/sh}$ are given in Section 5.5. The velocity constants $k^+_{sr,sh}$ can be expressed by an Arrhenius-type expression [99, 17, 204]:

$$k^+_s = k_{0,s} \cdot exp\left(\frac{-E_{act,s}}{RT}\right) \qquad \mathrm{mol/m^3 \cdot Pa^2 \cdot s} \tag{4.79}$$

with $s = sr, sh$.

Table 4.7.: Stoichiometric coefficients of the fuel gas components with regards to the steam reforming reaction $\nu_{sr,i}$ and the water-gas shift reaction $\nu_{sh,i}$.

species i	H_2	H_2O	CO	CO_2	CH_4	N_2
$\nu_{sr,i}$	3	-1	1	0	-1	0
$\nu_{sh,i}$	1	-1	-1	1	0	0

Based on the global rate expression approach, contains $k_{0,s}$ in Equation 4.79 as pre-exponential factor the catalysts surface area and texture characteristic, while $E_{act,s}$ denotes the regarding activation energy in kJ/mol. Values for k^+_{sr} are adopted from the work of H. Timmermann [204] and A. Kromp [206], both summarized in Section 5.5. For k^+_{sh} only a value for $T = 800$ °C was determined by Kromp et al. [17, 29] by means of impedance modeling and EIS measurements, recorded on cell Type-B. However, a value for $k_{0,sh}$ independent from surface area would be desirable to adopt the model framework to other cell types with different microstructural properties. Furthermore, in order to reproduce the WGS temperature dependency, a value for $E_{act,sh}$ is necessary. Hence, both values were determined within this work by means of combining gas conversion measurements and FEM modeling results, described in detail in Section 5.5.

4.4. Electrical Function Principle

The core feature of this model framework is to predict the cell voltage, U_{cell}, in V and corresponding current density, j_{cell}, in A/cm^{-2}, depending on the applied model geometry, material parameters and operating conditions. Presetting a load describes the simulation of *galvanostatic* operation, and it is equally possible to run a *potentiostatic* calculation by setting U_{cell} as default and receiving the resulting current response. Figure 4.10 illustrates the electric circuit schematic for *open circuit conditions* (OCC, Figure 4.10a) and for SOFC or SOEC operation under load (Figure 4.10b). The current flow is defined in the technical direction, which depends on whether power is generated (SOFC-mode) or used (SOEC-mode). In the following, the details displayed in Figure 4.10 are described and the model frameworks electrical function principle is explained with the aid of appropriate model equations.

4.4.1. Cell Voltage, Current Density and Loss Calculation

Under *open circuit conditions* (OCC, Figure 4.10a), no current is generated or used, and no overpotentials are created. The electrochemical potential difference between the electrodes, commonly denoted as *open circuit voltage* (OCV) or as U_{occ} in V, is defined by the electrochemical half-cell reactions (Eqs.2.3 and 2.2) in equilibrium, which depend on the corresponding electrode gas atmospheres and operating temperature. The OCV can be calculated using the Nernst-equation (Equation 2.8), which can be reformulated into:

$$U_{\text{occ}} = \Delta\Phi^{\text{TPB}}_{\text{eq,cat}}(p_i) - \Delta\Phi^{\text{TPB}}_{\text{eq,an}}(p_i) \qquad \text{V}. \tag{4.80}$$

Therein $\Delta\Phi_{\text{eq,an/cat}}(p_i)$ denote the corresponding electrodes half-cell potentials, each dependent on their individual gas composition, denoted by the partial pressure reference p_i. The following expression for the cathode half-cell potential is used in the Butler-Volmer approach (Section 4.3.3.1):

$$\Delta\Phi^{\text{TPB}}_{\text{eq,cat}} = -\frac{\Delta G_0}{2F} + \frac{RT}{4F}\ln\left(\frac{p\text{O}_{2,\text{cat}}}{p}\right) \qquad \text{V}. \tag{4.81}$$

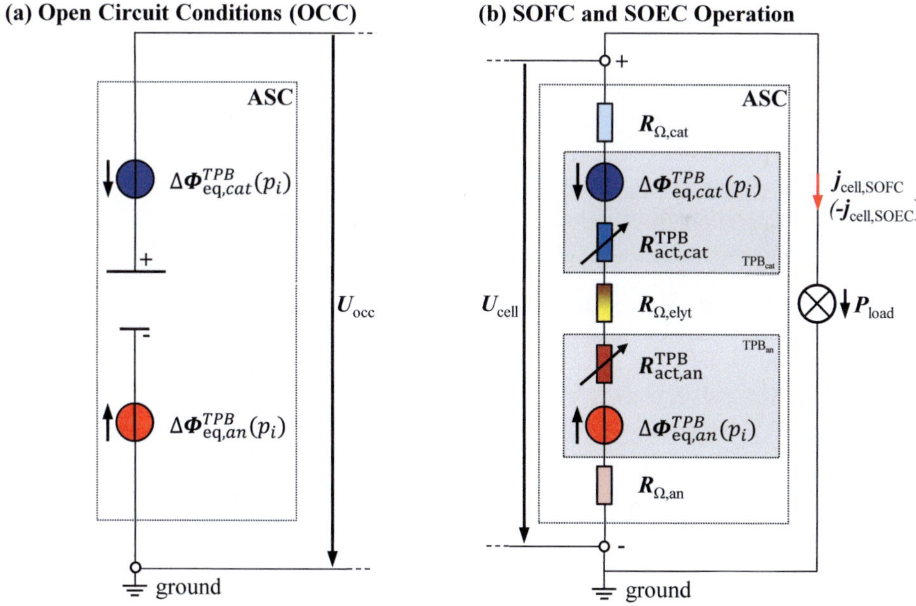

Figure 4.10.: Electrical circuit description in the model framework of an anode supported SOFC (ASC): (a) under open circuit conditions and (b) under load. Displayed are the partial pressure dependent electrodes half-cell potentials $\Delta\phi_{\text{eq,an/cat}}(p_i)$ and under load occurring ohmic ($R_{\Omega\text{i}}$) and polarization ($R_{\text{act,an/cat}}$) resistances, which govern (together with the half-cell potentials) the predicted cell voltage.

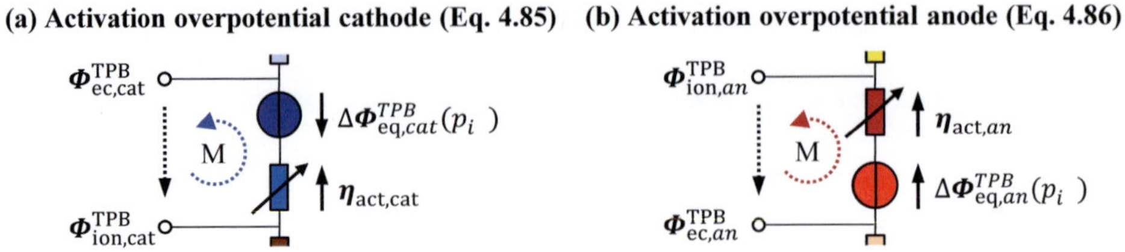

Figure 4.11.: Display of electrode activation polarization overpotential calculation by applying Kirchoff's 2nd Law of electrical circuits.

In the case of using the MIEC-cathode model (Section 4.3.3.2), $pO_{2,\mathrm{cat}}$ in Equation 4.81 is related to the oxygen ion concentration c_O in the bulk material: [65, 120]:

$$pO^{2-}_{\mathrm{eq}} = 10^{\frac{c_\mathrm{O}-C_2}{C_1}} \cdot p \qquad \mathrm{Pa}. \tag{4.82}$$

Therein C_1 and C_2 denote MIEC-material specific constants [65, 120].

The anode half-cell potential $\Delta\Phi_{\mathrm{eq,an}}$ solely depends on the electrochemically active fuel gas species present the anode. In hydrogen and mixed hydrogen/hydrocarbon operation (reformate, CH_4), only H_2 is electrochemically active [28], therefore $\Delta\Phi^{\mathrm{TPB}}_{\mathrm{eq,an}}$ is (according to Equation 2.3) given by:

$$\Delta\Phi^{\mathrm{TPB}}_{\mathrm{eq,an}} = \frac{RT}{2F} \ln\left(\frac{pH_2O_{\mathrm{an}}}{pH_{2,\mathrm{an}}}\right) \qquad \mathrm{V}. \tag{4.83}$$

In the case of mixed hydrocarbon operation, a calculation of U_{occ} is only possible in galvanostatic operation, because pH_2O_{an} and $pH_{2,\mathrm{an}}$ at the TPBs depend on possibly occurring gas conversion reactions in the gas channel (Section 4.3.4), if the applied fuel gas mixture is not in thermodynamic equilibrium.

Under load, Figure 4.10b illustrates the equivalent electrical circuit to calculate U_{cell} or j_{cell}. The current density, j_{cell}, is defined in the technical current direction scheme, pointing from positive to negative electrode. $R_{\Omega,\mathrm{an/cat}}$ describe the ohmic losses due to non-ideal charge transport in the electronic conduction electrode material phases, while $R_{\Omega,\mathrm{elyt}}$ covers ohmic loss in the electrolyte layer. Gas diffusion polarization is incorporated by the partial pressure dependency of $\Delta\Phi^{\mathrm{TPB}}_{\mathrm{eq,an/cat}}$ and charge transfer reaction loss at each electrode by $R_{\mathrm{act,an/cat}}$. Assuming galvanostatic operation, U_{cell} is governed by the half-cell potentials at their corresponding TPB and by the sum of current depending overpotentials $\eta_i = j \cdot R_i$ (Figure 4.10b):

$$U_{\mathrm{cell}} = \Delta\Phi^{\mathrm{TPB}}_{\mathrm{eq,cat}}(p_i) - \Delta\Phi^{\mathrm{TPB}}_{\mathrm{eq,an}}(p_i) - j_{\mathrm{cell}} \cdot \sum_i R_i \qquad \mathrm{V}. \tag{4.84}$$

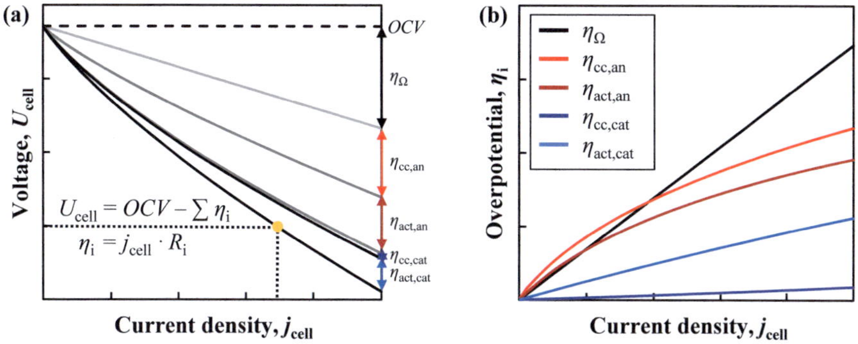

Figure 4.12.: Example display of (a) the OCV reduction by individual overpotentials η_i caused by occurring loss processes, resulting in the cell voltage U_{cell} at a certain current density j and (b) overpotential distribution calculated by the model framework dependent to j_{cell}.

Ohmic and gas diffusion losses result from the implemented transport equations, while the electrode activation overpotentials $\eta_{\text{act},el}$ remain unknown variables. However, in order to calculate the charge transfer current density, $j_{\text{ct,an/cat}}$ (Equations 4.59 and 4.72), a value for $\eta_{\text{act},el}$ is required. The solution is to introduce an electrical mesh concept based on Kirchhoff's 2nd Law at the corresponding electrode/electrolyte interfaces ($\text{IF}_{\text{an/cat,elyt}}$ as illustrated in Figure 4.11) and derive thereby the following expressions as boundary conditions:

$$\eta_{\text{act,cat}} = \Delta\Phi^{\text{TPB}}_{\text{eq,cat}}\left(pO^{\text{TPB}}_{2,\text{cat}}\right) - \Phi^{\text{TPB}}_{\text{ec,cat}} + \Phi^{\text{TPB}}_{\text{io,cat}} \qquad \text{V}, \qquad (4.85)$$

$$\eta_{\text{act,an}} = -\Delta\Phi^{\text{TPB}}_{\text{eq,an}}\left(pH^{\text{TPB}}_{2,\text{an}}, pH_2O^{\text{TPB}}_{\text{an}}\right) + \Phi^{\text{TPB}}_{\text{ec,an}} - \Phi^{\text{TPB}}_{\text{io,an}} \qquad \text{V}. \qquad (4.86)$$

In Equations 4.86 and 4.85 the corresponding variables from all implemented loss processes are coupled with each other. It is therefore vital to precisely follow Kirchhoff's 2nd Law and the therewith connected algebraic signs as depicted in Figure 4.11. It should be pointed out, that the expressions in Equations 4.85 and 4.86 differ from what is implemented in the COMSOL application mode for FuelCells&Batteries [125], where $\eta_{\text{act}} = \Phi_{\text{ec}} - \Phi_{\text{io}} - \Delta\Phi^{\text{TPB}}_{\text{eq,an/cat}}$ is implemented as a fixed expression for both electrode reactions. This is however not in agreement with Kirchhoff's 2nd Law and will lead to serious a miscalculation of $\eta_{\text{act},el}$ and consequently of U_{cell} or j_{cell}.

A stationary solution for U_{cell} is dependent on a preset j_{cell} or vice versa. A current/voltage characteristic (C/V) is generated by stepwise variation of one of the two parameters, exemplary shown in Figure 4.12a. Therein is illustrated how the OCV is decreased by certain overpotentials η_i, generated by the individual occurring processes with their specific loss R_i. η_i can be calculated individually using the formulations given in Section 4.4.3. Figure 4.12b displays the corresponding η_i distributions according to the C/V in Figure 4.12a, thus allowing a detailed analysis of performance limiting processes.

4.4.2. Differential Resistance Calculation

In order to compare measured area specific resistance (ASR) determined from impedance measurements (Section 2.3.2) with model predicted resistance, it is not necessary to calculate a complex impedance because only the real part is required at low frequencies. It is, however, not sufficient to simply calculate a stationary solution at the desired current density $j_{\text{EIS}} = 60$ mA/cm^{-2} and to solve R = U / j. In EIS measurements, the current amplitude is varied between $+/-j_{\text{EIS}} = 60$ mA/cm^{-2} at $f_{\text{EIS,min}}$, and the resulting ASR is actually the gradient at $j_{\text{EIS}} = 0$ mA/cm^{-2}. The value may deviate from the stationary solution at the same operating point, depending on the slope of the overpotential of interest [23]. Hence, the EIS procedure is emulated by calculation of stationary solutions at minimum of 4 current density values $j_{\text{EIS,min}} < j_{\text{cell}} < j_{\text{EIS,max}}$ mA/cm^{-2}. Thereafter, the desired overpotential, η_i, is plotted at each simulated j_{cell} and the ASR (the slope between η_i and j) can be deduced by means of linear approximation via MATLAB.

Figure 4.13a gives an example illustration of the described procedure, with several linear approximations for varying fuel gas humidity. The resulting gradients or ASR values are displayed in Figure 4.13b, plotted over the corresponding $p\text{H}_2\text{O}_{\text{an}}$ set in the model. By comparing the slopes in Figure 4.13a it becomes clear that a steep gradient is the equivalent of a high ASR value in Figure 4.13b.

Using the above described method, the required computational time is greatly reduced, but still allows a model comparison with highly precise EIS measurement results.

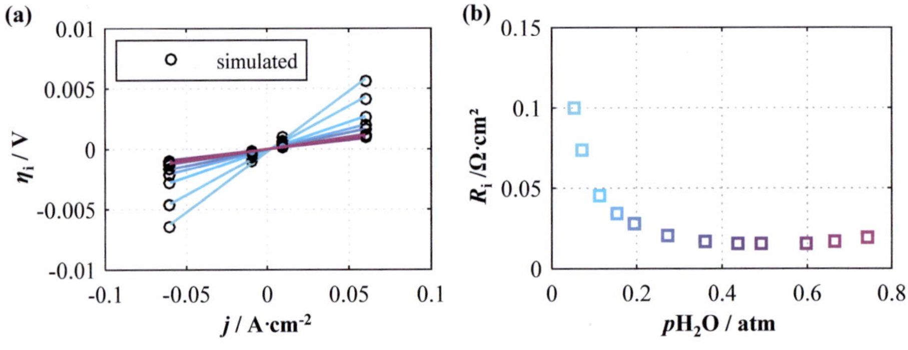

Figure 4.13.: Example of model predicted overpotentials η_i to calculated the slope in order to determine the differential resistance R_i of a process i under OCV.

4.4.3. Model-based Loss Process Quantification

The individual loss process contributions R_i to the overall loss R_{total} depend not only on the local species but also on the current density distribution. Activation ($R_{\text{act,an/cat}}$) and diffusion polarization ($R_{\text{cc,an/cat}}$) effect the transported current density at the TPBs, while ohmic loss (R_Ω) reduces the current flux locally in the conductive phases. The electrode activation overpotentials are calculated locally using Equations 4.85 and 4.86. The diffusion polarization is equally dependent on the local partial pressure distribution at the TPBs and are calculated by the following expressions (derived from the Nernst Equation) [9, 207]:

$$\eta_{\text{cc,an}}^{\text{TPB}} = \frac{RT}{2F}\ln\left(\frac{p\text{H}_2\text{O}_{\text{an}}^{\text{TPB}} \cdot p\text{H}_{2,\text{an}}}{p\text{H}_2\text{O}_{\text{an}} \cdot p\text{H}_{2,\text{an}}^{\text{TPB}}}\right) \qquad \text{V}, \tag{4.87}$$

$$\eta_{\text{cc,cat}}^{\text{TPB}} = \frac{RT}{4F}\ln\left(\frac{p\text{O}_{2,\text{cat}}}{p\text{O}_{2,\text{cat}}^{\text{TPB}}}\right) \qquad \text{V}. \tag{4.88}$$

The corresponding activation and diffusion losses, $R_{\text{act/cc},el}$, are now derived by calculating the arithmetic mean power loss based on the local current flux, $\vec{j}$, and the cell current density, j_{cell}. This is the only way that the influence of an inhomogeneous current distribution on the loss process evaluation may be regarded:

$$R_{\text{act/cc},el} = \frac{M\{\eta_{\text{act/cc},i}^{\text{TPB}} \cdot \vec{j}\}}{j_{\text{cell}}^2} \qquad \Omega \cdot \text{cm}^2. \tag{4.89}$$

The ohmic overpotential, $\eta_{\Omega,i}$, occurring in the electrode layers and electrolyte, is calculated by setting the mean of local electric field, $\vec{E}$, and current flux, $\vec{j}$, into relationship with the corresponding layer thickness, h_i, and j_{cell}:

$$\eta_{\Omega,i} = \frac{M\{\vec{E} \cdot \vec{j}\}}{h_i \cdot j_{\text{cell}}} \qquad \text{V} \tag{4.90}$$

and thereof resulting ohmic loss, $R_{\Omega,i}$, with

$$R_{\Omega,i} = \frac{M\{\vec{E} \cdot \vec{j}\}}{h_i \cdot j_{\text{cell}}^2} \qquad \Omega \cdot \text{cm}^2. \tag{4.91}$$

This completes the model framework description. The next chapter explains how to determine the material and modeling parameters required for the model equations introduced in this chapter.

5. Model Framework Parametrization

As well as the model equation description in the previous section, the significance of a model framework is based on a careful modeling parameter determination. The following section is therefore dedicated to describing methods to determine the required parameters.

5.1. Porous Electrode Microstructure Properties

The porous microstructure of SOFC electrodes or any porous structure in general is characterized by specific parameters, which strongly and individually influence the occurring loss processes. Consequently, the overall cell polarization loss is effected. The influences of especially sensitive parameters on predicted cell losses is further discussed in Section 6.5. The FIB/SEM tomography (Section 2.6) supplies a method by which the microstructure and related parameters can be quantified with very high precision. However, it is a very complex and time consuming process. In fact, it took more than two complete dissertations [120, 121] to develop the algorithms employed here. Nevertheless, the actual use of these tools is not completely autonomous and the gained results have to be evaluated carefully for their applicability before they are implemented in a model.

In the following, the microstructural parameters of the model framework presented here are briefly introduced and the determination procedure is described. These parameters are, in contrast to other literature models [144, 145, 147, 148, 177], all determined from cell samples on which the validation measurements are recorded. In this way, cell specific characteristics related to manufacturing and/or history of operation are captured within the parameters, thereby significantly raising the frameworks significance against other model approaches.

In addition to the tomography based determination, another method based on electrochemical impedance measurements can be employed, although limited to the pore phase characteristic parameter $\Psi_{\mathrm{por},el}$ [9, 208]. The results of both methods and their impact on FEM model-based SOFC performance predictions is discussed in Chapter 6.

5.1.1. Volume Fraction

A porous material is characterized by at least two phases: A material and a pore phase. The individual phases are detected in the segmentation process of the 3D reconstruction algorithm Section 2.6. The volume fraction, ε_i, is thus determined by calculating the sum, V_i, of voxels assigned to the corresponding material fraction and dividing it by the total number of voxels, V [120, 121]:

$$\varepsilon_\mathrm{i} = \frac{V_\mathrm{i}}{\sum_\mathrm{i} V_\mathrm{i}} = \frac{V_\mathrm{i}}{V}. \tag{5.1}$$

A possible error source for ε_i is erroneously segmented material fractions, which depend on the captured SEM image histogram properties and have to be individually assessed. Also, the reconstructed volume has to be large enough to be statistically representative (*n.b.* representative volume element, RVE) [120, 121]. This matter was extensively investigated by Joos *et al.* [120, 157] for the cell types used in this work and the reconstructed sample volumes used in the parameter determination are large enough. Furthermore, beyond all other microstructural parameters, ε_i, can be determined with the highest accuracy and is the least likely source of error.

The results for the individual cell type layers are summarized in Table 5.1 at the end of this section, together with all relevant microstructural parameters.

5.1.2. Tortuosity

The tortuosity, τ_i, is a measure of the twisted shape of the material phase, i. It effects the corresponding species transport properties (effective gas transport in the pores - Equation 5.7, effective electronic/ionic conduction in the material - Equation 5.10) because the transport pathways are elongated in a porous medium and may encounter thin bottlenecks in their transport cross section. An analytical solution to calculate τ is given in literature by the Brüggemann Relation [209], which is, however, based on the the assumption of closely packed spheres and therefore vastly underestimates τ [121]. Furthermore, different definitions of tortuosity exist in literature, causing a lot diffusion. In Ref. [210] the different definitions are discussed. In this work, the effective media definition is used, given here by the example of the effective conductivity [121, 120]:

$$\tau_\mathrm{i} = \frac{\sigma_\mathrm{bulk,i}}{\sigma_\mathrm{eff,i}} \cdot V_\mathrm{i}. \tag{5.2}$$

In this way, microstructure-inherent influences on the tortuosity (e.g. constrictivity) are considered.

To calculate all τ_i of the different phases, i, contained in the porous electrodes, an algorithm developed by M. Ender [121] is used. The algorithm is based on the finite volume method (FVM). Therein a numerical mesh is generated within MATLAB based on the discrete voxel information of the corresponding material phase, i, gathered from the 3D reconstruction of the corresponding porous electrode (Section 2.6), thus calculating with the highest possible resolution. Between top and bottom of the transport direction a potential difference is set as boundary conditions, with isolation at the sides, and the Poisson Equation (Equation 4.1) is solved to calculate the resulting effective conductivity $\sigma_{\mathrm{eff,i}}$. With this result (and the help of Equation 5.2) τ_{i} can be calculated for each material phase and transport direction. More detailed information about the algorithm is given in [121].

Large structures (10^9 voxels and larger) can be analyzed by the FVM algorithm in a much shorter time scale similar algorithm based on the FEM method developed by T. Carraro and J. Joos [159, 160, 211]. However, the achieved results can differ by up to 10% or more, depending on the structure [120]. The difference is related to the different discretization methods: even though in both methods the numerical mesh is composed of cubes based on the voxel information, the FEM uses 8 computational nodes for each cube (one per corner), whereas the FVM uses only one node in the middle of the cube. In the FVM a species flux is transported from phase to phase, while the FEM also allows diagonal transport. The result is called *staircase effect* [120], and causes the FVM algorithm from Ender to require a much higher numerical mesh resolution to reach the same result as the FEM algorithm from Carraro. However, the FVM mesh resolution is limited by the voxel information from the reconstruction and the possible errors must therefore be kept in mind.

As such, both error sources have to be regarded in the evaluation, especially in the validation of the anode gas diffusion transport. A deviation in the parameter estimation is much more likely to have an impact, due to longer transport pathways in the thick anode compared to the thin cathode. This is further discussed in the parameter sensitivity analysis of Section 6.1 and the model validation of Section 6.5.

Regardless of mentioned uncertainties, the results for anode pore phase tortuosity and the cathode material and pore phase tortuosities (as listed in Table 5.1) still show strong connections between layer thickness, volume fraction and tortuosity. This underlines the significance of a detailed microstructural cell analysis to produce data for model validation purposes. This feature clearly helps the work presented here to stand out from other models presented in literature [144, 145, 147, 148, 177, 154], where microstructural data (especially the tortuosity) is assumed or adopted from other sources without any knowledge of cell history or origin.

5.1.3. Stochastic Microstructure Generator

A unique tool to predict the pore and material phase tortuosity of real LSCF cathodes was developed by J. Joos *et al.* [120, 212]. Their stochastic *microstructure generator* was calibrated with reconstructions of real LSCF electrodes with varying pore fractions and thus reflecting the porous medium inherent parameter interdependence. This is shown in detail in Figure 5.1a for $\tau_{\mathrm{por/mat}}$ in dependence to the pore fraction. The calculated results predict that for low pore fractions a dramatic increase of τ_{por} will occur with τ_{mat} decreasing towards unity and vice versa for high pore fractions, which of course corresponds to low material fractions.

The following equations are the result of fitting exponential functions to the data shown in Figure 5.1a:

$$\tau_{\mathrm{por}} = 323.3 \cdot \exp(-16.14 \cdot \varepsilon_{\mathrm{por}}) + 2.54 \cdot \exp(-0.79 \cdot \varepsilon_{\mathrm{por}}), \tag{5.3}$$

$$\tau_{\mathrm{mat}} = 1.04 \cdot \exp(0.86 \cdot \varepsilon_{\mathrm{por}}) + 7.96 \cdot 10^{-4} \cdot \exp(11.9 \cdot \varepsilon_{\mathrm{por}}). \tag{5.4}$$

These functions are implemented in the model framework to reproduce realistic LSCF microstructural parameter interdependence in the calculations for the results presented in Section 7.2.3.

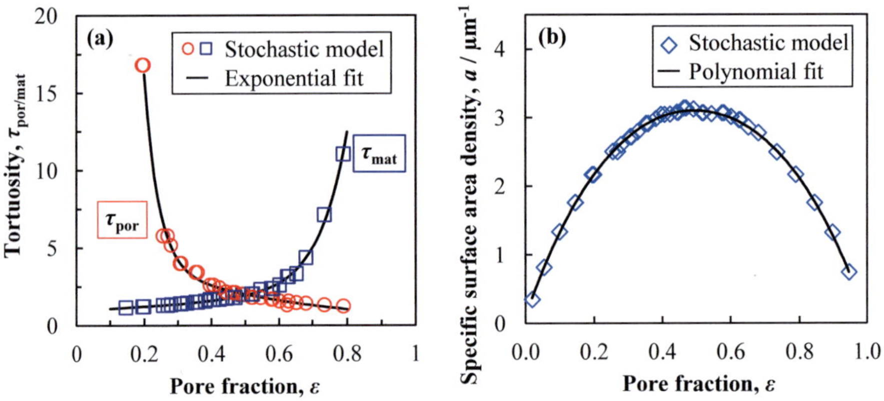

Figure 5.1.: (a) Pore and material phase tortuosity values and (b) specific surface area density, calculated (symbols) by stochastic LSCF microstructure-generator [120, 212] and exponential fits (lines) to numerical data.

5.1.4. Specific Surface Area

It is customary to express the surface area in relation to the total volume in order to gain a measure of comparison. The total surface area A is therefore normalized to the total reconstructed volume, resulting in a specific surface area, a_{i}, in $\mu m^2/\mu m^3$ or μm^{-1}.

The specific surface area within a porous electrode is a very important parameter because its size correlates directly to the kinetics of electrochemical and reforming reactions. For the model framework, the values of interest are (i) the pore/material surface area in the cathode, a_{MIEC}, which characterizes the surface area where the electrochemical oxygen reduction takes place and which is a key-parameter for the homogenized MIEC-cathode model (Section 4.3.3.2). And (ii) the Ni/pore surface area in the anode $a_{\mathrm{Ni/por}}$, where the reforming reactions occur and which is a key parameter to implement the reforming reactions into the model framework (Section 4.3.4).

The most straight forward way to calculate A is to count all voxel-faces between two different phases (material/pore or material/material) and multiply the sum by the actual voxel-face area. This will lead, however, to an overestimation of A of up to 50% because the electrode particle surface is rounded in various shapes and the voxel resolution is limited by the SEM resolution [121, 120]. A more precise method is given by the *marching cube* algorithm [213], which was implemented into a MATLAB code by M. Ender [121] to evaluate reconstructions of porous electrodes on voxel basis. The method is based on the triangulation of constant grey-values, thus increasing the resolution to capture the surface area of rounded particles. Still, an error up to 7% has to be expected, depending on the particle shape distribution within the reconstructed volume [121, 214]. More detailed information is given in the work of M. Ender [121].

Nevertheless, the determined specific surface area of interest for the model framework are listed in Table 5.1. Similar to the tortuosity, the specific surface area density in a porous electrode correlates with the porosity. The stochastic microstructure generator proposed by J. Joos [120] offers the option to calculate the pore fraction dependent surface area density, a_{MIEC}. The results, using the marching cube method in the determination process [121, 214], are displayed in Figure 5.1b. In order to regard the interdependence in later-performed parameter variations, the following polynomial function is implemented into the model framework:

$$a_{\mathrm{MIEC}} = -34.22 \cdot \varepsilon_{\mathrm{cat}}^4 + 83.41 \cdot \varepsilon_{\mathrm{cat}}^3 - 86.25 \cdot \varepsilon_{\mathrm{cat}}^2 + 41.15 \cdot \varepsilon_{\mathrm{cat}} - 4.2 \qquad \mu\mathrm{m}^{-1}. \tag{5.5}$$

5.1.5. Homogenization

In order to avoid the computational cost of a full microstructural model, but maintain the physical meaningful impact of on the implemented transport phenomena, an *effective* (or *homogenized*) modeling approach can be employed. Various techniques exist and are described in detail in Ref. [215]. In this work only two methods are used: *volume averaging* and *homogenization*.

The porous electrodes and semi-porous contact meshes are treated as homogenized volume elements, therefore electron and gas transport are described by effective transport parameters as a function of phase fraction and tortuosity, where species transport occurs. For this purpose, $\Psi_{\text{por/mat},el}$ functions as effective microstructural parameter for the pore or material phase of the corresponding electrode:

$$\Psi_{\text{por/mat},el} = \frac{V_{\text{por/mat},el}}{\tau_{\text{por/mat},el}}. \tag{5.6}$$

with $V_{\text{por/mat},el}$ denoting the volume fraction of pore or material phase of the corresponding electrode el. Hence, the in this work relevant *effective* species transport parameters are given by the following expressions:

$$D_{\text{eff},ik} = \Psi_{\text{por},el} \cdot D_{ik}, \tag{5.7}$$

$$D^{\text{eff}}_{Kn,i} = \Psi_{\text{por},el} \cdot D_{Kn,i}, \tag{5.8}$$

$$D^{\delta}_{\text{eff}} = \frac{1-\varepsilon_{el}}{\tau_{\text{mat},el}} \cdot D^{\delta}, \tag{5.9}$$

$$\sigma_{\text{eff},el} = \frac{1-\varepsilon_{el}}{\tau_{\text{mat},el}} \cdot \sigma_{\text{bulk},el}. \tag{5.10}$$

Furthermore, the surface area dependent electrochemical charge transfer reaction in the MIEC-cathode model (Section 4.3.3.2) and the chemical reforming reactions (Section 4.3.4) are transferred to volume related source terms by multiplication with the related, specific surface area a_{i}:

$$R^{\text{s}}_{\text{O}_2} = a_{\text{MIEC}} \cdot j_{\text{O}_2}, \tag{5.11a}$$

$$S_{\text{sh},k} = a_{\text{Ni-por}} \cdot j_k. \tag{5.11b}$$

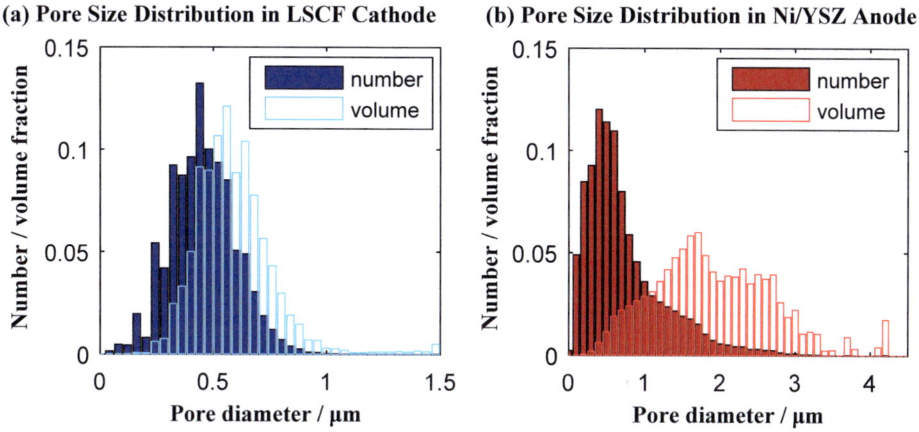

Figure 5.2.: Pore size distributions (PSD) in porous electrodes of an ASC Type-A (Table 5.1), calculated by number-weight and by volume-weight: (a) PSD in the LSCF cathode and (b) in the Ni/YSZ anode. Calculated mean values are listed in Table 5.1.

5.1.6. Mean Pore and Particle Size Diameter

A very important parameter of porous electrodes is $d_{\mathrm{por/mat}}$, the average *pore* or *particle diameter* respectively, denoted usually in µm. d_{por} plays a crucial role in the description of porous media gas transport, which is discussed in detail in Section 6.5. The most trivial approach in the determination of $d_{\mathrm{por/mat}}$ is to assume approximately spherical particles and pores to calculate a mean diameter by [121]:

$$\langle d_{\mathrm{i}} \rangle = \frac{6 \cdot \varepsilon_{\mathrm{i}}}{a_{\mathrm{i}}} \qquad \mu\mathrm{m} \tag{5.12}$$

wherein ε_{i} denotes the ith material phase volume fraction and a_{i} the corresponding volume specific surface area. This method is only accurate if the majority of particles or pores are equally sized spheres and not connected with each other. It is easy to understand that the ceramic electrodes are certainly not composed of uniform particles because the production powders are already inhomogeneous with a certain particle size distribution. Furthermore, in the high temperature production process, these particles form irregularly shaped, connected particles and pores, which are impossible to describe with a simple parameter [120]. However, it is possible to characterize such electrode structures with a method called *euclidean distance transform* (EDT), which was realized by M. Ender [121]. Although the algorithm was written for the characterization of Li-ion battery electrodes, it can be equally used for SOFC electrodes since its functional principle works with the same basic information. The interested reader can find detailed information in Refs. [120, 158, 216, 217].

Using the resulting size distribution from the EDT-analysis of the individual phases a mean particle or pore diameter is calculated using a number-weight ($d_{\mathrm{i}}^{\mathrm{n}}$) or volume-weight average

(d_i^v). The latter is based on the assumption that particle volume and particle size correlate with each other in the third dimension [121]:

$$d_i^v = \frac{\sum_{i=1}^{N} d_i \cdot d_i^3}{\sum_{i=1}^{N} d_i^3}. \tag{5.13}$$

The volume based weighting takes into account that species transport in more inhomogeneous distributions mainly occurs in larger particles or pores. It is important to note here that this differentiation is seldom made in literature, but has a notable effect on the Knudsen diffusion coefficient $D_{Kn,i}$ (Equation 4.31) and thus on the gaseous species transport properties.

An example for the ASC Type-A, with calculated pore size distribution using an average by number and by volume (Equation 5.13), is displayed in Figure 5.2. One can clearly see that the LSCF cathode (Figure 5.2a) is composed of much more homogeneously distributed pores over a smaller range of diameters. Nevertheless, a difference between number and volume average still exists: by a factor of ≈ 1.3.

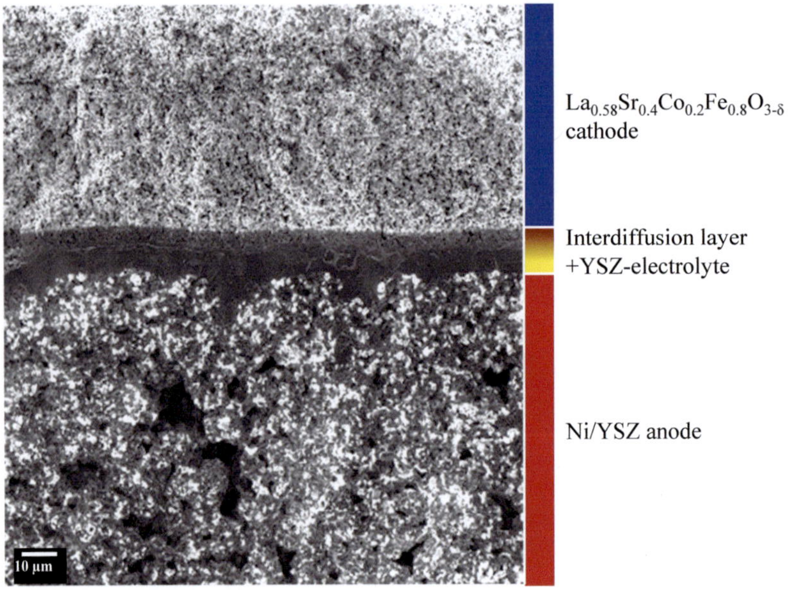

Figure 5.3.: Example display of inhomogeneous pore size distribution in the anode substrate layer of ASC Type-A section (pores colored black). SEM-image recorded with In-Lens-detector at 5k-magnification and EHT= 1.3 kV. [Cell Z2_275].

However, the Ni/YSZ anode in ASC Type-A has a more inhomogeneous microstructure compared to the cathode, which is demonstrated in the SEM-image (Figure 5.3). In the pore size distribution, this is reflected by the wider spread of detected pore sizes (Figure 5.2, right) compared to the LSCF cathode (Figure 5.2a). The number-weighing reveals that in the Ni/YSZ anode pores $< 1\mu m$ occur in much greater number, but a notable amount of larger pores exist, especially between $1 < d^{n}_{por} < 3\mu m$. The calculation by volume-weight shows that the small number of pores $> 3\mu m$ (which do not occur in the number-weighted calculation at all) occupy, due to their size, an appreciable amount of the volume fraction and influence the calculated average accordingly. Consequently, the volume and number average differ by a factor ≈ 2.6. The influence of this obvious difference will be discussed further in Section 6.5.

A disadvantage of the EDT-method is its dependence on the actual SEM image quality because the resulting voxels can be falsely segmented. It is very possible to detect multiple smaller pore/particle sizes instead of one larger if there is only one incorrectly assigned voxel within a pore or particle. The work of J. Joos determined a minimum amount of $10 - 15$ voxel per particle/pore and a volume of about $580\ \mu m^3$ is required to characterize a LSCF cathode with sufficient precision [120]. For Ni/YSZ-anodes, these values do not apply due to the much increased inhomogeneity. It was concluded by J. Joos that even a volume size of $8748\mu m^3$ is not sufficient, which is at the border of the possible reconstruction volume size using the FIB/SEM tomography. Larger volumes can be reconstructed using µ-CT, however due to a lower resolution negative consequences are to be expected regarding the accuracy of determined parameters. A combination of both tomography techniques might be an option for future investigations.

5.1.7. EIS-based Effective Microstructural Parameter

An alternative way to characterize the porous SOFC anode microstructure is to calculate a "microstructural paramcter" Ψ_{EIS} from electrochemical impedance measurements. The algorithm to determine Ψ_{EIS} was developed within A. Leonides dissertaion [9] and involves the application of a physically-based *equivalent circuit model* (ECM) to quantify recorded impedance data with the help of a *complex nonlinear least squares* (CNLS) fit (Section 2.3.2).

The *distribution of relaxation times* (DRT) is there a vital tool for identifying and separating the individual loss processes beforehand and supplying the user with start-parameters for the CNLS-fit (Section 2.3.2). Exemplary, Figure 5.4a displays DRTs calculated from impedance data recorded on an ASC Type-A for such a parameter variation: The hydrogen content in the fuel gas $pH_{2,an}$ is varied incrementally and substituted with inert nitrogen (N_2) while the steam content pH_2O_{an}, the cathode oxygen partial pressure $pO_{2,cat} = 1$ atm and the temperature $T = 800°C$ are kept constant. In this way, gas diffusion losses originating from the cathode are excluded and the impact on the cell impedance is solely related to the change of $pH_{2,an}$. The low frequency peak (P_{1A}) in the DRTs of Figure 5.4 decreases with increasing $pH_{2,an}$, showing typical Warburg-type behavior with an additional secondary peak at higher frequencies [18, 9, 29]. A comparatively low sensitivity towards the change in $pH_{2,an}$ can be identified in the high frequency range peaks (P_{2A} and P_{3A}), which are (according to Leonide) related to the anode fuel oxidation [9].

Consequently, in the ECM the anode gas diffusion polarization is regarded by a finite-length Warbug-type element (Figure 2.5), which is the equivalent circuit element of choice to describe continuous diffusion related processes based on Ficks law of diffusion [9, 18, 29]. Based on the Nernst-equation (Equation 2.5) and coupled with a generalized Ficks diffusion-model (Equation 4.22) the anode gas diffusion resistance $R_{cc,an}$ can be calculated by the following equation [9]:

$$R_{cc,an} = \left(\frac{RT}{2F}\right)^2 \frac{h_{an}}{p_0} \frac{1}{\Psi_{EIS,an}} \left(\frac{1}{D_{mol,H_2} \cdot xH_{2,an}} + \frac{1}{D_{m,H_2O} \cdot xH_2O_{an}}\right) \quad \Omega \cdot cm^2. \tag{5.14}$$

Here h_{an} denotes the anode substrate thickness in m (Table 5.1), $pH_{2,an}$ the hydrogen (H_2) and pH_2O_{an} the steam (H_2O) partial pressure in atm as present in the fuel gas composition. D_{mol,H_2} and D_{m,H_2O} denote in m/s^2 the corresponding molar diffusion coefficients, calculated by the Wilke-Bosanquet approach (4.3.2.3), thereby regarding intermolecular and Knudsen diffusion in the approach.

Combining Equation 5.14 with $R_{cc,an}$ obtained from CNLS-fitting, individual $\langle\Psi\rangle_{EIS,an}$ at every measured operating point can be determined. Inserting the resulting average $\Psi_{EIS,an}$ in Equation 5.14, a value for $R_{cc,an}$ can be predicted and compared with the measured data. The results of this procedure for a $pH_{2,an}$-variation recorded on an ASC Type-A (DRTs in Figure 5.4a) are displayed in Figure 5.4b. A high fit-accuracy (residuals < 2%) was achieved for all evaluated impedance spectra, shown by one representative CNLS-result, plotted in the upper right corner of Figure 5.4b. The deviation between individual and average microstructural parameters was below $\Delta\Psi_{EIS,an} < 0.015$. The results in Figure 5.4b therefore show, how Equation 5.14 parametrized with a microstructural parameter of $\Psi_{EIS,an} = 0.157$ reproduces the measured anode gas diffusion resistance with excellent precision.

The above described procedure was carried out on the same cell (Type-A, cell Z1-360) for a variation of pH_2O_{an}-variation. Again, with balance N_2 and constant $H_{2,an} = 0.4$ atm, $pO_{2,cat} = 1$ atm and $T = 800°C$. The resulting average microstructural parameter was calculated to $\Psi_{EIS,an} = 0.163$, also with a very low deviation of $\Delta\Psi_{EIS,an} < 0.01$. The achieved results are therefore in very good accordance with the first value. This further validates the applied method. The difference should be accounted to small deviations in the fitting process as well as in the calculation of the D_{mol,H_2} and D_{m,H_2O}, which is based on a linearized approximation for lower partial pressures of in a mixture containing species. The implemented generalized Fick model in Equation 5.14 only regards diffusive and no convective fluxes, as the static parameter p_0 in Equation 5.14 implies. This is important to recognize in the comparison of $\Psi_{EIS,an}$ obtained from the ECM-fitting process with $\Psi_{an,3D} = \varepsilon_{an}/\tau_{por,an}$, based on the microstructure information determined by the evaluation of 3D electrode reconstruction via FIB/SEM-techniques, as described in the previous sections.

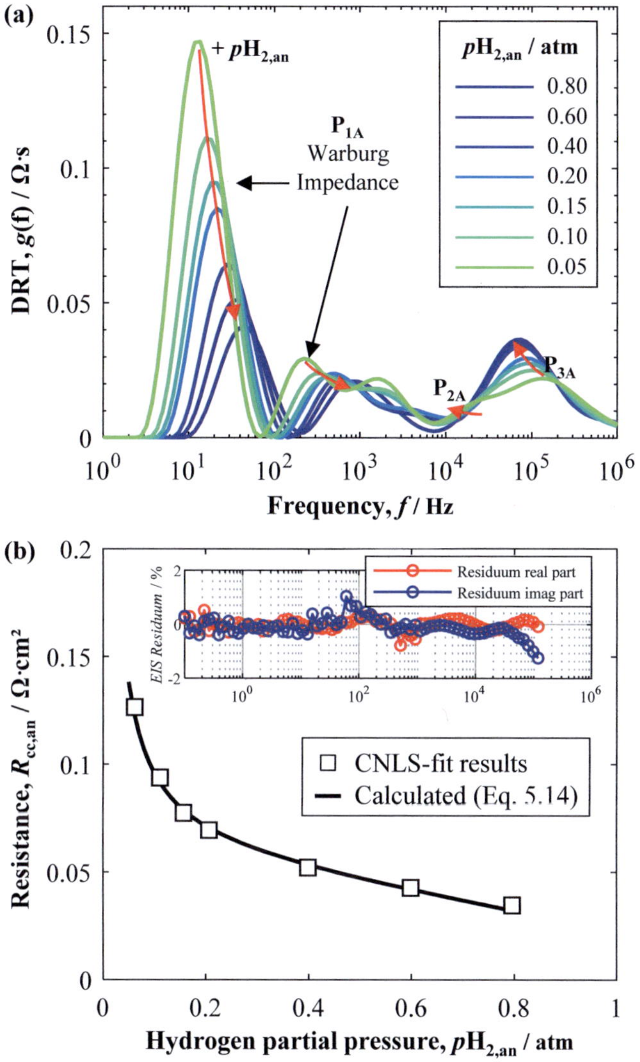

Figure 5.4.: (a) DRTs calculated from measured impedance data, recorded at varied hydrogen partial pressures in the fuel gas with simultaneous substitution of removed H_2 by an equal amount of nitrogen (N_2). Constant $pH_2O_{an} = 0.2$ atm, $pO_{2,cat} = 1$ atm and $T = 800°C$. (b) Anode gas diffusion resistance $R_{cc,an}$ comparison of measured (CNLS-fit results of measured impedance spectra, symbols) values with calculated data (line) using Equation 5.6 and $\Psi_{EIS,an} = 0.157$ (Table 5.1). [Type-A, cell Z1_360].

Furthermore, the influence of the Ni-mesh contact mesh on $R_{\mathrm{cc,an}}$ and thus on $\Psi_{\mathrm{EIS,an}}$ has to be taken into account when making a comparison. The matter is discussed further in detail in Section 6.5. Another point worth mentioning is the requirement for reliable data regarding the average pore diameter $d_{\mathrm{por,an}}$, which enters Equation 5.14 by the Knudsen diffusion coefficient $D^{\mathrm{eff}}_{\mathrm{Kn,i}}$ (Equation 4.31) in the calculation of the molar diffusion coefficients D_{i} (Equation 4.35). As discussed in Section 5.1.6, an error is incorporated in the determination algorithm, which of course influences the results here as well. However, single cells of ASC Type-B and C are also investigated by the method described above. All achieved results for $\Psi_{\mathrm{EIS,an}}$ are summarized in Table 5.1 at the end of this section.

It also should be mentioned that this work refrains from calculating a microstructural parameter for the cathode layer $\Psi_{\mathrm{EIS,cat}}$. Even though it was shown by A. Leonide that the cathode diffusion polarization exhibits a measurable resistance, $R_{\mathrm{cc,cat}}$, for $pO_{2,\mathrm{cat}} < 0.21$ atm [9], the resulting value is strongly influenced by the Au contact-mesh and a distinct value for the cathode gas diffusion resistance cannot be separated. Analytical approaches for calculating the mesh-influence can be found in literature by Gewiess [218] and Hayd [45], but nevertheless remain insufficient; certainly, an approach with spatial resolution is required to precisely regard the oxygen distribution beneath the flowfield rib.

5.1.8. Permeability

Permeability is the last material parameter required to adequately describe the species transport mechanisms regarded in this work. The parameter is fundamental to any porous medium, and depends on the medium porosity, pore shape and tortuosity. It is used in the field of fluid mechanics to characterize the ability of a fluid (or gas) to pass through the pore structure at a given pressure gradient. Traditionally, the parameter originates from the fields of petrol and geo-sciences, where it is use to characterize different rock formations. Nevertheless, permeability is required in any numerical fluid mechanic simulation where species transport in porous media occurs. It is also used in numerical SOFC modeling [177, 147], if pressure gradients are regarded in the species transport modeling. Typically, the Darcy equation (Equation 5.16) is used to model pressure induced viscous flux, wherein the permeability is denoted with κ_{el} in m^2 (indices: an for anode and cat for cathode, (Section 4.3.2.1). Hence, low permeability causes low permeation velocity $\vec{v}$ and consequently low viscous species flux $\vec{n}_{\mathrm{i}}$. In other words: a low permeability poses a high resistance against transported fluids.

Frequently cited SOFC modeling authors such as Suwanwarangkul *et. al* [169] or Virkar *et al.* [219] all together neglect viscous fluxes in their species transport modeling. They do so because they presume a low contribution to the total flux magnitude, and therefore they neglect the parameter. Other authors such as Tseronis *et al.* [148, 154] assume a certain value without further explanation, or blindly adopt values from literature sources. Finally, there are few sources in literature (e.g. Zhu *et al.* [177] and Bertei *et al.* [149, 147]), where analytical expressions are applied. Furthermore, their approaches differ.

Since this work aims to present a profound and physically correct model framework for homogenized SOFC FEM simulations, viscous flow has been included. Hence, reliable permeability data was required, the acquisition of which will be described in the following paragraphs.

Anode permeability

The best-case-scenario is to determine a permeability value from experiments conducted on the very same electrode samples as used in the validation measurements. However, elaborate experimental work with special equipment is required. So, unfortunately, supervised in-house experiments could not be carried out. Fortunately, experimental data on identical Forschungszentrum Jülich electrodes (published by Simwonis *et al.* [220, 221]) is available in literature where, among other parameters, the permeability of anode substrates was measured. In the sudies, Simwonis *et al.* used samples of varying porosity, which are similar to the ASCs used in this work (coat-mixing). In addition to that, detailed work on the matter was carried out in Jülich by Schafbauer [222] with anode samples of varying porosity. Both Simwonis and Schafbauer used the same method to determine κ_{an}: In specially designed test-setup, a defined air volume, V in m^3, is applied to a defined sample surface area, A in m^2, with the thickness, h, in m and the pressure drop, Δp in Pa, over the time, t in sec, was measured. With the help of the following expressions, based on DIN 51058 (basically a form of the Darcy equation Equation 4.12), a permeation number D_s (which is in fact equal to κ_{el}, therefore it is used in the following) is calculated [220, 222]:

$$\kappa_{\text{an}} = \frac{\eta_{\text{air}} h V}{A \Delta p t} \qquad \text{m}^2. \tag{5.15}$$

Therein, η_{air} denotes the dynamic viscosity of air in Pa·s. The experimental results from Simwonis and Schafbauer are depicted in Figure 5.5, denoted by squared symbols. A clear decreasing trend of κ_{an} with decreasing porosity ε_{an} can be observed. The samples used by Schafbauer are thereby comparable to the cells used in this work since both cell types originate from the same manufacturer and from the same development stage at Jülich. Furthermore, results of Simwonis and Schafbauer are in acceptable accordance with each other, thus delivering further evidence for regarding the data as reliable.

The experimental data from Jülich not only delivers excellent modeling parameters, but also offers the chance to test the validity of analytical approaches frequently used in literature. In the following, three common approaches used in the SOFC literature are introduced with a short discussion of any assumptions or simplifications. The required microstructure data are thereby adopted from the work of J. Joos [120] and determined by own-evaluations of 3D reconstructions based on the FIB/SEM algorithm (Section 2.6).

Starting with the approach of lowest complexity, an expression for κ_i can be derived by comparing the average fluid velocity in a porous media given by the Darcy velocity u_{Darcy} (Equation 5.16) and the velocity $u_{\text{capillary}}$ obtained by the Poiseuilles law applied for cylindrical capillaries (Equation 5.17) . The approach was first proposed by Kozeny in 1927 and later adopted and extended by other authors [223, 224]:

$$u_{\mathrm{Darcy}} = \frac{\varepsilon}{\tau_{\mathrm{por},el}} \cdot u = -\frac{\kappa}{\eta}\frac{dp}{dx} \qquad \mathrm{m/s}, \tag{5.16}$$

$$u_{\mathrm{capillary}} = -\frac{A}{\alpha}\frac{1}{\eta}\frac{dp}{dx} \qquad \mathrm{m/s}. \tag{5.17}$$

Both equations correlate velocity to a pressure gradient dp/dx with the fluid inherent viscosity η and a certain geometrical factor, characteristic for the perfused medium. In the case of Equation 5.16, this geometrical factor is the permeability, κ. In Equation 5.17 it is A/α with $A = \pi R^2$, denoting the area in m^2 for a cylindrical cross-section with the radius r, for which in a homogenized medium the average pore radius $r_{\mathrm{por,i}} = d_{\mathrm{por,i}}/2$ (Section 5.1.6) can be used. The value $\alpha = 8\pi$ is set as an empirical parameter for cylindrical tubes [225, 226, 227]. Comparing now Equations 5.16 and 5.17 with the mentioned considerations and regarding the characteristic microstructural effects by pore volume fraction $\varepsilon_{\mathrm{por},el}$ and tortuosity $\tau_{\mathrm{por},el}$ on the darcy velocity u_{Darcy}, the result delivers the following expression to calculate the permeability [223]:

$$\kappa_{el}^{\mathrm{K}} = \cdot \frac{\varepsilon_{el}}{\tau_{\mathrm{por},el}} \frac{d_{\mathrm{por},el}^2}{32} \qquad \mathrm{m}^2. \tag{5.18}$$

The approach given by Equation 5.18 is straight forward and has been used by Bertei *et al.* [147]. It does however require precise knowledge of the corresponding electrode microstructure properties. The wide distribution range of d_{por} (Figure 5.2) may negatively affect the prediction of κ_i^{K}.

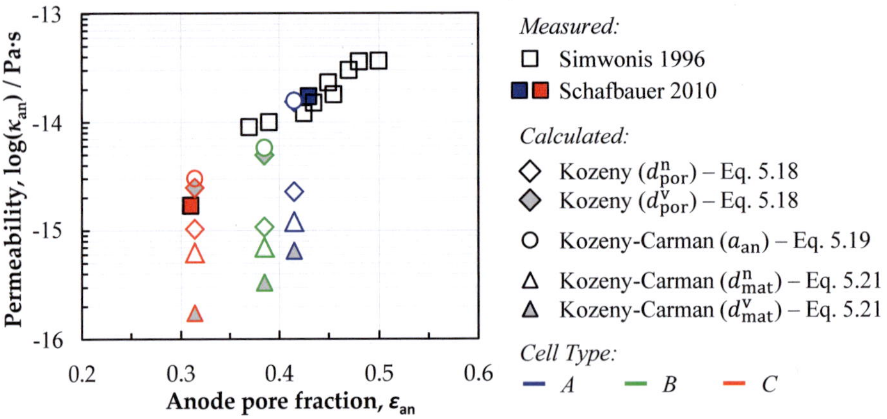

Figure 5.5.: Measured (squared symbols) anode substrate permeability values of samples with varying porosity [220, 221]. Calculated permeability values by Darcy-Poiseuille relation (Equation 5.18, diamond), by Kozeny-Carman relation based on surface density (Equation 5.19, circle) and by Kozeny-Carman relation based on average particle diameter (Equation 5.21, triangle).

The most common analytical approach found in the literature to calculate the permeability of a porous structure is the *Kozeny-Carman relation* [223, 228]. It is based on the Equation 5.18 and was further refined by Carman in 1937. Therein, a porous structure is assumed to be a bed of closely packed spheres, leaving the remaining inter-sphere spaces as open transport volume. Equation 5.18 is extended by replacing d_{por} by the hydraulic radius concept $R = \varepsilon_{el}/(a_{el} \cdot (1 - \varepsilon_{el}))$, wherein a_{el} denotes the surface area density between spheres and open volume. The resulting expression is hence given by [223, 228, 226]:

$$\kappa_i^{\text{KC}-\text{a}} = c \cdot \frac{\varepsilon_i}{\tau_{\text{por},i}} \cdot \frac{1}{8} \cdot \left(\frac{\varepsilon_i}{a_i \cdot (1 - \varepsilon_i)} \right)^2 \quad \text{m}^2 \tag{5.19}$$

wherein c is introduced as an empirical geometrical factor, known as the Kozeny constant (in the original derivation from 1927, Kozeny included the tortuosity as $c = C \cdot \tau$ [223]). In fact, the value c is basically a parameter to fit experimental data, even though it is seldom explained as such. Probably due to difficulties in determining a_{el}, it is commonly expressed by relating it to the sphere diameter d_{sphere}:

$$a_{\text{sphere}} = \frac{A_{\text{sphere}}}{V_{\text{sphere}}} = \frac{4 \cdot \pi \cdot r_{\text{sphere}}^2}{\frac{4}{3} \cdot \pi \cdot r_{\text{sphere}}^3} = \frac{6}{d_{\text{sphere}}} \tag{5.20}$$

and by inserting Equation 5.20 into Equation 5.19 results in

$$\kappa_i^{\text{KC}-\text{d}} = c \cdot \frac{\varepsilon_i}{\tau_{\text{por},i}} \cdot \frac{1}{288} \cdot \frac{d_{\text{mat},i}^2 \cdot \varepsilon_i^2}{(1 - \varepsilon_i)^2} \quad \text{m}^2. \tag{5.21}$$

Please note, in Equation 5.21 the average particle diameter $d_{\text{mat},el}$ is regarded and not the pore diameter $d_{\text{por},el}$, which is often confused in modeling approaches found in literature. Additionally, the Konzeny parameter c is assumed differently, causing multiple forms of the KC-equations to be circulating in literature. However, in SOFC literature Equation 5.21 with $c = 4$ has been used by Zhu *et al.* [177], for example.

Numerous, far more complex approaches for calculating permeability can be found in literature, generally fueled by the quest to capture the irregular shape of a porous medium or due to the incapability to determine the tortuosity. They are, however, accomplished by far more complex mathematical formulations: more importantly, these require additional parameters, which are either hard to come by, bear no physical meaning or are pure fitting-parameters to begin with. These models are therefore not addressed here.

In order to verify the applicability of aforementioned permeability-approaches, each model is parametrized with microstructure data from three different ASCs, determined from FIB/SEM analyzed anode substrates (Table 5.1). The results are plotted together with the measured values from Jülich in Figure 5.5. Analyzing this, it can be concluded that the simple approach by Kozeny (Equation 5.18), parametrized with a volume-averaged $d_{\text{por}}^{\text{v}}$, and the more complex KC-approach based on pore surface area, both deliver almost equal results for a Kozeny-constant $c = 3.25$. Furthermore, the results from both models are in very good accordance with the measured data. In contrast to these good results, the Kozeny-model parametrized with the number-averaged $d_{\text{por}}^{\text{n}}$ underestimates measured data of all cell types by half a decade and the KC-model based on the material phase diameter d_{mat} underestimates even up to decade or more, regardless of the used pore-diameter average. Such errors are cause by the model assumption to represent the porous microstructure as closely packed spheres and thereby overestimate the surface area by a factor ~ 5 [121]. Although the surface-based KC-model is derived from similar assumptions, the parametrization with surface density data obtained from FIB/SEM-analysis clearly makes a difference.

Cathode permeability

The thinness of the cathode layer prevents experimental determination of its permeability, κ_{cat}, due to material stability reasons. Nevertheless, it is doubly important to validate the presented model, both to implement a reliably value for mode validation purposes, and to incorporate a physically correct dependence on microstructure determining parameters such as porosity. Only in this way can one secure a parameter variation that calculates the implemented species transport properties in a physically correct way.

Hence, in the frame of a supervised master thesis [229], fluid mechanic FEM simulations were performed on various real and generated 3D LSCF cathode microstructures with varying porosity with the goal to determine their permeability. Figure 5.6 illustrates the applied workflow: (i) A stack of continuous 2D images (Figure 5.6a) is generated by FIB/SEM tomography of real LSCF cathode microstructures or by the LSCF microstructure generator [120]. The benefit of artificial structures is generation of desired microstructure properties such as a specific porosity. (ii) The 2D raw data is imported into the commercial software Synopsys' Simpleware^{TM} ScanIP to generate a 3D structure and produce a numerical mesh of the pore volume (Figure 5.6b). (iii) In the last step, the generated 3D mesh is imported into the commercial FEM software COMSOL Multiphysics v4.3b [125] and the Navier-Stokes equation (Equation 4.9) is solved. The result is a spatial velocity distribution, $\vec{u}$, within the pore volume (Figure 5.6c). In a final step, the Darcy Equation 4.12 is rearranged, whereby an isotropic permeability distribution is assumed and only the predetermined transport direction is regarded:

$$\kappa_{xx} = -\varepsilon \cdot \eta \cdot u_{D,x} \cdot \left(\frac{\delta p}{\delta x}\right)^{-1} \qquad \mathrm{m}^2. \tag{5.22}$$

Hereby, in Equation 5.22, the character ε denotes the pore fraction, η the fluid viscosity and u_{Dx} the Darcy-velocity, which is calculated by averaging the local velocity, $\vec{u}$, in the relevant transport direction:

$$u_{D,x} = \frac{1}{V_p} \int_{V_p} u_x dV \qquad \mathrm{m/s}. \tag{5.23}$$

Besides extensive work in finding the representative volume element (RVE) and investigations regarding mesh resolution and computational effort, the main result of the master thesis is displayed in Figure 5.7. It shows κ_{cat} calculated from 3D structures with varied porosity, $\varepsilon_{\mathrm{cat}}$, and calculated results from analytic models (lines), fed with microstructure data from the artificial structures.

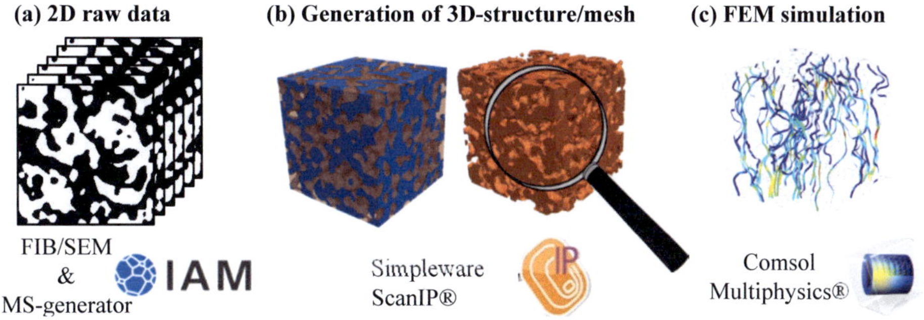

Figure 5.6.: Workflow to determine κ_{cat} based on 3D microstructure simulations: (a) Gather 2D raw data via FIB/SEM or stochastic microstructure generator, (b) generate the 2D a 3D structure and a numerical mesh and (c) perform FEM simulation within COMSOL to solve a pressure drop problem by the Navier-Stokes equation (Eq.4.9).

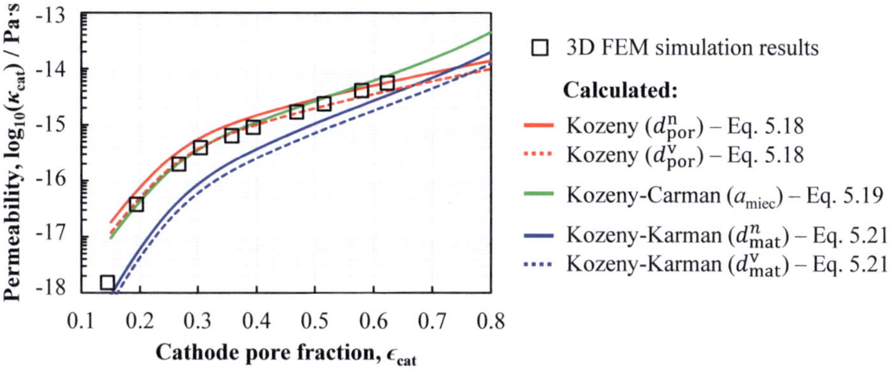

Figure 5.7.: Permeability values for a LSCF cathode with varying porosity, predicted by 3D FEM simulations (symbols) and calculated via various analytic approaches (lines).

The results displayed in Figure 5.7 clearly show that the surface area dependent Kozeny-Carman-model (Equation 5.19) is the best at reproducing the simulated results over the widest porosity range. The Konzeny-model (Equation 5.18) overestimates, and the material diameter dependent Kozeny-Carman-model (Equation 5.21) underestimates the simulated data. The Kozeny-constant was kept constant at $c = 3.25$, similar as in the anode permeability calculation. Comparing the cathode permeability results with the results obtained for the anodes (Figure 5.5), a smaller deviation between the pore/particle diameter averaging method is observed, which is attributed to the much more homogeneous cathode microstructure.

Table 5.1.: Microstructural parameters of anode supported SOFCs (ASCs) manufactured at Forschungszentrum Jülich, which have been used in this work. The ASCs differ in their Ni/YSZ-anode substrate layer-thickness but are equipped with identical LSCF-cathodes. All parameters have been determined by FIB/SEM-analysis of samples taken from measured ASCs, unless noted otherwise.

		Type-A	**Type-B**	**Type-C**	
Anode (substrate)	Parameter		value		unit
Layer thickness	h_{an}	1500	1000	500	µm
Pore fraction	$\varepsilon_{\mathrm{an}}$	41.5	38.5	30.8	%
Tortuosity (por)	$\tau_{\mathrm{por,an}}$	3.03	3.52	7.1	-
Surface area (por-mat)	$a_{\mathrm{an}}^{\mathrm{por-mat}}$	1.33	1.74	1.22	μm^{-1}
Surface area (por-Ni)	$a_{\mathrm{an}}^{\mathrm{por-Ni}}$	0.40	0.59	0.43	μm^{-1}
Avg. pore diameter	$d_{\mathrm{por,an}}^{\mathrm{n}}$	729	559	677	nm
Avg. pore diameter (vol.)	$d_{\mathrm{por,an}}^{\mathrm{v}}$	1910	1820	1133	nm
Avg. particle diameter	$d_{\mathrm{mat,an}}^{\mathrm{n}}$	912	826	1043	nm
Avg. particle diameter (vol.)	$d_{\mathrm{mat,an}}^{\mathrm{v}}$	1244	1214	1536	nm
Permeability[1] [222]	$\kappa_{\mathrm{an}}^{\mathrm{m}} \cdot 10^{-14}$	1.73	-	0.17	m^2
Permeability[2] (Equation 5.18)	$\kappa_{\mathrm{an}}^{\mathrm{K}} \cdot 10^{-14}$	1.53	1.35	0.17	m^2
Eff. microstructural parameter[3]	$\Psi_{\mathrm{EIS,an}}$	0.157	0.133	0.06	-
Reconstructed volume size	-	8748	14600	8700	μm^3
Cathode	Parameter		value		
Active electrode area	A_1 & A_{16}	1 & 16	1	1	cm^2
Layer thickness	h_{cat}		45		µm
Pore fraction	$\varepsilon_{\mathrm{cat}}$		44.5		%
Tortuosity (por)	$\tau_{\mathrm{por,cat}}$		2.04		-
Tortuosity (mat)	$\tau_{\mathrm{mat,cat}}$		1.68		-
Surface area (por-mat)	a_{MIEC}		2.88		μm^{-1}
Avg. pore diameter	$d_{\mathrm{por,an}}^{\mathrm{n}}$		454		nm
Avg. pore diameter (vol.)	$d_{\mathrm{por,an}}^{\mathrm{v}}$		590		nm
Avg. particle diameter	$d_{\mathrm{mat,an}}^{\mathrm{n}}$		470		nm
Avg. particle diameter (vol.)	$d_{\mathrm{mat,an}}^{\mathrm{v}}$		556		nm
Permeability[2] (Equation 5.18)	$\kappa_{\mathrm{cat}}^{\mathrm{K}} \cdot 10^{-14}$		0.15		m^2
Reconstructed volume size	-		30816		μm^3

[1]measured at Forschungszentrum Jülich [2]calculated [3]EIS measurements

5.2. Electrochemical Charge Transfer Kinetics

In Section 4.3.3, two approaches were proposed to implement electrochemical charge transfer reaction in the model framework: The *Butler-Volmer*-approach (BV), applicable for both anode and cathode reactions and secondly, the homogenized *mixed-ionic-electronic conductive* (MIEC) approach (for only the cathode charge transfer reaction).

Required kinetic parameters for both approaches are experimentally determined from impedance data. For this purpose, impedance measurements are conducted as described in Section 3.3 on anode supported single SOFCs (ASCs) with an active electrode area of 1 cm^2, thus ensuring a homogeneous gas and temperature distribution and well-defined operating parameters. The ASCs differ in their anode support layer thickness, but are equipped with a comparable 8YSZ-electrolyte, CGO interdiffusion-layer and a LSCF-cathode layer (Table 5.1).

5.2.1. Butler-Volmer Model (BVM)

For the BV-parameter determination of each cell type, roughly ~120 impedance spectra have to be analyzed with the help of the DRT (Section 2.3.2) and evaluated by a CNLS-fit of the equivalent circuit model (ECM), described in Section 2.3.2. Measurement details and evaluation algorithms are further described in Refs. [9, 28].

Table 5.2.: Kinetic parameters for Butler-Volmer approach (Equations 4.59 and 4.60) to model anodic and cathodic electrochemical charge transfer reactions. From literature adopted values as cited.

	Ni/YSZ-anode		
Operating Gas	H_2/H_2O	reformate	CO/CO_2
Exponent a	-0.10	-0.06	-0.04
Exponent b	0.33	0.34	0.25
Pre-exp. factor γ_{an} / $10^6\cdot$ A·m^{-2}	$1.83 \cdot T$	$1.83 \cdot T$	$4.56 \cdot T$
Act. Energy E_{act} / eV	1.09	1.10	1.23
Charge transfer coeff. α	0.59	0.59	0.62
Reference	[9]	[28]	[16]

	LSCF-cathode	
ASC Type	Type - A [Cell Z2_275]	Type - B [Cell Z1_188]
Operating Gas	ambient air	ambient air
Exponent m	0.13	0.22
Pre-exp. factor γ_{cat} / $10^8\cdot$ A·m^{-2}	$2.35 \cdot T$	$1.52 \cdot T$
Act. Energy E_{act} / eV	1.42	1.45
Charge transfer coeff. α	0.65	0.65
Reference	this work	[9]

The high reproducibility of ASCs fabricated by Forschungszentrum Jülich makes them ideal. Further, they have uniform anode support-layers (thickness and microstructure properties), and are equipped with comparable remaining layers (Ni/8YSZ-AFL, 8YSZ-electrolyte, CGO interdiffusion-layer and LSCF-cahtode). Therefore this work adopts the BV electrode kinetics for H_2/H_2O- and CO/CO_2-operation determined by A. Leonide [9, 16] and for reformate-operation determined by Kromp [28]. The cathode activation energy, $E_{\mathrm{act,cat}}$, has been determined for ASC Type-A, because A. Leonide performed all impedance measurements under ambient air, which causes an ongoing degradation of the cathode charge transfer kinetics [31]. Table 5.2 lists all parameters required in the model framework for Equations 4.60, 4.60a and 4.60b.

5.2.2. MIEC-Cathode Model

Before discussing the kinetic parameters k^{δ} and D^{δ}, it should be mentioned that different notations exist in literature, often inconsistently used. In [45], a clear breakdown is given, related to individual measurement procedure: (i) Electrical conductivity measurement: k^{Q} and D^{Q}, (ii) Tracer experiments: k^* and D^* and (iii) Chemical experiment: k^{δ} and D^{δ}. Details regarding the different measurement methods can be found in [45]. Furthermore, values for k^{δ} and D^{δ} vary a great deal across literature, especially for D^{δ}, where data can be found ranging over several orders of magnitude. Multiple reasons can be found for this: (i) Varying measurement methodology with different accuracy aspects, (ii) dense or porous sample fabrication method, requiring different sinter-temperatures (impact on grain size, chemical composition) and (iii) varying sample history with respect to temperature and unrelated prior operations. However, operating temperature and time have a tremendous impact on the activation resistance ($R_{2C} = R_{\mathrm{chem}}$) and time constant ($t_{\mathrm{chem}}$) of perowskite-type cathodes. For a LSCF-cathode this has been discussed in detail in Refs. [9, 31, 69].

Most recent results by L. Almar *et al.* [231] and J. Szasz [232] point out how cathode characterization in ambient air is greatly affected by surface segregation [233, 234], secondary phase formations at the cathode/electrolyte interface [72] and degradation due to gas contamination (CO_2, SO_2, Cr, Si, etc.) [230, 235, 236]. Furthermore, it was shown by Almar and Szasz that the impedance response of a MIEC electrode operated in synthetic air does not show a Gerischer-type behavior and therefore cannot be modeled with the according Gerischer-element in the ECM. In order to reproduce real stack operation conditions in a model, a more application-oriented approach would be to conduct characterizing and parameter determination in ambient air. C. Endler-Schuck *et al.* have also shown that at $T = 750$ °C after approx. 300h a saturation of R_{chem} is reached and further degradation proceeds at a much lower rate. Hence, it can be regarded as constant [31, 69]. Leonide and Joos confirmed how the impedance responds after the initial degradation phase can be satisfactorily described by a Gerischer-element and models parametrized with accordingly determined values reproduce the measured results with high precision [9, 120].

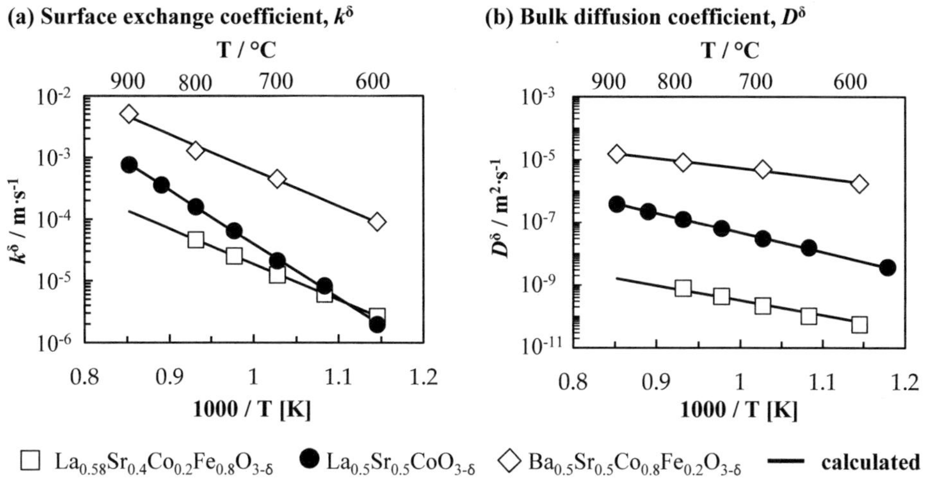

Figure 5.8.: MIEC parameters, temperature dependency: (a) surface exchange coefficient k^δ and (b) bulk diffusion coefficient D^δ in air for different MIEC materials between 600 °C and 900 °C. Values for LSCF (squares) were determined in this work from impedance measurements, based on the method proposed by J. Joos *et al.* [31, 120]. Values for LSC are taken from Ref. [230] and for BSCF from Ref. [231]. Symbols indicate measured values and lines calculated values using Equations 5.24 and 5.25 with data from Table 5.3.

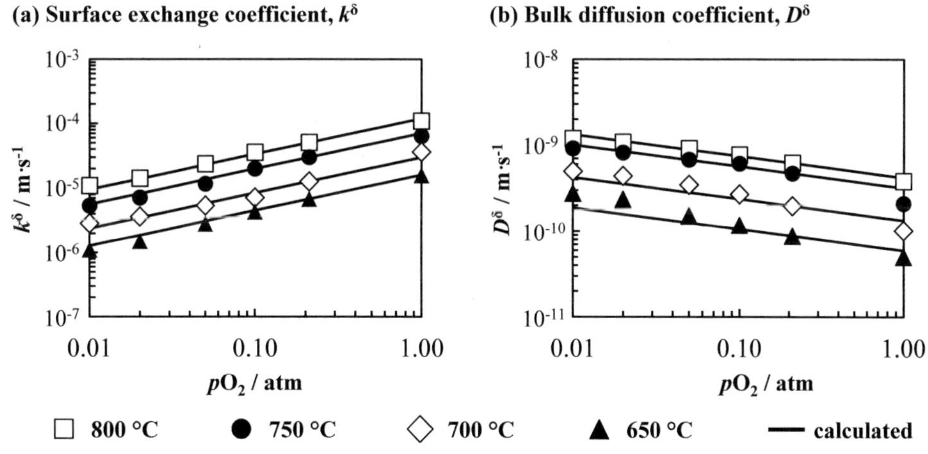

Figure 5.9.: MIEC parameters (LSCF), oxygen partial pressure dependency at various temperatures: (a) surface exchange coefficient k^δ and (b) bulk diffusion coefficient D^δ at $pO_{2,\mathrm{cat}} = 1 \ldots 0.01$ atm between 650 °C and 800 °C. The data was determined in this work from impedance measurements, based on the method proposed by J. Joos *et al.* [31, 120]. Symbols indicate measured values and lines calculated values using Equations 5.24 and 5.25 with data from Table 5.3.

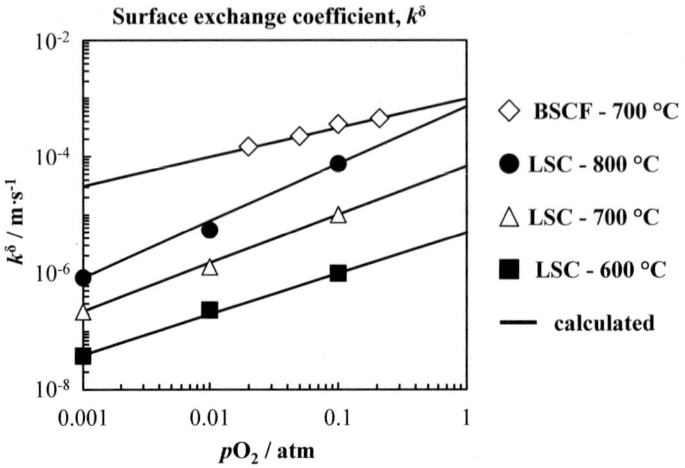

Figure 5.10.: Surface exchange coefficient k^δ at various oxygen partial pressures $pO_2 = 1...0.001$ atm for different MIEC materials between 600 °C and 800 °C (values for LSC from Ref. [237] and for BSCF from Ref. [231]). Symbols indicate measured values and lines indicate values calculated by Equations 5.24 and 5.25 with data from Table 5.3.

Bearing all this in mind, k^δ and D^δ for various operating temperatures and oxygen partial pressures are determined from impedance measurements conducted at an ASC Type-A with a LSCF-cathode, after the mentioned initial 300h degradation phase in ambient air. The exact method is adopted from Refs. [9, 31, 120]. The results for an operating temperature ranging from $T = 900 \ldots 600$ °C and $pO_{2,\mathrm{cat}} = 0.21$ atm are plotted in Figure 5.8. In addition, values for lathanium doped strontium cobaltide ($La_{0.5}Sr_{0.5}CoO_{3-\delta}$) and barium doped strontium cobalt ferrite ($Ba_{0.5}Sr_{0.5}Co_{0.8}Fe_{0.2}O_{3-\delta}$) are included in Figure 5.8. Similar LSCF, both materials are likewise perowskite-oxides and are discussed in the literature as promising cathode materials due to their high oxygen kinetics. This is clearly shown in Figure 5.8 by the faster (higher value) surface exchange and bulk diffusion coefficients, especially of BSCF and for both materials at lower temperatures.

Naturally, these oxygen-ion conducting materials react to their surrounding atmosphere, so their dependence to change in pO_2 should be taken into account. It is discussed by Almar [231] that for BSCF the limiting process is the surface exchange and therefore no values for the bulk diffusion coefficient D^δ can be determined. The same seems to be true for LSC, as no data regarding a pO_2-dependence of D^δ can be found in the literature. However, for LSCF a pO_2-dependence is detected in the evaluation of measured impedance data. The results found for k^δ and D^δ in the temperature range of $T = 800 \ldots 600$ °C are shown in Figure 5.9, while Figure 5.10 depicts the pO_2-dependence of k^δ for LSC and BSCF (only at $T = 700$ °C) as found in literature [231, 230]. A decreasing trend for k^δ of all materials is given for lower pO_2, while D^δ increases for LSCF. The slowdown of surface exchange kinetics can be simply explained by the decreasing amount of oxygen molecules available in the gas atmosphere. This causes a decreased oxygen concentration in the lattice structure, which is consequently the reason for the increased D^δ-values as more oxygen vacancies are left open for transport to take place.

The depicted lines in Figures 5.8 to 5.10 are calculated k^δ and D^δ values using the following Arrhenius-type expressions [120]:

$$k^\delta = \exp\left(A_\mathrm{k}\right) \cdot \exp\left(\frac{-E_\mathrm{act,k}}{RT}\right) \cdot \left(\frac{pO_2}{pO_{2,ref}}\right)^{\alpha_\mathrm{k}} \qquad \mathrm{m/s}, \tag{5.24}$$

$$D^\delta = \exp\left(A_\mathrm{D}\right) \cdot \exp\left(\frac{-E_\mathrm{act,D}}{RT}\right) \cdot \left(\frac{pO_2}{pO_{2,ref}}\right)^{\alpha_\mathrm{D}} \qquad \mathrm{m/s^2}. \tag{5.25}$$

Herein $A_\mathrm{k/D}$ denotes the pre-exponential factors, $E_\mathrm{act,k/D}$ the according activation energies in kJ/mol by which the reference-values for k^δ and D^δ at $pO_{2,ref} = 0.21$ atm have been determined. The partial pressure dependency is realized by the second fraction in Equations 5.24 and 5.25, where the exponent $\alpha_\mathrm{k/D}$ determines the de- or increasing trend. All values for the various cathode materials are listed in Table 5.3. It can be deduced from Figures 5.8 and 5.9 that the measured values can be reproduced with excellent precision for the temperature variation (for LSCF $\Delta_\mathrm{error} < 5$ %) and with satisfactory precision for the partial pressure dependency (for LSCF $\Delta_\mathrm{error} < 12$ % at 800 °C while at 650 °C $\Delta_\mathrm{error} < 33$ % due an temperature dependent, increasing error in the calculation method). Finally, Joos showed in his work [120] that for an ideally contacted cathode with sufficient gas supply, the sensitivity of $\alpha_\mathrm{k/D}$ is negligible and constant oxygen kinetics may be assumed. However, if the gas supply is non-ideal (as in the stack operation), the results later presented will show that it is vital to include the partial pressure dependency in the modeling.

In summary, the presented in this section parameters are vital for the presented modeling approach. It enables the model to regard the oxygen kinetics of different cathode materials, thus comparison of their performance under variable conditions. The results for different temperatures, cell and contact designs are presented and discussed in Section 7.2.

Table 5.3.: Parameters for calculating oxygen ion (O^{2-}) surface exchange (k^δ) and bulk-diffusion (D^δ) coefficients in dependence of operating temperature T and oxygen partial pressure $pO_{2,\mathrm{cat}}$, using Equations 5.24 and 5.25.

Material	A_k m/s	$E_\mathrm{act,k}$ kJ/mol	α_k	A_D m²/s	$E_\mathrm{act,D}$ kJ/mol	α_D
$La_{0.58}Sr_{0.4}Co_{0.2}Fe_{0.8}O_{3-\delta}$	2.48	111.2	0.55	-8.19	115.4	-0.25
$La_{0.5}Sr_{0.5}CoO_{3-\delta}$ [237]	10.05	167.8	0.83	-2.72	117.6	-
$Ba_{0.5}Sr_{0.5}Co_{0.8}Fe_{0.2}O_{3-\delta}$ [231]	6.02	111.4	0.49	-4.92	60.2	-

5.3. Electronic and Ionic Conductivities

Temperature dependent, effective electronic/ionic charge transport is described in the model framework by Ohm's law (Equation 4.1), wherein the required material parameters electronic conductivity $\sigma_{\mathrm{an/cat}}$ and $\sigma_{\mathrm{elyt}}^{\mathrm{eff}}$ govern the charge transport. In contrast to existing models in literature, only experimentally determined cell parameters are used in this work.

5.3.1. Electronic Conductivity of Ni/8YSZ-Anodes

Electronic conductivity in the cermet-anode is supplied by the Ni-phase and is here adopted from the work of M. Kornely [37, 68]. Kornely carried out constant current conductivity measurements on Ni/8YSZ-anode samples similar to the cell types used in this work, therefore it is appropriate to adopt the data. In the relevant SOFC operating temperature range, the measured effective electronic Ni/8YSZ-anode conductivity, $\sigma_{\mathrm{an}}^{\mathrm{eff}}(T)$, is reproduced with high precision by the following Arrhenius-type expression:

$$\sigma_{\mathrm{Ni/8YSZ}}^{\mathrm{eff}}(T) = 82643 \cdot \exp\frac{5.349[\mathrm{kJ/mol}]}{R \cdot T} \qquad \mathrm{S/m}. \tag{5.26}$$

For example, at $T = 800°\mathrm{C}$ an effective conductivity of $\sigma_{\mathrm{an}}^{\mathrm{eff}} = 1505.1$ S/cm is calculated.

5.3.2. Electronic Conductivity of various Cathode Materials

As discussed in the fundamentals of Section 2.4.1, the electronic conductivity in a cathode is dependent on its working principle. In a layered or composite electrode, electron transport only occurs in the containing metallic phase, whereas in a MIEC-electrode electronic conduction takes place in the entire material phase. The underlying transport mechanism is further discussed by J. Hayd [45]. In any case, effective electronic conductivity data can be acquired by constant current experiments carried out by regarding electrode samples. For LSCF, constant current 4-point measurements were carried out at the Institute for Applied Materials (IAM-WET) [37]. The used samples are produced from the same LSCF-paste and under comparable manufacturing conditions as applied for the standard Jülich cathode-layers [120]. The following expression for $\sigma_{\mathrm{LSCF}}^{\mathrm{eff}}$ is gained by fitting a polynomial function to the experimental results [37]:

$$\sigma_{\mathrm{LSCF}}^{\mathrm{eff}} = 4.176 \cdot 10^{-8} \cdot T_{\mathrm{c}}^{3} - 3.575 \cdot 10^{-4} \cdot T_{\mathrm{c}}^{2} + 0.373 \cdot T_{\mathrm{c}} - 2.707 \qquad \mathrm{S/cm}. \tag{5.27}$$

Due to the fact that production-paste and manufacturing conditions of conductivity experiment samples and ASC cathode layers are equal, similar microstructural parameters and electron conduction abilities are assumed. Consequently, an electronic bulk-conductivity of LSCF $\sigma_{\mathrm{LSCF}}^{\mathrm{bulk}}$ is calculated using Equation 5.10 and Equation 5.27 together with microstructural data from Table 5.1. These alterations to the model framework allow it to consider changes in the cathode microstructure via Equation 5.10, which is a key feature for the results discussed in Section 7.2.2.

Table 5.4.: Electronic conductivities (effective values for LSCF, LSM, LSM+YSZ) of various cathode materials at different SOFC operating temperatures, listed from top to bottom with decreasing conductivity.

Cathode material	Unit	Temperature, T / °C		
		600	700	800
$La_{0.5}Sr_{0.5}CoO_{3-\delta}$ (LSC) [237]		1858	1546	1231
$La_{0.58}Sr_{0.4}Co_{0.2}Fe_{0.8}O_{3-\delta}$ (LSCF)		101.5[1]	97.7[1]	88.4[1]
$La_{0.8}Sr_{0.2}MnO_{3-\delta}$ (LSM) [68, 37]		17.2[1]	17.9[1]	18.4[1]
$Ba_{0.5}Sr_{0.5}Co_{0.8}Fe_{0.2}O_{3-\delta}$ (BSCF) [202]	$S{\cdot}cm^{-1}$	36	36	36
LSM+YSZ [37, 68]		-	-	1.9[1]
$SrTi_{1-x}Fe_xO_{3-y}$ (STF) [238]		-	-	$0.99 - 1.8$
$Pr_{0.005}Ce_{0.995}O_{2-\delta}$ (PCO) [239, 240]		$10^{-2.5}$	$10^{-2.5}$	$10^{-2.5}$

[1]In-house measured effective electronic conductivity.

Table 5.4 lists the conductivity-data of various cathode materials, which are already applied in SOFC stacks (LSCF, LSM+LSM/YSZ) and which are discussed as promising future materials due to their high electrochemical kinetics (PCO, STF, LSC, BSCF). The data for LSM, LSM/YSZ and LSCF are effective conductivity values, whereby the microstructures of LSM and LSM/YSZ samples are unknown. Nevertheless, the data was measured by M. Kornely in-house [37, 68] with samples which were fabricated by Forschungszentrum Jülich as reference cathodes for SOFC stacks (at that time). For LSC and BSCF, literature provides reliable bulk-conductivity data in Ref.[237] for LSC and in Ref. [202] for BSCF, whereby the data for LSC is extrapolated for $T > 700$ °C.

Analyzing the data in Table 5.4, it becomes clear that σ_{cat} can range over several orders of magnitude depending on the applied material. The according impact on SOFC stack performance and consequences regarding layer thickness and interconnector design are evaluated and discussed further in Section 7.2.2.

For the sake of completeness, it should be mentioned that σ_{cat} has a specific pO_2-dependency [237, 202], as the conductivity decreases for lower pO_2. The effect is related to the formation of more oxygen vacancies and thereby a change in the local charge compensation and charge transport. At this point the effect is not implemented into the model framework. However, the impact on SOFC stack performance is considered to be low, but should be kept in mind for future model extension. The same applies to time dependent degradation mechanisms (e.g. Chromium poisoning [118]).

5.3.3. Effective Ionic Conductivity of Electrolyte/Interdiffusion-Layer

The ionic charge transport in the electrolyte and interdiffusion-layer is modeled in this work by an effective approach. The individual layers are not resolved but combined into a single layer with a resulting effective ionic conductivity $\sigma_{\mathrm{elyt}}^{\mathrm{eff}}$. The approach is advantageous, because $\sigma_{\mathrm{elyt}}^{\mathrm{eff}}$ can be determined from impedance measurements with high precision, thus including manufacturing related decrease of ionic conductivity in the model (e.g. secondary phase formation due to Sr-diffusion [72]). It has to be pointed out, however that ongoing degradation due to as-yet-unknown mechanisms, the determined $\sigma_{\mathrm{elyt}}^{\mathrm{eff}}$ is only valid for a certain time window and will decrease further. Also, the spatial charge distribution in both layers should not be considered as precise, because in reality GCD has a lower conductivity compared to 8YSZ. The resulting overall loss in the electrolyte/interdiffusion-layer due to non-ideal charge transport is maintained in the approach and therefore serves the required purpose of the model framework [9]:

$$\sigma_{\mathrm{elyt}}^{\mathrm{eff}} = \left[\frac{T}{B_{\mathrm{elyt}}} \cdot \exp\left(\frac{E_{\mathrm{act,elyt}}}{RT} \right) \right]^{-1} \cdot h_{\mathrm{elyt}} \qquad \mathrm{S/cm}. \tag{5.28}$$

The effective ionic conductivity, $\sigma_{\mathrm{elyt}}^{\mathrm{eff}}$, is calculated by Equation 5.28, wherein the required parameters are the activation energy, $E_{\mathrm{act,el}}$, a pre-factor, B_{el}, and the combined electrolyte/interdiffusion-layer thickness, h_{elyt}. The latter is determined from SEM-image analysis, whereas the others result from impedance measurements at various temperatures. The process involves fitting the equivalent circuit described in Section 2.3.2 to measured impedance spectra and evaluating R_0 from the ohmic resistance element. Plotting R_0 over $T/1000$ in an Arrhenius-plot allows the determination of $E_{\mathrm{act,elyt}}$ and B_{elyt} by fitting Equation 5.28 to the measured data. Table 5.5 lists all parameters needed to calculate $\sigma_{\mathrm{elyt}}^{\mathrm{eff}}$ by Equation 5.28. The deviation between the determined values for both cell types is accounted to small manufacturing differences as well as to a difference in relative measurement point in time. In total however, both cell types have a very low ohmic resistance with only small differences to each other.

Table 5.5.: Required parameters to calculated via Eqation 5.28 the effective ionic conductivity $\sigma_{\mathrm{elyt}}^{\mathrm{eff}}$ of the combined 8YSZ-electrolyte/GDC-interdiffusion-layer.

Description	Parameter	Unit	Cell Type-A	Cell Type-B[1]
Pre-factor	B_{elyt}	S·K/m^2	$8.745 \cdot 10^{12}$	$4.1879 \cdot 10^{12}$
Activation energy	$E_{\mathrm{act,elyt}}$	kJ/mol	96.05	90.31
Electrolyte/GDC-layer thickness	h_{elyt}	µm	~14	~17

[1]Adopted from Ref. [9].

5.4. Metallic Contact Mesh

Metallic contact meshes are integrated into the ideal test-setup between the active electrode areas and the flowfield gas channels to ensure ideal electrical contacting and sufficient, homogeneously-distributed gas supply to the porous electrodes surface (Section 2.5.2). The feature distinguishes ideal and stack contact design at the cathode side (Section 2.5.3). Consequently, detailed informations regarding the size and design are required in order to reproduce the test-bench setup in the model.

A schematic layout is illustrated by a mesh-section on the left in Figure 5.11, while geometric details are given on the right. In the test setup (Section 3.2.1), two single meshes are stacked above each another, and a thicker wire is welded on to each side for electrical contact and stability. It is assumed that both meshes are geometrically perfectly aligned. The porosity is estimated by considering the geometric dimensions given in Figure 5.11 in 2D and a total mesh density of 1024 meshes/cm^2 (adopted from the Heraeus data sheet). A similar value as listed in Table 5.6 was also used by J. Hayd [45].

Tortuosity values for the pore and material phase are somewhat difficult to directly calculate. From the mesh design (Figure 5.11 on the right) it is possible to comprehend, however that due to the crossing of wires a more tortuous pore phase results in the x-direction compared to the wide open spaces in y-direction.

Table 5.6 lists the required parameters implemented into the model framework for the mesh description. Admittedly, the geometric and semi-porous microstructure data (especially the tortuosities) are not universally applicable, but need to be carefully reviewed after the measurement by comparison of predicted and measured gas diffusion resistances at lower cathode oxygen partial pressures. The matter is further examined in Section 6.5. The use of noble metals (Ni at the anode and Au at the cathode side) ensures the required high electrical conductivity and a long-lasting sustainability against degradation in the reducing/oxidizing gas atmosphere conditions at high temperatures.

Figure 5.11.: Schematic display of single, metallic contact mesh layer on the left. Geometric information of mesh-grid on the right.

Table 5.6.: Material, geometric and microstructural parameters of metallic contact meshes.

Parameter description	variable	value	Unit
Mesh layer thickness	h_{mh}	240	µm
Porosity	$\varepsilon_{\mathrm{mh}}$	0.64	
Pore tortuosity	$\tau_{\mathrm{por,x}}^{\mathrm{mh}}$	1.1	
	$\tau_{\mathrm{por,y}}^{\mathrm{mh}}$	1	
Material tortuosity	$\tau_{\mathrm{mh-mat}}$	1	
Electr. conductivity	σ_{Au}	$4.46 \cdot 10^7$ [166]	S/m
	σ_{Ni}	$1.39 \cdot 10^8$ [166]	S /m

5.5. Chemical Reforming Reaction Kinetics

Methane-steam-reforming (SR) and the Watergas-Shift reaction (WGS) are regarded in the model framework to predict SOFC performance not only under H_2-, but also for hydrocarbon-operation. Section 4.3.4 provides the model expressions required by Equation 4.74 as SR reaction rate $R_{\mathrm{sr,i}}$ and Equation 4.75 as WGS reation rate $R_{\mathrm{sh,i}}$, which are both incorporated into the corresponding i^{th} species balance equation Equation 4.28. The rate expressions are thereby dependent on forward reforming reaction velocity constant $k_{\mathrm{sr/sh}}^{+}$ (Equation 4.79) in combination with the equilibrium constants $K_{\mathrm{sr/sh}}$ (Equation 4.76). Hence, in this section suitable expressions from literature for mentioned parameters are provided and an experimental determination process is presented, where previously no, or only insufficient, formulations were to be found in literature.

Methane-Steam Reforming (SR) Reaction

Within the scope of H. Timmermanns dissertaion [103], a suitable rate expression for $S_{sr,i}$ (Equation 4.74) was developed, based on extensive experimental work carried out on Ni/8YSZ-anodes from Jülich. It is obvious to incorporate the expression into this works model framework, as the same ASC-type is regarded. The required equilibrium constant, K_{sr}, may be calculated by an exponential fit to experimental data provided by Rostrop-Nielson [105]. However, higher resolution with respect to temperature steps may be acquired from the thermodynamic database MALT [205]. The following expression reproduces the MALT-data with very high precision:

$$K_{\mathrm{eq,SR}} = exp\left(30.52 - \frac{27350\mathrm{K}}{T}\right) \cdot 10^{10} \cdot \mathrm{Pa}^2. \tag{5.29}$$

In the model framework, the result from Equation 5.29 is combined with Equation 4.76 for each species balance (Equation 4.26), leaving k_{sr}^{+} with Equation 4.79 as remaining unknown parameter. Therein, values for the rate constant $k_{0,\mathrm{sr}}$ and activation energy $E_{\mathrm{act,sr}}$ are required. To find these, a routine was developed within the diploma-thesis of A. Kromp [26], which is described in deteail in this next section because it is applied to determine the parameters for the WGS-reaction rate. Briefly summarized, simulated and in test-bench B (Section 3.2.2) experimentally recorded gas distributions along the gas channel length are compared, leading to the following expression for k^+_{sr} as result [26]:

$$k^+_{\mathrm{sr}} = 30.14 \left[\frac{\mathrm{mol}}{\mathrm{m}^3 \cdot \mathrm{Pa} \cdot \mathrm{s}}\right] \cdot exp\left(\frac{-61\frac{\mathrm{kj}}{\mathrm{mol}}}{RT}\right) \qquad \mathrm{mol/m^3 \cdot s \cdot Pa^2}. \tag{5.30}$$

Results presented in Sections 6.2 and 6.3 will demonstrate an accurate reproduction of measured gas distributions along the anode gas channel under OCV and load, thus implying a physically meaningful methane conversion due to properly determined SR kinetic parameters.

Water-gas Shift (WGS) Reaction

The water-gas shift reaction rate $S_{\mathrm{sh,i}}$ in mol/m^3/s (or $s_{\mathrm{sh,i}}$ in kg/m^3/s as in a mass-based calculation) describes species conversion of CO and H_2O to H_2 and CO_2 or vice versa via Equation 4.75. It thereby depends on the local thermodynamic equilibrium, defined by temperature and present species partial pressures; the available Ni-catalyst surface area is equally important. From the modeling point of view, it is desirable to acquire an expression *independent* of surface area and temperature, in order to adapt a model to different cell types and operating conditions. The following section describes how such an expression was derived in a supervised master thesis [241].

First of all, the rate expression Equation 4.75 provided by Lehnert et al. [99] is extended to a surface area dependent expression:

$$S_{\mathrm{sh,i}} = \nu_{\mathrm{sh,i}} \cdot a_{\mathrm{Ni,ASC}} \cdot \hat{k}^+_{\mathrm{sh}} \cdot p^2 \cdot \left(x_{\mathrm{CO}} \cdot x_{\mathrm{H_2O}} - \frac{1}{K_{\mathrm{eq,sh}}} \cdot x_{\mathrm{H_2}} \cdot x_{\mathrm{CO_2}}\right) \qquad \mathrm{mol/m^3 \cdot s} \tag{5.31}$$

with $\hat{k}^+_{\mathrm{sh,ASC}}$ in mol/m^2·s · Pa^2 as modified reaction velocity constant and $a_{\mathrm{Ni,ASC}}$ as specific surface area density in µm^2/µm^3. Temperature dependent data for $K_{\mathrm{eq,sh}}$ can be calculated by the thermodynamic software package MALT [205]. The following expression reproduces the data supplied by MALT with excellent precision:

$$K_{\mathrm{eq,sh}} = exp\left(-4.235 + \frac{4521}{T \cdot \frac{1}{\mathrm{K}}}\right) + 0.0999. \tag{5.32}$$

In order to calculate k^+_{sh} of Equation 5.31 temperature-dependently, the Arrhenius-type expression of Equation 4.79 is employed. The parameters $k_{0,\mathrm{sh}}$ and $E_{\mathrm{act,sh}}$ are determined by a combined numerical/experimental method, which is explained in the following.

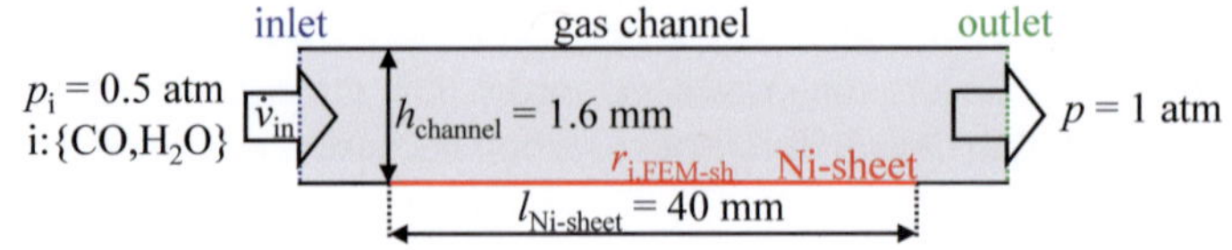

Figure 5.12.: Schematic display 2D FEM Model for WGS reaction kinetic determination. Geometric details and boundary conditions.

The idea is to fit simulated reformate species concentration gradients in a numerical model to experimental data by a systematic variation of $\hat{k}^{+}_{\text{sh}}$. For this purpose, a 50:50 CO/H_2O gas-mixture is applied in test-bench A4 to a thin Ni-sheet at various temperatures at a fixed gas flow rate $\dot{v} = 600$ sccm. The mixture composition is thereby determined at certain gas extraction points (Section 3.2.2). A 2D FEM-model with appropriate momentum and species balance equations is set up, regarding conservation of momentum by Equation 4.9 and of species by Equation 4.25. The models geometry resembles the longitudinal gas channel cross-section of test bench B (Figure 3.6). The anode flowfield was removed for the experiment in order to simplify the procedure. In this way, the catalytic gas conversion can take place over the entire Ni-sheet surface area A_s without hindrance from the flowfield contact ribs. In Figure 5.12 geometric details and boundary conditions of the FEM-model are given schematically, whereby the red line denotes the catalytically active Ni-surface with the modified WGS reaction rate expression of Equation 5.31.

The entrance velocity, $\vec{v}_{\text{in}}$, of the reformate gas mixture entering the test setup was set to laminar inflow and calculated by setting the entrance area $A_{\text{in}} = (40 \cdot 1.6)\ \text{mm}^2$ into relation with $\vec{v}_{\text{in}}$ and T:

$$\vec{v}_{\text{in}} = \dot{v}_{\text{in}} \cdot \frac{T}{273.15\ \text{K}} \cdot A_{\text{in}}^{-1} \qquad \text{m/s}. \tag{5.33}$$

The parameter $\hat{k}^{+}_{\text{sh}}$ is systematically varied in the model until the best agreement between simulated and measured species partial pressure gradients is found. The process is repeated for each measured temperature step. Example results achieved for $T = 800/650$ °C are shown in Figure 5.13. Therein, measured (symbols) and simulated (lines) partial pressure distributions are plotted over the gas channel length y in mm. Additionally, equilibrium concentrations calculated by MALT [205] are plotted (dotted lines). The results illustrated in Figure 5.13 demonstrate acceptable agreement between the measured and model-predicted data at high and low temperatures. The small deviation at $y = 20\,mm$ in Figure 5.13b is attributed to gas leakage at the corresponding gas extraction point in the test setup. A comparison of the results in Figure 5.13 also reveal a faster gas conversion rate due to the temperature activated WGS reaction. The gradients at 650 °C (Figure 5.13b) progress less steeply and with a more linear trend, whereby at 800 °C (Figure 5.13a) a non-linear progress due to the faster conversion rate is observable. However, even at 800 °C, the MALT-calculated thermodynamic equilibrium is not reached for H_2/H_2O for the set gas inflow rate of $\dot{v}_{\text{in}} = 600\,\text{sccm}$.

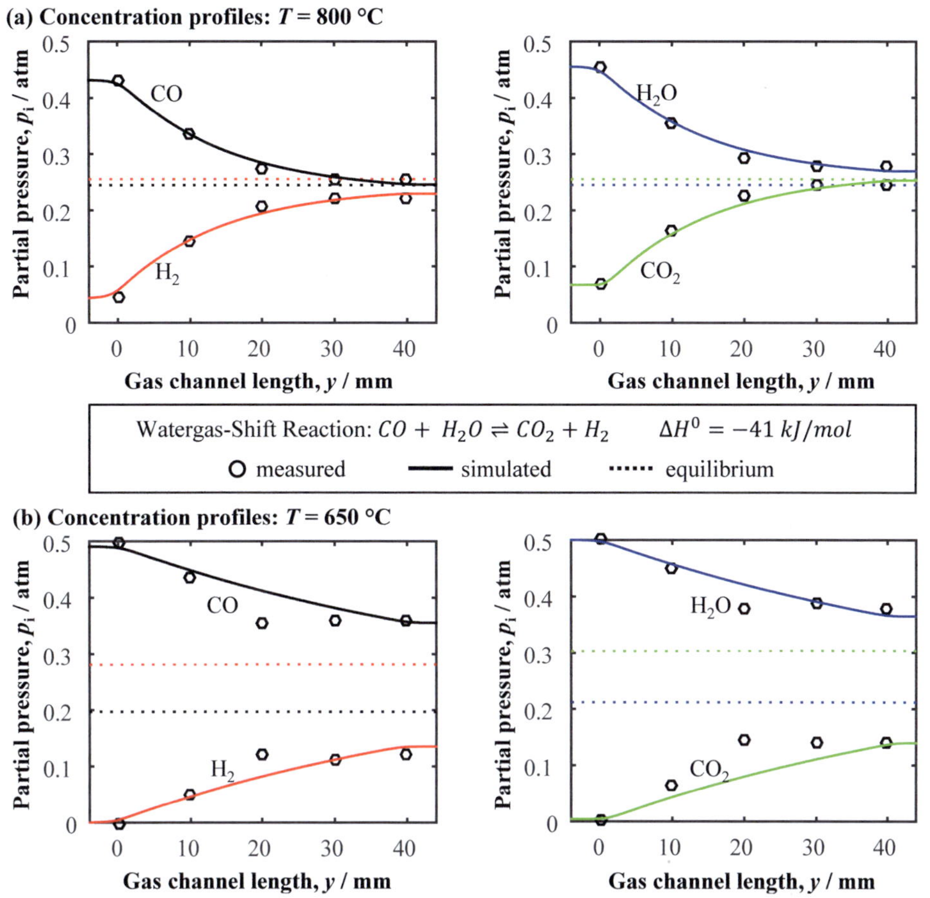

Figure 5.13.: Measured (symbols) and by FEM model predicted (lines) species concentrations along the anode fuel gas flow direction of test bench B (without flowfield). $\dot{v}_{\mathrm{in}} = 600\,\mathrm{sccm}$, $p\mathrm{CO}_{\mathrm{in}} = p\mathrm{H_2O}_{\mathrm{in}} \approx 0.5\,\mathrm{atm}$, $1 = p\mathrm{H}_{2,\mathrm{in}} + p\mathrm{CO}_{2,\mathrm{in}} + p\mathrm{CO}_{\mathrm{in}} + p\mathrm{H_2O}_{\mathrm{in}} + p\mathrm{N}_{2,\mathrm{in}}$. (a) $T = 800\,°\mathrm{C}$, (b) $T = 650\,°\mathrm{C}$.

The Arrhenius-type behavior of k^{+}_{sh} (Equation 4.79) allows to calculate $\hat{k}_{0,\mathrm{sh}}$ and $E_{\mathrm{act,sh}}$ from the simulated data of $\hat{k}^{+}_{\mathrm{sh}}$. $E_{\mathrm{act,sh}} = 82.25$ kj/mol is determined directly from the exponential slope, as it is independent from the catalyst geometry. $\hat{k}_{0,\mathrm{sh}}$ however incorporates the catalysts geometry. In an ASC, the catalyst surface area will not change significantly during a relatively short experiment time (24h), as it is embedded in a rigid YSZ skeleton. On the other hand, the thin Ni-sheet surface A_{s} changes during the gas conversion experiment due to the high operating temperature. This is shown by the SEM images in Figure 5.14 taken before (Figure 5.14a on the left) and after the experiment (Figure 5.14b on the right). Before exposing the Ni-sheet to heat ($T = 25...800$ °C), the surface appears very smooth with no grain boundaries visible, while after the experiment significant grain boundary formation is observable.

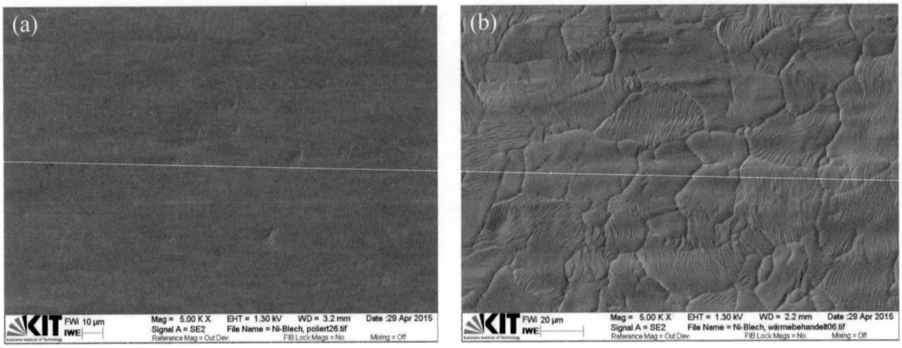

Figure 5.14.: (a) Polished Ni-sheet, prepared for WGS gas conversion experiment. (b) Ni-sheet after WGS reaction gas conversion experiment.

A rougher, and therefore greater surface area results, which has been quantified by laser microscopy at 10 different positions across the Ni-sheet surface. A mean surface area increase by the factor $a_{\mathrm{Ni}} = 1.22$ was calculated. With this the pre-factor results to

$$\hat{k}_{0,\mathrm{sh}} = 9.52 \times 10^{-7} \qquad \mathrm{mol/m^2 \cdot s}. \tag{5.34}$$

Indisputably, the changed surface texture has an influence on how the adsorption and desorption of reactants proceed, as well as on the reaction mechanism as a whole. This effect, however, is not captured by the simplistic global reaction approach chosen here as all effects are condensed within the pre-exponential factor $\hat{k}_{0,\mathrm{sh}}$. Further revelation of underlying chemical reactive mechanisms must be left open for future investigations with a more detailed approach. An option could be the Langmuir-Hinshelwood mechanism [204], where the reactants surface coverage is regarded. In any case, the temperature impact on the surface is considerably higher at high temperatures as it progresses faster, wherefore the data to be used in the parameter determination was recorded in chronological order and from high to low temperatures in 50 K steps. This strategy was chosen to minimize the temperature induced surface change. In fact, it is assumed that the surface only changes up to $T = 800$ °C and remains constant from there on down to lower temperatures.

Combining now the gained results with Equation 4.79 delivers the desired expression for the WGS reaction forward velocity constant $\hat{k}^{+}_{\mathrm{sh}}$, applicable to any temperature and Ni-catalyst surface area density:

$$\hat{k}^{+}_{\mathrm{sh}} = 9.52 \times 10^{-7} \cdot exp\left(\frac{-82.25\mathrm{kJ/mol}}{RT}\right) \qquad \mathrm{mol/m^2 \cdot s \cdot Pa^2}. \tag{5.35}$$

For cell Type-B (same cell Type-As used by Kromp et al.) with $a_{\mathrm{Ni-AFL}} = 0.38$ 1/µm [120], a velocity constant of

$$k^{+}_{\mathrm{sh}} = 3.59 \times 10^{-5} \qquad \mathrm{mol/m^3 \cdot s \cdot Pa^2}. \tag{5.36}$$

at $T = 800$ °C is calculated. The value for k^{+}_{sh} is here significantly larger compared to the value determined by Kromp et al. ($k^{+}_{\mathrm{sh}} = 1.6 \times 10^{-6}$ $\mathrm{mol/m^3 \cdot s \cdot Pa^2}$) [17, 29, 171]. The deviation could be attributed to a certain margin of error in the gas chromatograph accuracy ($\approx 10\%$), and to the fact that two totally different methods were applied in the determination process (EIS measurements by Kromp et al.). However, the impact of parameter variance is tested in section 6.1 by a sensitivity analysis, where the difference between both determined values was found to be negligible. The results in Section 6.2 and Section 6.3 will demonstrate excellent reformate operation model predictions at various temperatures under OCV and load, thus supporting the expression for the WGS kinetic found here.

5.6. Contact Resistance

The use of interconnectors (ICs) is essential to electrically interconnect the various layers of a full stack unit (Section 2.5.3) in order to increase the overall power output. While at higher operating temperatures ceramic materials are applied, ferritic steels may be used at lower temperatures. The advantage of metallic components are similar expansion coefficients, good durability and maybe most importantly they are more economical [242, 243]. A standard material for example is Crofer22APU [83] or more generally ferritic steals with a chromium content between 18 and 25% [244]. The electric resistance caused by such metals is negligible (in theory) due to their high electronic conductivity ($\sigma_{\mathrm{Crofer22APU}} = 8.7 \cdot 10^3$ S·cm^{-1} at 800 °C). However, at the elevated operating temperature corrosion occurs at the ICs surface exposed to the cathode air atmosphere causing a continuously growing oxide scale, which has very low electronic conductivity. Chromium contained in the metal slows down the oxide scale growth, but it leads to increased degradation due to the *chromium poisoning* of the cathode [118]. In recent years different coatings have been developed, which significantly decrease the chromium evaporation into the cathode air atmosphere, but which themselves exhibit a comparatively low electronic conductivity. The overall resistance caused by these individual layers is condensed into the expression ASR_{contact}. This surface related entity is denoted by $\Omega \cdot \mathrm{cm}^2$ and generally called *contact resistance*.

At Forschungszentrum Jülich the spinel $MnCo_{1.9}Fe_{0.1}O_4$ (MCF) was developed as protective coating with a thermal coefficient fitting to Crofer22APU [245]. The coating is deposited on top of the IC surface using atmospheric plasma spraying (APS), which has been proven to be an advantageous technique for operation in SOFC stacks [246, 247]. Table 5.7 lists specific values at certain temperatures, including the expectable tolerance, measured at Forschungszentrum Jülich between an IC with MFC coating and a LSCF cathode layer. The data was gathered by measuring 13 different samples, hence a sufficient reproducibility can be assumed. An exponential fit to the measured data is implemented into the model framework and set as an electrical boundary condition at $\mathrm{IF}_{\mathrm{IC-el}}$ (Figure 4.3) via Equation 4.4.

An even more cost efficient technique is to apply a metallic coating, which will oxidize during cell operation and form the protective layer. Promising candidates with very high conductivities are $Cu_{1.3}Mn_{1.7}O_4$ or $Mn_{1.3}Co_2O_4$. Using these coatings an $ASR_{\mathrm{contact}} \approx 10\ \mathrm{m}\Omega \cdot \mathrm{cm}^2$ at 600 °C is possible [248]. However, according to the authors of Ref. [248], the main contact resistance contribution is caused by the Cr_2O_3 layer, which increases over time and may become the main loss in SOFC stack operation. Hence, the influence of ASR_{contact} on an optimal IC flowfield design is investigated numerically in Section 7.3.2 in order to provide useful information for stack developers.

Table 5.7.: Area specific contact resistance values, ASR_{contact}, in $\Omega \cdot \mathrm{cm}^2$ at certain temperatures T, measured at Forschungszentrum Jülich between a metallic interconnector coated with $MnCo_{1.9}Fe_{0.1}O_4$ (MCF) via atmospheric plasma spraying (APS) and a LSCF cathode layer. The data is kindly supplied by K. Sick and N. Menzler and it can also be found in Refs. [85, 247].

T / °C	850	800	750	700	650	600
ASR_{contact} / $\Omega \cdot \mathrm{cm}^2$,	25 ± 5	29 ± 5	38 ± 6	56 ± 7	95 ± 10	175 ± 15

6. Validation and Modeling Approach Assessment

The content of this chapter forms part of the results obtained in this work. It is separately present in order to emphasize its importance. The information given in the previous chapters is now combined and the first actual simulation results are presented and compared with measured data. The underlying idea is to analyze and validate the model-framework behavior in the technically relevant operating condition range, and beyond. Only in this way can one out carry trustworthy geometric and material parameter variations, which are presented in the next chapter, and reliably maintain a physically correct representation of implemented loss processes.

In the first section, each relevant modeling parameter is investigated in a sensitivity analysis. The objective is to identify very sensitive (and therefore critical) modeling parameters and assess their importance in the following validation process. The main part of this chapter presents the model validation. The procedure has basically three parts, wherein simulated results are compared to measured data:

(i) Area specific resistances (ASR) of individually occurring loss processes,

(ii) Current/voltage (C/V) characteristics,

(iii) Gas conversion measurements along the fuel gas channel.

The validation is carried out individually at various operation temperatures, fuel gas mixtures and cathode oxygen partial pressures. It is differentiated between pure hydrogen (H_2) operation with different amounts of steam (H_2O) and inert nitrogen (N_2) in the fuel gas mixture and operation with fuel containing additional hydrocarbons (in the latter called reformate operation). This artificial syngas is composed of $H_2/H_2O/N_2$ and additional carbon monoxide and dioxide (CO/CO_2).

The experimental ASR were obtained using electrochemical impedance spectroscopy (EIS) to record spectra from anode supported SOFCs (ASC) with an active electrode area of 1 cm^2 (A_1-cells) and a subsequent algorithm. Therein, complex non-linear least squares (CLNS) fitting of the equivalent circuit model (ECM) for hydrogen [27] and reformate operation [28] (Figure 2.5) to experimental spectra was used to quantify the individual ASR under OCV. The C/Vs were recorded at A_1-cells and A_{16} (Figure 3.1). Gas conversion measurements were carried out at A_{16}. The experimental methods are described in Ch.3. For all measurements, ideal cell contacting is assured in the laboratory test-bench via metallic contact-meshes (Section 2.5).

Mainly, the model framework is applied to the 2D-RPU geometry (Section 4.2), because the majority of presented results in Chapter 7 are produced by that process. The molar-based Dusty-Gas Model (DGM-c) (Section 4.3.2.3) functions as gas transport model, equipped for the anode with $\Psi_{\mathrm{EIS,an}}$ as microstructural parameter (Section 5.1.7), while for the cathode microstructural data determined from 3D-reconstructions is used (Section 5.1). In both transport models, the volume-averaged pore diameter $d_{\mathrm{por}}^{\mathrm{vol}}$ is implemented (Section 5.1.6). If not noted otherwise, the Butler-Volmer approach (Section 4.3.3.1) is used to model the electrode electrochemical charge transfer reactions. For reformate operation simulations the 2D-GCh geometry (Section 4.2) is used to regard gas conversion due to occurring reforming reactions and the DGM-c is extended to five gases accordingly.

In the final part of this chapter, accuracy aspects regarding the various presented porous media gas transport models (Section 4.3.2.3) and methods of microstructure parameter determination are discussed. Furthermore, the need to use a 2D or 3D model geometry and electrochemical cathode model is elaborated on, with respect to intended purpose and computational effort.

6.1. Sensitivity Analysis

The purpose of performing a parameter sensitivity analysis is to identify individual modeling parameters, which are exceptionally sensitive or even critical for deviations in the calculated solution. Hence, all relevant modeling parameters are varied $\pm 20\%$ under the following standard operating conditions: $U_{\mathrm{cell}} = 0.7$ V, $T = 800/600$ °C, $p\mathrm{H_2O_{an}} = 0.6$ atm with $p\mathrm{H_{2an}} = 1 - p\mathrm{H_2O_{an}}$ and $p\mathrm{O_{2,cat}} = 0.21$ atm. Gas conversion is excluded by assuming high gas-flow rates.

The resulting current density difference, Δj, between reference value j_{ref} and calculated result j is set into percentage relation for each varied value. Parameters with a sensitivity of $\Delta j < 1\%$ are listed in Table 6.1 and their significance discussed shortly thereafter.

To begin with, Figure 6.1a shows the results of parameters with a sensitivity $\Delta j > 10\%$. In detail, the electrodes charge transfer activation energies, $E_{\mathrm{act,an/cat}}$ (Equation 4.60), exhibit a higher sensitivity, if the parameter is decreased. A significant temperature impact is thereby given for an increased $E_{\mathrm{act,cat}}$. A still greater sensitivity is determined for $E_{\mathrm{act,elyt}}$, the activation energy required to calculate the effective ionic conductivity related to the combined electrolyte/interdiffusion-layer (Equation 5.28). Again, a stronger temperature dependence is given if $E_{\mathrm{act,elyt}}$ is increased. The pre-exponential factor, B_{el} of Equation 5.28 still shows a strong sensitivity, however it is by far the lowest in this group. An almost symmetrical dependence towards its variation is coupled with a non-significant temperature dependence.

Displayed in Figure 6.1b are sensitivity analysis results of parameters also associated with the electrode charge transfer reaction, modeled by the Butler-Volmer (BVM) approach (Equation 4.59). It also shows activation energies, required to calculate reforming reaction velocity constants (Equation 4.79). In contrast to previous parameters, none of these exhibit a sensitivity $> 5\%$. The BVM-coefficients, $\alpha_{\mathrm{an/cat}}$, and exchange current density pre-factors, $\gamma_{\mathrm{an/cat}}$, show an increased sensitivity at lower temperatures, whereby the impact of α_{cat} is lower compared to α_{cat}. Similar results are given for $\gamma_{\mathrm{an/cat}}$, combined with an almost symmetrical sensitivity at low and high temperatures. The cathodic oxygen partial pressure exponent shows only very limited sensitivity, but compared to the anode exponents (a and b) it is still noticeable and presumably higher at lower $pO_{2,\mathrm{cat}}$. The reforming reaction's activation energies, $E_{\mathrm{act,sr/sh}}$, exhibit a similar behavior as that found for the electrodes activation energies. Higher sensitivity for decreased values, coupled with almost no temperature dependence. In summary, all discussed parameters require a similar precision in their determination, whereby their individual impact on the overall loss ranges is considerably lower compared to the parameters displayed in Figure 6.1a.

Thirdly and finally, parameters regarding the electrode microstructure are grouped together with material constants connected to the mixed-ionic-electronic conductive (MIEC) cathode model. Corresponding calculated sensitivity results are displayed in Figure 6.1c. Similar to results in Figure 6.1b, the deviation of varied parameters in Figure 6.1c remains below $< 5\%$. The electrode pore fractions $\varepsilon_{\mathrm{an/cat}}$ are identified as the most sensitive candidates in this group. Both parameters have a direct impact on gas species and electronic charge carrier transport due to the homogenization approach (Section 5.1.5 and Equation 5.6). At the cathode, the ionic charge carrier approach is coupled to the aforementioned processes with the homogenized surface exchange reaction approach (Section 4.3.3.2). However, the sensitivity characteristic is opposing for the electrodes pore fraction parameter. An increased $\varepsilon_{\mathrm{an}}$ causes an increasing current density, j, while a decrease is predicted for increasing $\varepsilon_{\mathrm{cat}}$. Furthermore, at lower temperatures the impact of $\varepsilon_{\mathrm{an}}$ almost disappears while it increases by a factor of two for $\varepsilon_{\mathrm{cat}}$. These contradicting results can be explained by the individual electrode characteristic and modeling approach. Gas transport losses are pronounced in the thick anode layer, while electrochemical charge transfer is not modeled with variable space dependency but as fixed boundary condition at the anode/electrolyte interface (Section 4.3.3.1). Consequently, this increases the anode pore fraction gas transport properties positively and charge transfer reactions remain unaffected. On the other hand, gas transport losses in the thin cathode layer are insignificant and therefore unaffected by changes of microstructural parameters. The homogenized MIEC-cathode approach, however, is affected by $\varepsilon_{\mathrm{cat}}$ as the results displayed in Figure 6.1c prove. Here, an increase in active material phase (lower $\varepsilon_{\mathrm{cat}}$) causes an increased electrochemical activity, which is more pronounced at lower temperatures due to a stronger temperature activation compared to gas transport [9]. This assertion is confirmed by the dependence of anode pore and cathode material phase tortuosity $\tau_{\mathrm{por,an}}$ and $\tau_{\mathrm{mat,cat}}$, which also bear similar opposing temperature dependencies, only now in reverse fashion due the inverse influence in Equation 5.6. The line of argument is continued by the results for the mean anode pore diameter $d_{\mathrm{por,an}}$, which only reveals a noticeable impact at higher temperatures, where temperature activated loss processes are less dominant and gas transport losses become more noticeable. The active surface area, a_{MIEC}, surface exchange and bulk-diffusion coefficient, k^{δ} and D^{δ}, exhibit a similar characteristic with increasing performance by increasing the parameter and a higher impact at lower temperatures.

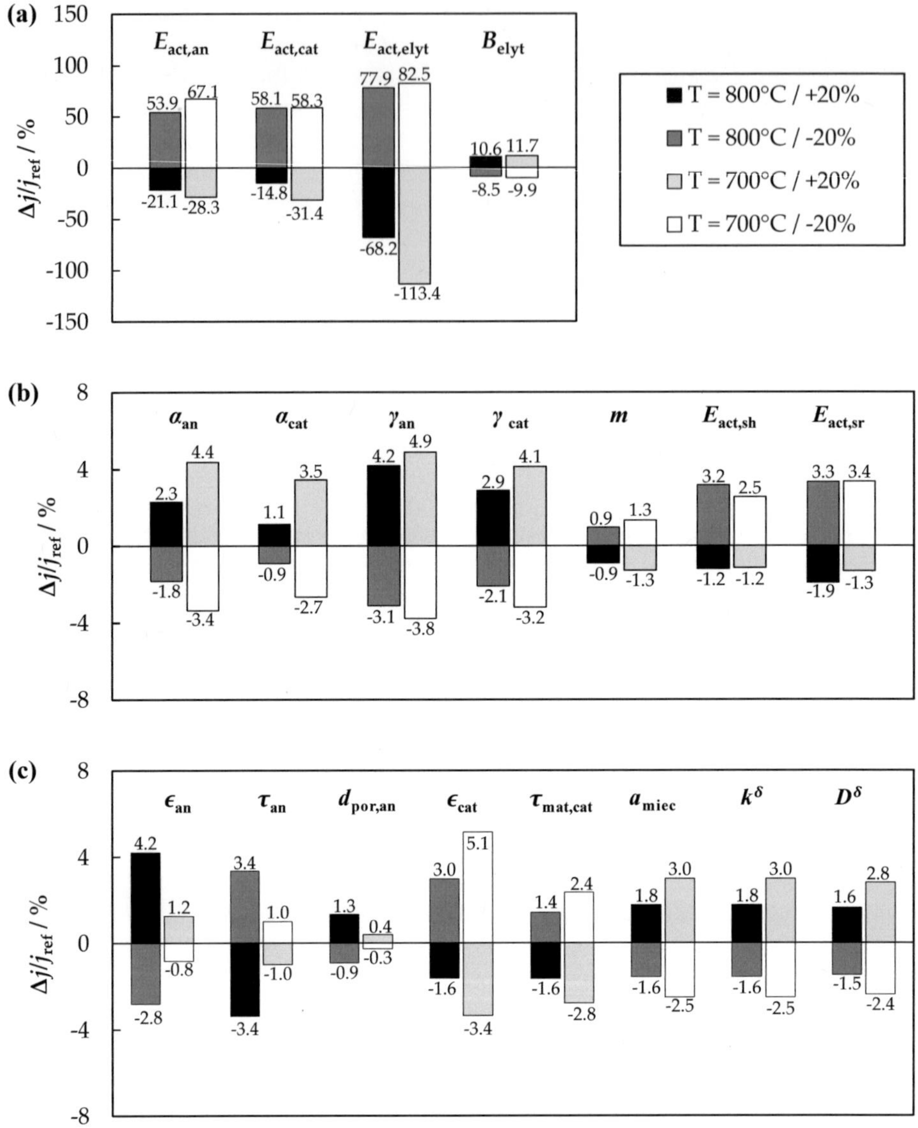

Figure 6.1.: Results of sensitivity analysis: Percentage current deviation from the reference value for ± 20 % varied parameters. (a) Electrode activation energies ($E_{\mathrm{act,an/cat}}$) and activation energy with pre-exponential factor of effective ionic electrolyte conductivity ($E_{\mathrm{act,elyt}}$, B_{elyt}). (b) Butler-Volmer parameters ($\alpha_{\mathrm{an/cat}}$, $\gamma_{\mathrm{an/cat}}$), cathode oxygen partial pressure dependency exponent (m) and reforming reactions activation energies ($E_{\mathrm{act,sh/sr}}$). (c) Pore fractions ($\varepsilon_{\mathrm{an/cat}}$), anode pore tortuosity (τ_{an}), anode mean pore diameter ($d_{\mathrm{por,an}}$), cathode material phase tortuosity ($\tau_{\mathrm{mat,cat}}$), cathode specific surface area (A_{miec}), surface exchange coefficient of O_2-reduction reaction (k^{δ}) and O^{2-} bulk-diffusion coefficient (D^{δ}).

Table 6.1.: Parameters with an impact of < 1 % deviation on the simulated current density for an ideally contacted SOFC in the laboratory. At low $pO_{2,\mathrm{cat}} < 0.1$ atm or in stack operation with non-ideal contacting (where O_2 transport limitations are encountered, Section 7.1), the sensitivity of cathode parameters related to species transport increases significantly.

Parameter	Description
a, b	Anode partial pressure exponents (Equation 4.60a)
$\tau_{\mathrm{por,cat}}$	Cathode pore tortuosity (Section 5.1.2)
$d_{\mathrm{por,cat}}$	Mean cathode pore diameter (Section 5.1.6)
$\Psi_{\mathrm{mesh,x/y}}$	Contact mesh microstructure parameter (Section 5.4)
$\kappa_{\mathrm{an/cat}}$	Anode / cathode permeability (Section 5.1.8)
$\sigma^{\mathrm{eff}}_{\mathrm{an/cat}}$	Electronic conductivities (Section 5.3)
$\dot{v}_{\mathrm{an/cat}}$	Gas channel inlet flow rates (Equation 4.58)
$k_{0,\mathrm{sr/sh}}$	Frequency factors of reforming reactions (Section 5.5)

Reviewing Table 6.1 and the list of parameters with an impact below $< 1\%$, certain assumptions may be drawn from the results. Starting in Table 6.1 top to bottom, the exact anode gas composition seems less important for calculating the exchange current density via Equation 4.60a with sufficient precision because the partial pressure dependency exponents a and b have no noticeable impact on the calculated solution. However, the low sensitivity can be explained by the chosen high water-vapor content in the anode gas composition. As shown by Ref. [9], at lower fuel gas humidities, a and b have a noticeable impact on the resulting current density and should therefore not be selected randomly. Furthermore, the mean cathode pore diameter $d_{\mathrm{por,cat}}$ and tortuosity $\tau_{\mathrm{por,cat}}$ exhibit a very low sensitivity, signifying a negligible contribution of the cathode gaseous species transport losses to the overall cell ASR. As mentioned in the previous paragraph this effect is to be expected, considering the cathodes thickness ($l_{cat} = 45$ µm) with a highly-permeable contact mesh positioned in the laboratory test-bench (Section 2.5). However, it is very important to understand for the results in Chapter 7 that cell design and non-ideal contacting (e.g. stack contacting) have a massive impact on the cathode gaseous species transport properties. Equal considerations account for the cathode electronic conductivity $\sigma^{\mathrm{eff}}_{\mathrm{cat}}$, which is not a critical parameter, if the electrode is contacted by a high electronically conductive mesh as in the ideal-contact situation in the laboratory test-bench (Section 2.5.2). The work of Kornely has extensively proven that electronic in-plane conduction limitations in the cathode affects SOFC-stack performance [37]. Results presented in Section 7.1 will show (as a continuation of Kornely's work) that SOFC stack performance is governed not only by the material inherent electronic conductivity but also by the electrode microstructural characteristic. In contrast to this is the anode effective electronic conductivity $\sigma^{\mathrm{eff}}_{\mathrm{an}}$. It is not a critical parameter as it is significantly higher compared to any contemplable cathode material, which has also been proven by M. Kornely experimentally and by simulation results [37]. The contact mesh microstructure represented by $\Psi_{\mathrm{mesh,x/y}}$ has, in the chosen standard operation conditions, no significant impact on the calculated solution. However, in the case of $pO_{2,\mathrm{cat}} < 0.1$ atm, the sensitivity increases due to the increased domination of cathode gas transport losses. Similar considerations account for the electrode permeabilities, $\kappa_{\mathrm{an/cat}}$, which are indeed dependent on the chosen gas transport model. Further details are discussed in Section 6.3. Also listed in Table 6.1 and therefore without an impact are the gas channel inflow rates, $\dot{v}_{\mathrm{an/cat}}$. This result confirms setting high gas-flow rates in the laboratory test-bench, which are set intentionally high to exclude any influence on measured results. The matter was further investigated in a supervised master thesis [151, 152]. Therein, a 2D-FEM model with anode gas channel and substrate was generated and used to simulate the transient gas transport impedance response for a incrementally lowered anode inlet gas-flow rate, $\dot{v}_{\mathrm{an}}$. A noticeable impact on the resulting simulated impedance spectra was detected below $\dot{v}_{\mathrm{an}} < 150$ sccm due to changes in the gas composition along the gas channel length. Consequently, in an application-oriented case with high fuel utilization, a greater sensitivity of $\dot{v}_{\mathrm{an}}$ should be considered. The last parameters in Table 6.1 are the frequency factors for calculating the reforming reaction velocity constants using Equation 4.79. The results show that in the chosen variation range, the activation energies, $E_{\mathrm{act,sr/sh}}$, in Equations 4.74 and 4.75 have a much greater impact, regardless of the chosen operating temperature. Nevertheless, an exact determination is scientifically desirable.

To summarize this section, the sensitivity analysis revealed that the most critical parameters are the activation energies used to calculate the charge transfer reaction at anode and cathode (Equation 4.60), together with the activation energy used to calculate the effective conductivity of electrolyte and interdiffusion-layer (Equation 5.28). All parameters exhibited a strong temperature dependence in their sensitivity, therefore the model validation at various temperatures bears a significant importance. All other parameters seem to individually have only minor impacts on the performance calculation at the regarded operating conditions. Comparable results were found by Bertei [147], whose analysis is however limited to microstructure properties and TPB length. The following section will discuss how under certain conditions, where associated processes become limiting, the significance of precise microstructure data increases. Furthermore, the above presented results are based on the 2D RPU lab-model with ideal contacting as used in the validation measurements in order to judge the parameters' importance in the following validation process. In the stack contacting (non-ideal) cathode, related parameters will be more significance, which will be discussed later in Chapter 7.

6.2. Temperature Dependence

The prior-presented sensitivity analysis revealed the activation energies for the electrochemical charge transfer reactions ($E_{\mathrm{act,an/cat)}}$) and for the ionic transport property in the electrolyte ($E_{\mathrm{act,elyt}}$) to be the most critical parameters with the highest impact on the predicted performance, especially at lower operation temperature. Hence, it is of major importance to demonstrate high model prediction accuracy at various temperatures, but with constant gas mixture supply in order to exclude gas transport influences. To begin with, simulated and measured results for hydrogen operation are compared, followed by results obtained under operation with hydrocarbon containing fuel (reformate).

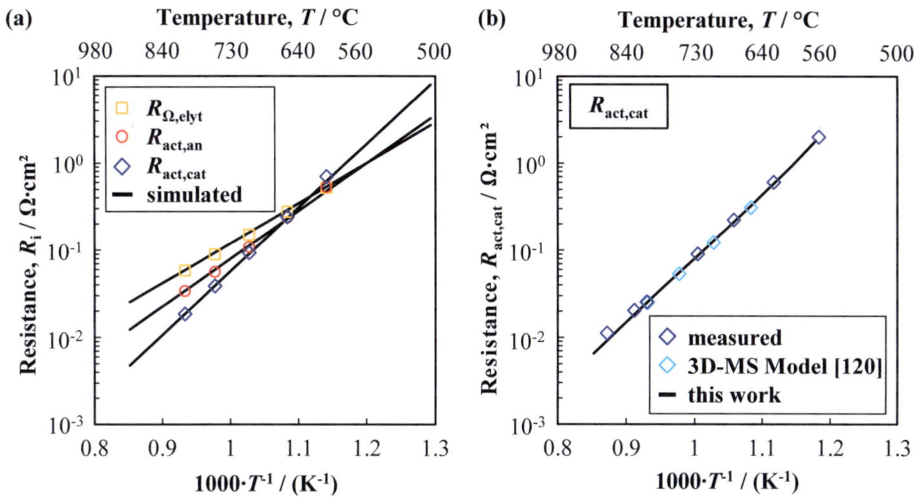

Figure 6.2.: Measured and predicted values, R_i, of ohmic and electrode polarization resistance for varying $T = 900\ldots600$ °C and constant fuel and oxidant gas compositions with 250 sccm gas-flow rate ($pH_{2,an} = 0.4$ atm, $pH_2O_{an} = 0.6$ atm, $pO_{2,cat} = 0.21$ atm). (a) Butler-Volmer approach and [Cell Z2_275]. (b) MIEC-cathode approach (dark blue diamonds), real microstructure (3D-MS) model (light blue diamonds, adopted from [120]) and [Cell Z1_188].

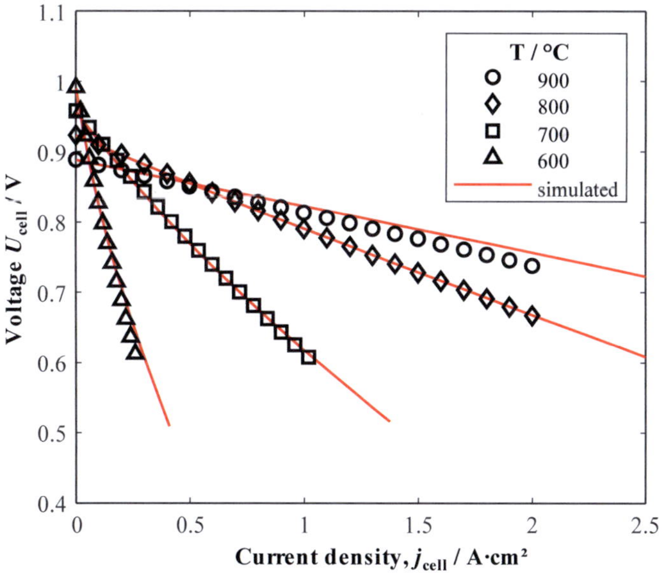

Figure 6.3.: Comparison of measured and predicted C/Vs recorded at varying $T = 900\ldots600$ °C and constant fuel and oxidant gas compositions with 250 sccm gas-flow rate. $pH_{2,an} = 0.4$ atm, $pH_2O_{an} = 0.6$ atm, $pO_{2,cat} = 0.21$ atm. [Cell Z2_275].

Hydrogen Operation

At higher temperatures, faster electrochemical reaction rates lead to lower polarization losses, $R_{\mathrm{act,an/cat}}$, at the corresponding electrode. A similar temperature dependence is expected for the ohmic loss, $R_{\Omega,\mathrm{elyt}}$, due to the Arrhenius-type behavior (Equation 5.28). Figure 6.2a displays measured (symbols) and simulated resistances (lines) over a broad temperature range. Measured resistances ($R_{\mathbf{ohm}}$ and $R_{\mathrm{act,an/cat}}$) are determined from impedance measurements recorded for cell Type-A. By comparing the results displayed in Figure 6.2a, it can be concluded that simulated and experimental data are in good agreement. Figure 6.2b displays the cathode electrochemical polarization resistance ($R_{\mathrm{act,cat}}$, dark blue diamond), determined from impedance measurements on cell Type-B (adopted from the work of A. Leonide [9]), together with simulated R_{cat} results predicted by a real 3D microstructure FEM-model (adopted from the work of J. Joos [120]; light blue diamonds). The results predicted by this work's homogenized MIEC-cathode approach (Section 4.3.3.2) are depicted by lines. Again, measured and simulated data are in excellent agreement over the total regarded temperature range. These first results justify the adaptation of the electrochemical and ohmic modeling parameters of cell Type-B for for simulating the results for cell Type-A. Both ASCs are manufactured by Forschungszentrum Jülich with the same materials and production methods and only differ in the anode support layer thickness.

Figure 6.3 displays at cell Type-A measured C/Vs (symbols) for the same operating conditions as used above, together with corresponding model predictions (lines). It can be deduced that higher operating temperatures yield lower open current voltages (Equation 2.8). Furthermore, the model predicts the temperature dependent behavior of measured curves with high precision within the regarded temperature range (900 . . . 600 °C). The deviation between measured and predicted data lies within the range of $0 < \delta < 2.4$ % up to high current densities of $j = 2$ A/m^2, as the calculated results in Figure 6.4 show. The deviation can be ascribed to small manufacturing related differences in the cell production.

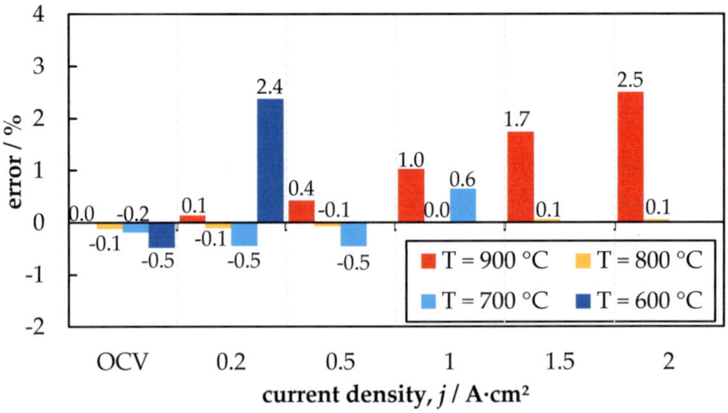

Figure 6.4.: Calculated error in % between measured and predicted data displayed in Figure 6.3 using the Butler-Volmer approach as charge transfer model at both electrodes.

In summary, the above presented results imply that the chosen, temperature dependent modeling approaches are in principle correct and can reproduce measured C/Vs under hydrogen operation over a broad range of temperatures ($T = 900 \ldots 600$ °C). Furthermore, the results prove that a 2D-RPU model geometry is sufficient to precisely predict C/Vs for high gas-flow rates and that the determination of used material parameters ($E_{\mathrm{act,elyt}}$, $E_{\mathrm{act,an/cat}}$), B_{elyt}, $\gamma_{\mathrm{an/cat}}$,k^{δ} and D^{δ}) is precise. In the next subsection, the fuel gas is changed to a mixture containing hydrocarbons (reformate).

Reformate Operation

To validate the model for temperature dependence in SOFC operation with fuel containing hydrocarbons, an "artificial reformate" was used in the measurements. Due to gas conversion via the water-gas shift reaction occurring along the gas channel length at the anode's catalytically active Ni surface (Section 4.3.4), the model framework is applied to the 2D Gas Channel (2D-GCh) model geometry (Section 4.2). Appropriate kinetic parameters regarding reformate operation were adopted (Table 5.2) for the Butler-Volmer approach (Equation 4.59). The molar-based Dusty-Gas Model (DGM-c) (Section 4.3.2.3) is used as gas transport model at both electrodes, extended to five gases at the anode to account for additional carbon-monoxide (CO) and carbon-dioxide (CO_2) alongside the H_2, H_2O and N_2 in the fuel gas.

Figure 6.5 illustrates simulated and measured C/Vs for varied temperatures with constant fuel gas mixture at the anode and constant oxygen partial pressure at the cathode (details given in Figure 6.5). The measurement was recorded at cell Type-B (Table 5.1) with an active electrode area of 1 cm^2 (A_1-cell). Comparing measured (symbols) and simulated (lines) model prediction, a very good agreement is found for all *OCVs*. This means that the same thermodynamic equilibrium is present in the anode fuel gas-filled compartment in both measurement and simulation, which implies a correct implementation of the water-gas shift reaction kinetics (Equation 4.75).

Furthermore, it strengthens the Kromp *et al.* [17, 28] result: the anode half-cell potential is governed by the electrochemical hydrogen oxidation (Equation 2.3) and not additionally by CO-electro-oxidation as it is suggested by some works [249]. For increasing load an excellent agreement between measured and predicted operating voltage, U_{cell}, is given at higher temperatures $T > 780$ °C and minor deviations are found at lower temperature. A possible explanation could be minor differences in the electrolyte/interdiffusion-layer thickness due to manufacturing variations among the individual ASCs used in the parameter determination and actual C/V-measurement. Coking can be excluded however, because at $T = 680$ °C and for the given gas composition the thermodynamic software MALT [205] only predicts a ratio of $6.888 \cdot 10^{-32}$ atm for solid carbon. In principle it can be concluded that the model framework reproduces the measured temperature dependence above $T > 700$ °C for A_1-cells under reformate operation with sufficient precision.

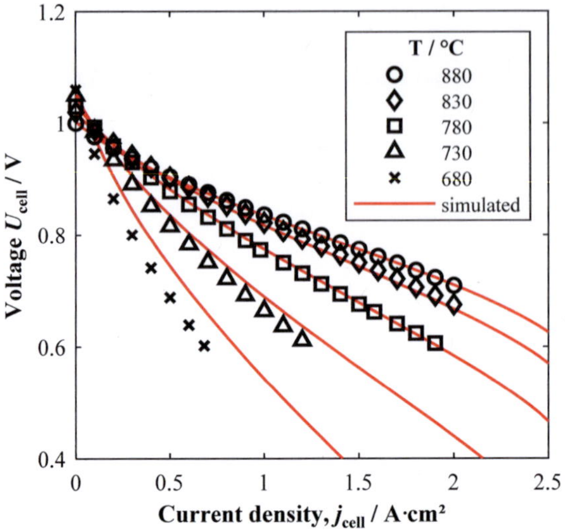

Figure 6.5.: Comparison of simulated and measured C/Vs recorded at varied temperature $T = 880 \ldots 680\ ^\circ\mathrm{C}$ with constant fuel and oxidant gas mixtures on an ASC with 1 cm^2 active electrode area (A_1-cell). Artificial reformate as fuel ($p\mathrm{H}_{2,\mathrm{an}} = 0.2$ atm, $p\mathrm{CO} = 0.22$ atm, $p\mathrm{H_2O}_{\mathrm{an}} = p\mathrm{CO}_2 = 0.03$ atm, $p\mathrm{N}_2 = 0.52$ atm, $\dot{v}_{\mathrm{an}} = 187.5$ sccm) and ambient air as oxidant ($p\mathrm{O}_{2,\mathrm{cat}} = 0.21$ atm, $\dot{v}_{\mathrm{cat}} = 250$ sccm). [Cell Z1_190].

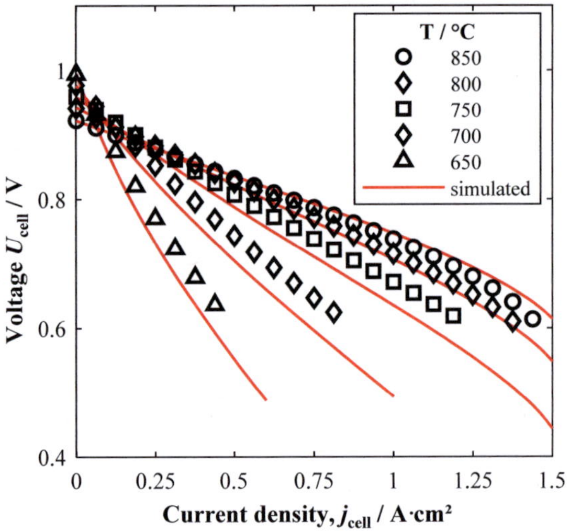

Figure 6.6.: Comparison of simulated and measured C/Vs recorded at varied temperature $T = 880 \ldots 680\ ^\circ\mathrm{C}$ with constant fuel and oxidant gas mixtures on an ASC with 16 cm^2 active electrode area (A_{16}-cell). Reformate from Diesel-ATR as fuel ($p\mathrm{H}_{2,\mathrm{an}} = 0.2$ atm, $p\mathrm{CO} = 0.22$ atm, $p\mathrm{H_2O}_{\mathrm{an}} = p\mathrm{CO}_2 = 0.03$ atm, $p\mathrm{N}_2 = 0.52$ atm, $\dot{v}_{\mathrm{an}} = 187.5$ sccm) and ambient air as oxidant ($p\mathrm{O}_{2,\mathrm{cat}} = 0.21$ atm, $\dot{v}_{\mathrm{cat}} = 250$ sccm). [Cell Z4_194].

The results shown in Figure 6.6 show simulated and measured results recorded for cell Type-A at varied temperatures ($T = 850 \ldots 650$ °C), but with a larger active electrode area of 16 cm^2 (A_{16}-cell). The fuel gas composition was composed of an application oriented reformate composition, typical for Diesel-ATR (Section 2.4.3). While the OCV is precisely reproduced by the model for all measured temperatures and the measured C/V-course is followed with acceptable precision at higher temperatures, a certain error is given at lower temperatures $T \leq 750$ °C. However, larger temperature gradients have to be expected across the increased active electrode area at higher current densities, which cannot be regarded by the isothermal model.

In summary, the results presented in this section show a good model validity regarding temperature dependency for reformate operation, even for ASCs with larger active electrode area.

6.3. Anode Gas Composition Dependence

This subsection is dedicated to validating the model framework regarding gaseous species transport prediction in the anode pore structure.

Hydrogen Operation

Gas Transport Resistance

Depicted in Figure 6.7 are the measured ASRs obtained from two independent experiments: Firstly (denoted by diamond shaped symbols), are those from CNLS-fits resulting $R_{1\mathrm{A}}$-values of a hydrogen partial pressure variation ($p\mathrm{H}_{2,\mathrm{an}}$), with constant steam partial pressure ($p\mathrm{H}_2\mathrm{O}_{\mathrm{an}}$). The fuel gas mixture was balanced with N_2 in order to keep a constant total gas-flow throughout the measurement. Hence, the increase in $R_{1\mathrm{A}}$ can be linked directly to the decrease in $p\mathrm{H}_{2,\mathrm{an}}$. Secondly (denoted by circle shaped symbols), are the resulting $R_{1\mathrm{A}}$ of a fuel gas humidity-variation ($p\mathrm{H}_{2,\mathrm{an}}/p\mathrm{H}_2\mathrm{O}_{\mathrm{an}}$). Unlike in the first experiment, $p\mathrm{N}_2$ is attributed to gas leakage and assumed as constant with $p\mathrm{N}_2 = 0.01$ atm. The resulting $R_{1\mathrm{A}}$-dependence depicted in Figure 6.7 shows the characteristic dependence on the changes in the fuel gas humidity: decrease of either H_2 or H_2O results in an increase of $R_{1\mathrm{A}}$, thus giving the curve its characteristic basin-shape. Furthermore, the experiment may be regarded as more application oriented (compared to changes of a single gas component), because it resembles the fuel conversion in SOFC/SOEC-operation.

Instead of producing a complete numerical EIS spectrum for each operating condition and deducing thereby the real part intersection as $R_{1\mathrm{A}}$, a much faster stationary algorithm, described in Section 4.4.2, was used to obtain the simulation results depicted in Figure 6.7. The algorithm allows one to resolve $R_{1\mathrm{A}}$ with increased detail under less computational time. Comparing experimental results and numerical predictions in Figure 6.7, it is evident that both are in very good agreement.

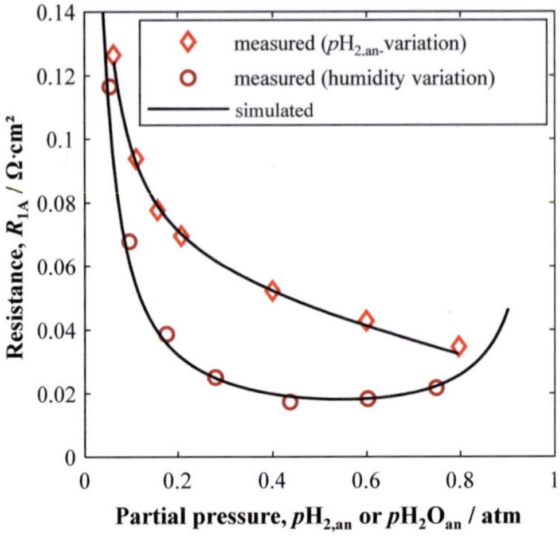

Figure 6.7.: Measured (symbols) and simulated (lines) anode gas diffusion resistances (R_{1A}) under varied hydrogen operating conditions: (i) $pH_{2,an}$-variation (diamonds) with constant $pH_2O_{an} = 0.2$ atm and balance N_2. (ii) Fuel gas humidity variation (circles) with $pN_2 = 0.01$ atm resulting from leakage. In both experiments: $pO_{2,cat} = 0.21$ atm, $T = 800$ °C and $\dot{v} = 250$ sccm. [Cells Z2_275, Z1_360].

Hence, at this point it can be concluded that the model framework using the molar-based DGM and parametrized with $\Psi_{EIS,an}$ and d^{V}_{por} is capable of reproducing the characteristic Warburg-type diffusion impedance and the anode gas diffusion loss R_{1A} as measured under OCV with excellent precision over a very broad range of fuel gas compositions composed of hydrogen and steam.

Current/Voltage Characteristics

The above presented results achieved under OCV are already promising, but a model behavior evaluation under load is still required. At higher current densities the fuel conversion increases and causes the formation of larger concentration gradients between the electrochemical active zones (TPBs) and the gas channel supply because gas species transport occurs non-ideally. Likewie, opposing species fluxes consequently increase to supply/remove reaction reactants/products and they affect each other in doing so. Under certain conditions, diffusion polarization will become predominant in the overall loss. In SOFC operation, these extreme conditions are only reached at $pH_{2,an} < 0.2$ atm and $j > 2$ A/cm^2, as the experimental results (symbols) in Figure 6.8 demonstrate. In the corresponding measurement recorded on cell Type-A, the hydrogen content in the fuel gas was incrementally lowered and balanced with N_2 with constant $pH_2O_{an} = 0.2$ atm, temperature $T = 800$ °C and $pO_{2,cat} = 0.21$ atm.

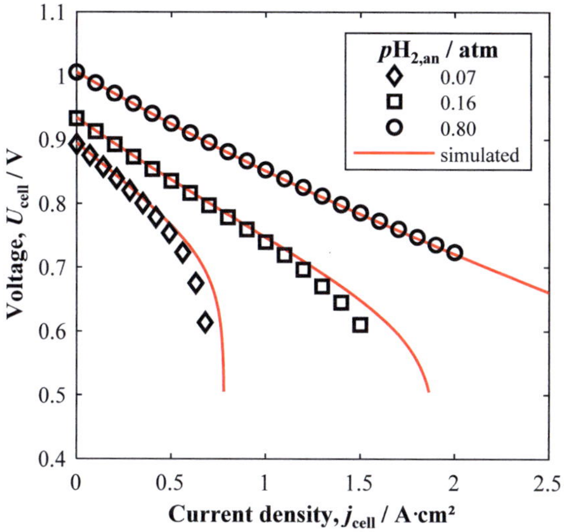

Figure 6.8.: Simulated (using the 2D-RPU model geometry; lines) and measured (symbols) C/Vs, recorded at cell Type-A at different hydrogen and water partial pressures in the fuel gas (pH_2O_{an} : pH_2O_{an} = [(0.16 : 0.18), (0.80 : 0.20)] atm, balance N_2), constant $pO_{2,cat}$ = 0.21 atm, T = 800 °C and $\dot{v}$ = 250 sccm. [Cell Z2_275].

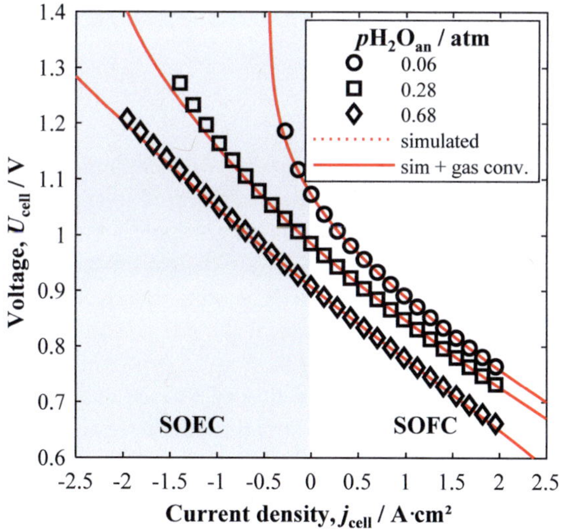

Figure 6.9.: Simulated (using the 2D-RPU model geometry; lines) and measured (symbols) C/Vs, recorded at cell Type-A at three different fuel gas humidities (pH_2O_{an} = 0.68 . . . 0.06 atm, constant pN_2 = 0.01 atm from leakage and $pH_2O_{an} = 1 - pH_2O_{an} - pN_2$ = 0.01 atm) constant $pO_{2,cat}$ = 0.21 atm, T = 800 °C and $\dot{v}$ = 250 sccm. [Cell Z2_275].

Comparing the measurement results in Figure 6.8 (symbols) with depicted model predictions (lines), it is evident that the measurement and simulation results are in very good agreement. Hence, the model framework seems to reproduce in SOFC mode the occurring gaseous species transport fluxes and therewith the related diffusion overpotential with sufficient precision.

Further limiting operating conditions regarding the gas transport properties are given under SOEC operation and low pH_2O_{an}. Figure 6.9 illustrates simulated and recorded C/Vs for varied fuel gas humidities ($pH_2O_{an} = 0.68 \ldots 0.06$ atm, $pN_2 = 0.01$ atm from leakage, constant temperature $T = 800$ °C and $pO_{2,cat} = 0.21$ atm). Again, by comparing simulation and measurement results, an excellent agreement is given. Only a slight deviation between simulation and measurement is given for $j < -1$ A/cm^2 in SOEC mode, where additional temperature induced losses occur [250], which are not yet regarded in the isothermal modeling approach.

Based on the results presented in this section, the anode gas transport model can be regarded as validated. The results clearly show how the implemented, molar-based, Dusty-Gas Model (DGM-c) together with the microstructure parameter, $\Psi_{EIS,an}$, and the mean anode pore diameter $d^{v}_{\mathbf{por}}$, reproduce occurring gas transport losses and the connected limiting current densities with sufficient precision under hydrogen operation.

Before the anode gas transport model validation is extended to reformate operation, another important fact should be recognized: The results presented above in Figure 6.9 also show that the model is able to reproduce measured C/Vs in SOEC operation (at $T = 800$ °C).

Reformate Operation

As mentioned above, the 2D-GCh model geometry with implemented reaction kinetics and appropriate electrochemical kinetic parameters is required to simulate SOFC performance under reformate operation. In addition, the applied gas species transport model is required to handle additional hydrocarbon species (CO/CO_2) in the gas mixture. Due to their presence in the fuel gas, the water-gas shift reaction (WGSR) occurs via Equation 2.17 at the catalytically active Ni-surface within the porous substrate, if the gas mixture is not in thermodynamic equilibrium. This can be the case for the gas mixture as it enters the gas channel, which causes a reaction rate, S_{sh}, according to Equation 4.75 at the anode surface. Additionally, if a current is drawn from the cell, electrochemical fuel conversion via Equation 2.1 occurs in the electrochemical active zones (TPBs), thus causing the mixture to change from its thermodynamic WGSR-disequilibrium, and an additional WGSR occurs according to S_{sh} of Equation 4.75. The consequence is a complex coupling of gaseous species transport, electrochemical and reforming reactions, schematically displayed in Figure 6.10. The imposed high demand on the implemented gas transport model (molar-based DGM-c) is analyzed in the following.

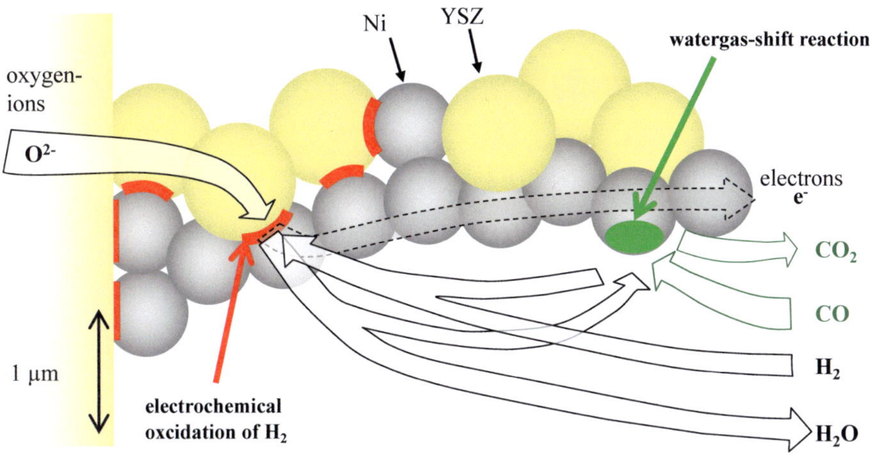

Figure 6.10.: Illustration of the reaction and transport processes for Ni/YSZ anodes in anode supported SOFCs operated with reformate fuels [17, 29, 171].

Gas Transport Resistance

Figure 6.11 shows simulated and measured ASR, associated with gas diffusion polarization under reformate operation [17, 29, 171]. The operating conditions are stated in the caption. The good agreement with the measured data demonstrates the satisfactory applicability of the DGM-c gas transport model approach in combination with implemented microstructure parameters ($\Psi_{\mathrm{EIS,an}}$, d^{V}_{por}). Similar to the results found by Kromp *et al.* [17, 29, 171], an overestimation of model-predicted losses at very low $p\mathrm{H_2O}_{\mathrm{an}}$ becomes noticeable. Nevertheless, the error using this works DGM-c approach is considerably lower and no additional fitting parameters are required. Another advantage of this work's approach is the ability to simulate C/Vs as gas conversion along the gas channel can be considered. This is presented in the next subsection.

Current/Voltage Characteristics

Figure 6.12 illustrates simulated and measured C/Vs (lines & symbols, respectively) for a varied anode fuel gas composition. The fuel gas inlet concentrations were chosen to simulate a certain "fuel utilization" (*f.u.*) (see Table 6.2). Accordingly, a lower OCV was observable in the measurement for higher *f.u.*, which is perfectly reproduced by the model. As in the already-mentioned temperature dependent validation, this implies a similar thermodynamic condition in measurement and simulation that hydrogen is the preferred electrochemical active species and that the reforming kinetic approach with its experimentally determined parameters is correct. Furthermore, under increasing current density, a limitation due to predominant diffusion polarization was observable at 69% *f.u.*

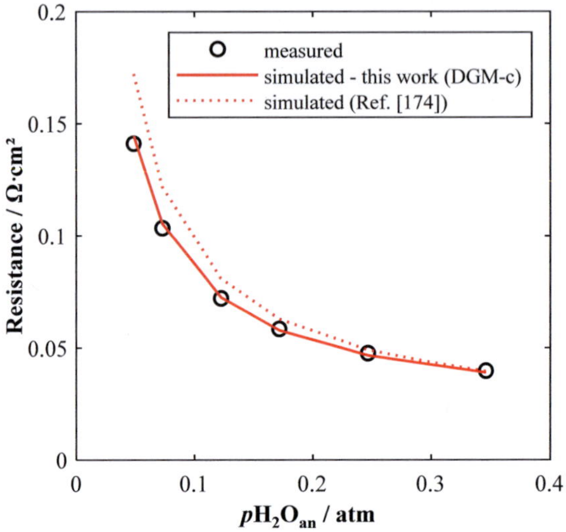

Figure 6.11.: ASRs of the measured (R_{1A} + R_{ref}) and simulated anode diffusion polarization process for varying pH_2O_{an} in the anode with reformate as fuel gas ($pH_2O_{an} = 0.15$ atm, $pH_2O_{an} = 0.07 \ldots 0.35$ atm, $pCO = 0.15$ atm, $pCO_2 = 0.07 \ldots 0.35$ atm, balance N_2; $T = 800$ °C). [Cell Z2_190].

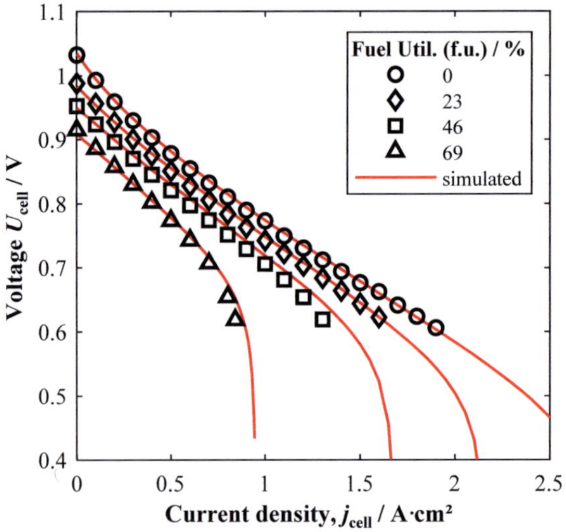

Figure 6.12.: Simulated (2D-GCh model geometry; symbols) and measured C/Vs for varied fuel utilizations (0%, 23%, 46% and 69%) recorded at cell Type-B with an active electrode area of $A = 1$ cm^2. Fuel gas inlet composition see Table 6.2, constant $\dot{v}_{an} = 187.5$ sccm, $T = 800$ °C, $pO_{2,cat} = 0.21$ atm. [Cell Z2_190].

The excellent agreement of simulation and measurement results in Figure 6.12 demonstrates the model's ability to reproduce the performance reduction due to fuel gas transport limitation.

Hence, the above presented results imply that the chosen gas transport model (DGM-c for five gases, parameterized with $\Psi_{\mathrm{EIS,an}}$ and $d^{\mathrm{v}}_{\mathrm{por}}$) is able to accurately predict the gas transport in the anode gas channel and porous substrate. However, these results regard the smaller A_1-cells, where gas conversion gradients along the gas channel are less pronounced, and so a more application-oriented validation with longer gas channels is desirable to strengthen the transport model validation. This concern is addressed in the next section.

Table 6.2.: Operating conditions set for measurement of C/Vs at various fuel utilizations (*f.u.*) at A_1-cell Type-B, results are shown in Figure 6.12. Constant $pO_{2,\mathrm{cat}} = 0.21$ atm

f.u.	T	$\dot{v}_{\mathrm{an}}$	$pH_{2,\mathrm{an}}$	pCO	pH_2O_{an}	pCO_2	pN_2
0%	780 °C	187.5 sccm	0.200	0.220	0.030	0.030	0.52
23%			0.159	0.165	0.071	0.085	0.52
46%			0.114	0.144	0.136	0.116	0.52
69%			0.067	0.065	0.163	0.185	0.52

Gas Conversion Measurements

In order to further validate the implemented gas transport model, gas conversion measurements under reformate operation were carried out at the test-bench for A_{16}-cells (Section 3.2.2). The setup allows tracking of gas conversion along the gas channel length via gas extraction probes at various positions with a subsequent concentration determination via gas chromatograph (GC). The measurements were performed at A_{16}-cells with a gas channel length $y = 40$ mm under load of $j = 1$ A·cm^{-2} at various fuel utilizations (*f.u.*). The fuel gas composition, which is typically supplied by a commercial Diesel-ATR, gives an application-oriented output to the presented results.

Figure 6.13 illustrates measured and simulated (symbols &lines, respectively) partial pressure distributions p_{i} along the gas channel length, where $y = 0$ mm denotes the channel's entrance. In SOFC operation mode H_2 is converted to H_2O at the TPBs, whereby the amount of H_2 decreases and H_2O increases towards the channels exit ($y = 40$ mm). In Figure 6.13a, at low fuel utilization (20%), the measured gradients follow a low and linear trend. Similar observations can be made for CO/CO_2 as CO is consumed and CO_2 produced via the WGSR. With increasing fuel utilization (Figure 6.13b), significantly larger species gradients of H_2/H_2O and moderately increased species gradients of CO/CO_2 occur. At 60% *f.u.* (Figure 6.13c) additionally to further increased gradients of all species, a non-linear trend is observable for H_2/H_2O. Comparing measured and simulated results in Figure 6.13, it is evident that the model predictions are in good agreement. Even the non-linear H_2/H_2O gradients are reproduced with only minor deviations, which are ascribed to a certain measurement error in the GC.

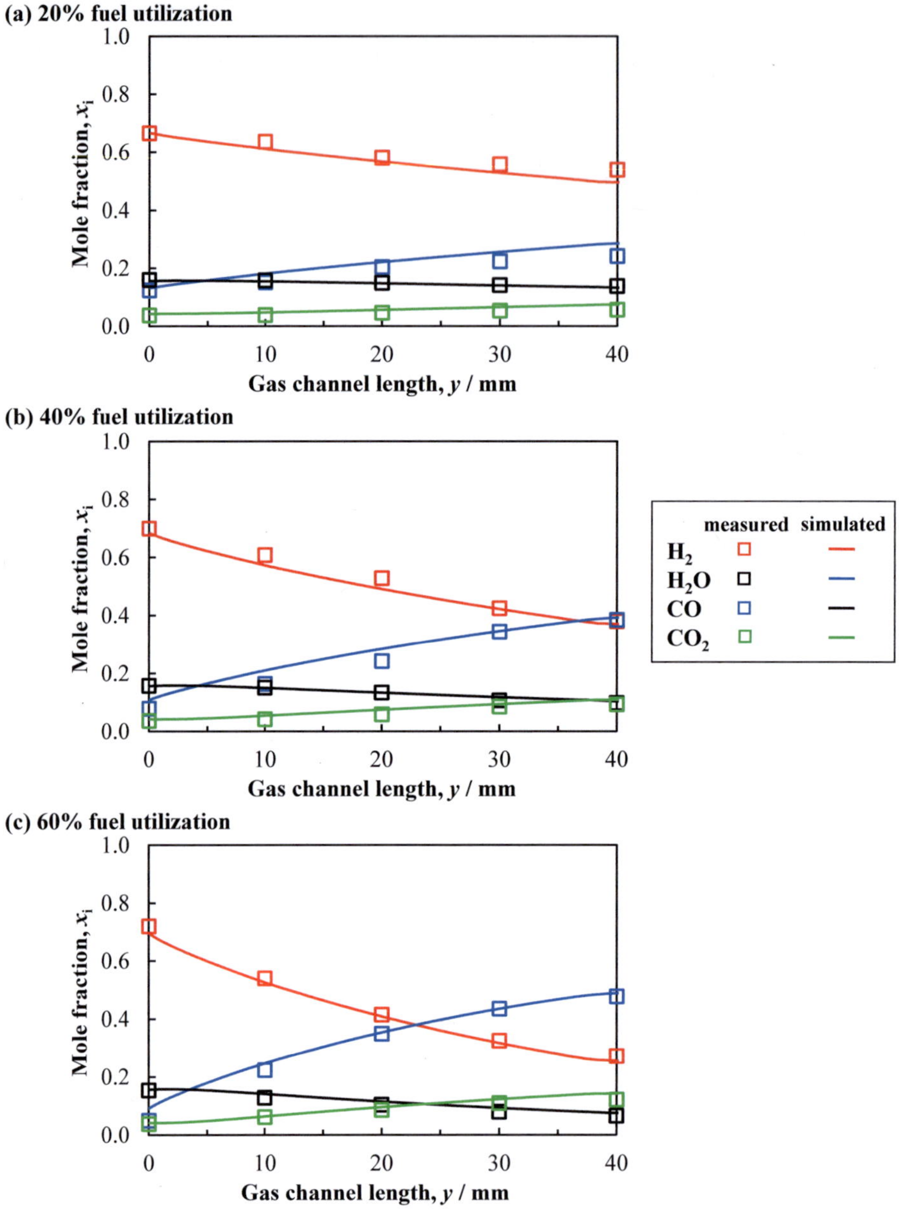

Figure 6.13.: Simulated (2D-GCh Model, Type-A A_{16}-cell, DGM-c, $\Psi_{\mathrm{EIS,an}}$, $d^{\mathrm{V}}_{\mathbf{por}}$; lines) and measured (test-bench B; symbols) molar concentrations along the anode gas channel length for varied anode gas-flow rates of a Diesel-ATR reformate fuel gas mixture at $j = 1$ A·cm^{-2} resulting different fuel utilizations: (a) f.u = 20% and $\dot{v}_{\mathrm{an}} = 689$ sccm, (b) f.u = 40% and $\dot{v}_{\mathrm{an}} = 347$ sccm, (c) f.u = 60% and $\dot{v}_{\mathrm{an}} = 238$ sccm. Constant $p\mathrm{O}_{2,\mathrm{cat}} = 0.21$ atm, $T = 810$ °C. [Cell Z4_194].

However, the simulated cell voltage was for all *f.u.* ~4% higher than in the measurement. Bearing in mind the good agreement of the results illustrated in Figure 6.13 regarding the anode species transport, the additional overpotential must be accounted to one or both of the electrode charge transfer reactions or to the ionic charge transport in the electrolyte. Nevertheless, the error lies within a tolerable range considering the isothermal approach, ongoing cathode degradation in ambient air operation [69] and the fact that several thermal cycles were performed before the gas conversion measurements were recorded.

To conclude the anode gas transport validation section, the above presented results achieved for the larger A_{16}-cell confirm the findings presented earlier for the smaller A_1-cell: The applied molar-based Dusty-Gas Model (DGM-c) equipped with the mentioned microstructure parameters reproduces all measured results with good, or even excellent, agreement. Therefore, it is perfectly suited to describe the multi-component gas species transport fluxes in anodes at SOFC/SOEC operation and is thus considered as validated. In the next section, this matter will be analyzed for the cathodic gas transport.

6.4. Cathode Oxidant Gas Composition Dependence

To analyze and validate the model framework with respect to gas composition variation at the cathode, simulated and measured ASR values obtained from impedance measurements under OCV and recorded C/Vs are compared at various, decreasing oxygen partial pressures $p\mathrm{O}_{2,\mathrm{cat}} = 0.21...0.01$ atm (21 % O_2 is considered as standard at ambient air composition). The anode gas composition ($p\mathrm{H}_{2,\mathrm{an}} = 0.4$ atm, $p\mathrm{H}_2\mathrm{O}_{\mathrm{an}} = 1 - p\mathrm{H}_{2,\mathrm{an}} - p\mathrm{N}_{2,\mathrm{leak}}$) and operating temperature $T = 800$ °C were kept constant. gas-flow rates for both electrode gas compositions were also kept constant at 250 sccm.

The experimental data presented in the following is based on part of the work of A. Leonide [9], who used ASCs of Type-B for his experimental results (Table 5.1). The microstructure of the applied LSCF cathode layer is considered to be identical to the cell Type-A cathode, therefore the results are regarded as applicable for all simulations in this work. Hence, as general model geometry, the 2D-RPU model geometry for ideal contact (Figure 4.3) is applied with geometrical and micostructural parameters for cell Type-B (Table 5.1). The molar-based Dusty-Gas Model (DGM-c, Section 4.3.2.3) as gas transport model and electronic conductivity, σ_{cat}, is regarded as constant over the considered $p\mathrm{O}_2$-range. The Butler-Volmer model (BVM, Section 4.3.3.1 and Section 5.2.1) is used as standard charge transfer model at anode and cathode. Additionally, as one of this work's features, the mixed-ionic-electronic cathode model is applied (MIEC, Section 4.3.3.2 and Section 5.2.2) and is therefore also a subject for validation. Kinetic parameters regarding the surface exchange reaction (k^{δ}) and oxygen ion bulk diffusion (D^{δ}) at $T = 800$ °C are adopted from the work of J. Joos [120] for ASC Type-B and were determined at various temperatures for ASC Type-A within this work (Section 5.2.2).

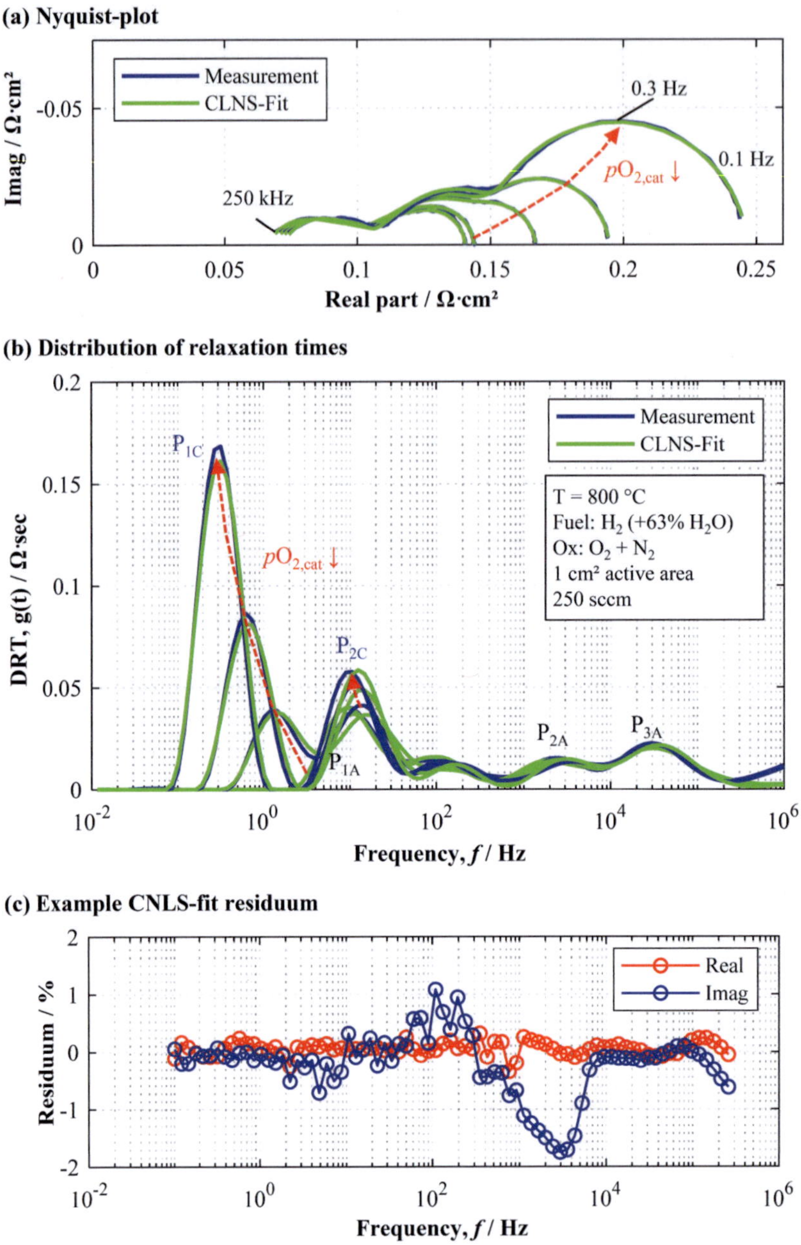

Figure 6.14.: Measured and CNLNS-fitted EIS spectra (a), calculated DRT results (b) and example fit residuals. Operating conditions: $T = 800$ °C, Hydrogen with 60 % steam as fuel and varied oxygen partial pressure $pO_{2,\mathrm{cat}} = 0.21 \ldots 0.01$ atm with balance N_2. Constant gas-flow rate of $\dot{v}_{\mathrm{an/cat}} = 250$ sccm. [Cell Z1_188].

Figure 6.14 displays experimental impedance results of cell Type-B used to determine the cathode gas diffusion, R_{1C}, and activation resistance, R_{2C}, which are used among results from ASC Type-B in the following FEM model validation process. The Nyquist-plot (Figure 6.14a) displays measured EIS spectra and results of fitting the equivalent circuit of 2.5 using a CNLS-fit method. It is shown how with decreasing $pO_{2,\mathrm{cat}}$ the total ASR increases, mainly due to an additional arc formation in the low frequency range. The DRTs (Figure 6.14b) show that two processes are responsible for the ASR increase. P_{1C} is associated with the gas diffusion polarization, visible only below $pO_{2,\mathrm{cat}} = 0.1$ atm [9], and P_{2C} as the cathode activation polarization. Figure 6.14c displays example CNLS-fit residuals, demonstrating sufficient precision with residuals below 2 %.

With this setup, the analysis starts by comparing predicted and measured cathode charge transfer activation loss R_{2C}.

Charge Transfer Activation Loss

Figure 6.15 shows experimentally determined resistances (R_{2C}, symbols) at varied $pO_{2,\mathrm{cat}}$ for cell Type-A (this work) and for cell Type-B (experimental data adopted from Ref. [9]), representing the LSCF-cathode activation polarization loss. In the equivalent circuit model a Gericher-Element was used (Section 2.3.2), thus determining R_{2C} by applying the CNLS-fit method to measured electrochemical impedance spectra [9, 36]. At lower $pO_{2,\mathrm{cat}}$ the surface exchange reaction rate is hindered due to decreased oxygen molecule availability in the gas phase, which results in an increases of the associated loss, R_{2C}. This behavior is with satisfactorily agreement met by the BVM prediction as the results in Figure 6.15 demonstrate (blue and yellow line). Using the MIEC-model with pO_2-independent $k\&D$-values (violet lines, Equations 5.24 and 5.25, $\alpha^{\mathrm{k/D}} = 0$) results, unsurprisingly, in a constant, pO_2-independent resistance prediction and consequently in an increasing deviation for decreasing $pO_{2,\mathrm{cat}}$. However, at standard conditions ($pO_{2,\mathrm{cat}} = 0.21$ atm), where $k\&D$ were determined, an excellent agreement of simulation and measurement results is given.

The results in Figure 6.15 serve as a perfect example of why it is essential to use $pO_{2,\mathrm{cat}}$-dependent kinetic data to model cathode activation loss, if the work is to be considered reliable for cathode loss prediction. Using Equations 4.63 and 4.67 with the $\alpha^{\mathrm{k/D}}$-values given in Table 5.3 deliver excellent reproduction of measured data by the MIEC-model (red line for ASC Type-A and green line for Type-B).

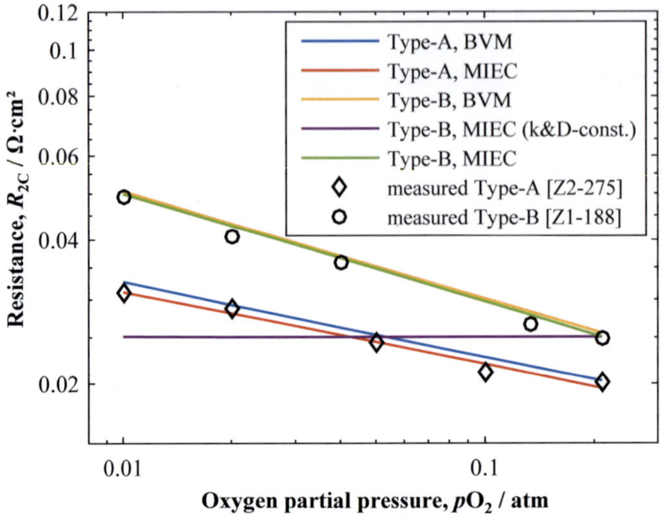

Figure 6.15.: Simulated and measured cathode activation loss (R_{2C}) of ASCs Type-A and Type-B. Model predictions using the Butler-Volmer approach (BVM, blue/yellow lines): Equations 4.59 and 4.60b with kinetic data from Table 5.2. MIEC-approach (red/green/violet lines) with Equations 5.24 and 5.25 and parameters from Ref. [120] and in this work determined values for LSCF (Table 5.3). $T = 800$ °C. Fuel: $H_2 + 60$ % steam. Oxidant: $pO_{2,\text{cat}} = 0.21 \ldots 0.01$ atm with balance N_2. $\dot{v}_{\text{an/cat}} = 250$ sccm. $\dot{v}_{\text{an/cat}} = 250$ sccm. [Cells Z1_188, Z2_275].

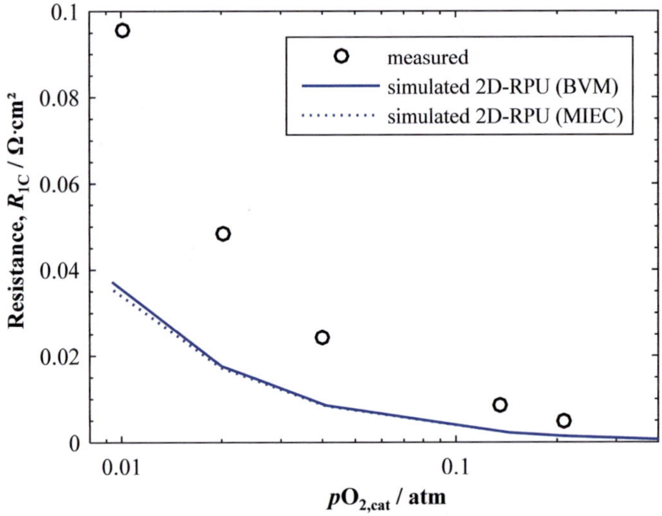

Figure 6.16.: Simulated and measured cathode diffusion loss (R_{1C}) of ASC Type-B. Predicted values obtained using the Butler-Volmer (BVM, solid line) approach with kinetic data from Ref.[9] (Table 5.2) and by the MIEC-cathode approach (dotted line) with data from Ref.[120], Table 5.3). $T = 800$ °C. Fuel: $H_2 + 60$ % steam. Oxidant: $pO_{2,\text{cat}} = 0.21 \ldots 0.01$ atm with balance N_2. $\dot{v}_{\text{an/cat}} = 250$ sccm. [Cell Z1_188].

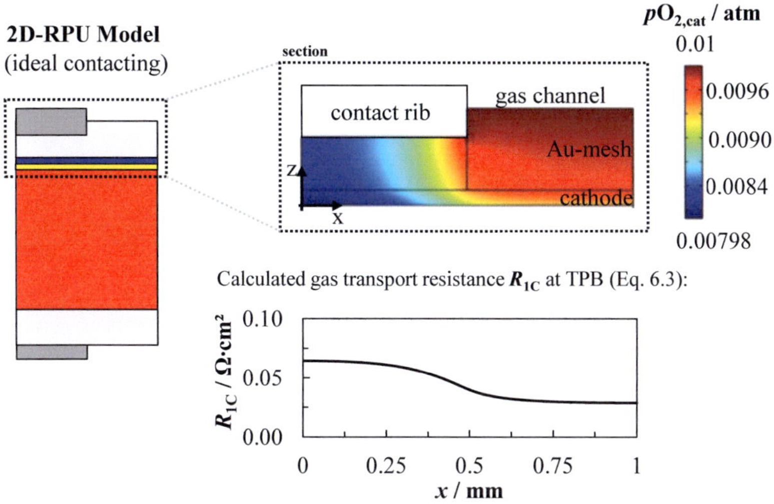

Figure 6.17.: Predicted $pO_{2,\mathrm{cat}}$-distribution in the cathode layer and adjacent Au-contact mesh at $pO_{2,\mathrm{cat}} = 0.01$ atm, $j_{\mathrm{EIS}} = 60$ A/cm^{-2}, $T = 800$ °C. Corresponding $R_{1\mathrm{C}}$-distribution at IF$_{\mathrm{cat-elyt}}$, calculated using Equation 6.3.

Gas Diffusion Resistance

Figure 6.16 illustrates measured values (symbols), BV-model predicted values (solid line) and MIEC-model predicted values (dashed line) over corresponding $pO_{2,\mathrm{cat}}$. The measured $R_{1\mathrm{C}}$ for decreasing $pO_{2,\mathrm{cat}}$ increases due to limited gas transport properties, partly within the Au-mesh, but mostly in the cathode layer. The result is the formation of an O_2 concentration gradient between gas channel ($pO^{\mathrm{in}}_{2,\mathrm{cat}}$) and electrochemical active zone ($pO^{\mathrm{TPB}}_{2,\mathrm{cat}}$). From the O_2-gradient the cathode gas diffusion overpotential $\eta_{\mathrm{cc,cat}}$ can be calculated by Equation 4.88, derived from the Nernst-equation [9], and given here again for better comprehension:

$$\eta_{\mathrm{cc,cat}} = \frac{RT}{4F} \ln\left(\frac{pO^{\mathrm{in}}_{2,\mathrm{cat}}}{pO^{\mathrm{TPB}}_{2,\mathrm{cat}}}\right) \quad \mathrm{V}. \tag{6.1}$$

In contrast to the calculation method for gas diffusion losses in the anode (Equation 5.14, Section 5.1.7), $R_{1\mathrm{C}}$ cannot be used to calculate a microstructure parameter $\Psi_{\mathrm{EIS,cat}}$ with sufficient accuracy due to two reasons. Firstly, the Au-mesh impact on $\eta_{\mathrm{cc,cat}}$ is not negligible in contrast to the mesh influence on $\eta_{\mathrm{cc,an}}$, where regarding partial pressure gradients in the thick anode substrate are considerably larger than in the thin cathode layer. And secondly, the pO_2-gradient evolves not only in z-direction from gas channel to TPBs but also in x-direction to electrochemically active zones beneath the flowfields contact ribs. This is illustrated by a FEM-simulation result shown in Figure 6.17, which shows the pO_2-distribution at $pO_{2,\mathrm{cat}} = 0.01$ atm ($j_{\mathrm{EIS}} = 60$ A/cm^{-2}, $T = 800$ °C). The calculated $R_{1\mathrm{C}}$-distribution in Figure 6.17 shows how the result varies by a factor of two. Even with the help of geometric estimations for l_{cat} as suggested in Ref. [45], it is simply not possible to use a non-spatially dependent expression such as the following Equation 6.2 to calculate $R_{1\mathrm{C}}$ reliably for a two-dimensional $p\mathrm{O}_2$-gradient [9, 45]:

$$R_{1\mathrm{C}} = \left(\frac{RT}{4F}\right)^2 l_{\mathrm{cat}} \frac{1}{\Psi_{\mathrm{EIS,cat}} \cdot p\, D_{\mathrm{mol,O_2}}} \frac{1}{\Psi_{\mathrm{EIS,cat}}} \left(\frac{1}{p\mathrm{O}_{2,\mathrm{cat}}} - 1\right) \quad \Omega \cdot \mathrm{cm}^2. \tag{6.2}$$

$$R_{1\mathrm{C}}^{\mathrm{FEM}} = M \left\{ \frac{\frac{RT}{4F} \ln\left(\frac{p\mathrm{O}_{2,\mathrm{cat}}^{\mathrm{in}}}{p\mathrm{O}_{2,\mathrm{cat}}^{\mathrm{TPB}}}\right) \cdot j_{\mathrm{ct,cat}}}{j_{\mathrm{cell}}^2} \right\} \quad \Omega \cdot \mathrm{cm}^2. \tag{6.3}$$

However, using the 2D-RPU model and Equation 6.3 to predict the mean value of $R_{1\mathrm{C}}^{\mathrm{FEM}}$ (Section 4.3) with the BVM (solid line) or MIEC-model (dashed line) results in a nearly equal progression of simulated $R_{1\mathrm{C}}$-values, as the results in Figure 6.16 show. Both model predictions underestimate the measured results under standard conditions ($p\mathrm{O}_{2,\mathrm{cat}} = 0.21$ atm). The deviation is ~2-3 mΩ·cm^2, which is well within the measurable tolerance considering a total $ASR = 130$ mΩ·cm^2 at the given operating conditions. Nevertheless, the deviation between measured and predicted $R_{1\mathrm{C}}$ increases further towards lower $p\mathrm{O}_{2,\mathrm{cat}}$ up to an total error of ~60 %. For this mismatch we must consider that the numerical results were obtained assuming a constant $p\mathrm{O}_{2,\mathrm{cat}}$ distribution at the cathode/gas channel interface ($\mathrm{IF}_{\mathrm{GC,cat}}$, Figure 4.3), which is apparently not the case at $p\mathrm{O}_{2,\mathrm{cat}} < 0.10$ atm. This even accounts for the high applied gas-flow rates $\dot{v}_{\mathrm{cat}} = 250$ sccm and small EIS current stimulus $j_{\mathrm{EIS}} = 60$ A/cm^2. The results presented in Section 6.5 will demonstrate, how an improved gas diffusion loss prediction with the 2D-RPU model is possible by considering a current dependent gas conversion function at $\mathrm{IF}_{\mathrm{GC,cat}}$.

In summary, the results in Figure 6.15 demonstrate a precise reproduction of electrochemical charge transfer losses in the cathode, using the BV-model or MIEC-model with $p\mathrm{O}_2$-dependent surface exchange kinetics (k^δ with Equation 4.63). Results in Figure 6.16 show that without considering gas conversion, gas diffusion polarization at the cathode is underestimated by the 2D-RPU model. However, the deviation accounts ~ 5 mΩ·cm^2 at standard conditions ($p\mathrm{O}_{2,\mathrm{cat}} = 0.21$ atm), which is within the measurable tolerance.

Current/Voltage Predictions

Measured C/Vs at various, decreasing $p\mathrm{O}_{2,\mathrm{cat}}$ and at constant anode gas composition ($p\mathrm{H}_{2,\mathrm{an}} = 0.4$ atm, $p\mathrm{H}_2\mathrm{O}_{\mathrm{an}} = 1 - p\mathrm{H}_{2,\mathrm{an}} - p\mathrm{N}_{2,\mathrm{leak}}$) and operating temperature $T = 800$ °C are displayed by symbols in Figure 6.18. From literature it is well known that the limiting current density decreases at lower $p\mathrm{O}_{2,\mathrm{cat}}$ [9, 184, 251]. Limited diffusion properties cause an increase of $\eta_{\mathrm{cc,cat}}$ accompanied by a decreased electrochemical activity or increased loss $\eta_{\mathrm{act,cat}}$. In sum, both the non-linear current dependent loss mechanisms dominate the overall loss, whereby gas diffusion becomes the predominant loss at lower $p\mathrm{O}_{2,\mathrm{cat}}$. This behavior is demonstrated by the measured results in Figure 6.18. Furthermore, a lower OCV at lower $p\mathrm{O}_{2,\mathrm{cat}}$ results from the partial pressure dependent cathode half-cell potential, which is represented in the model framework by Equation 4.81.

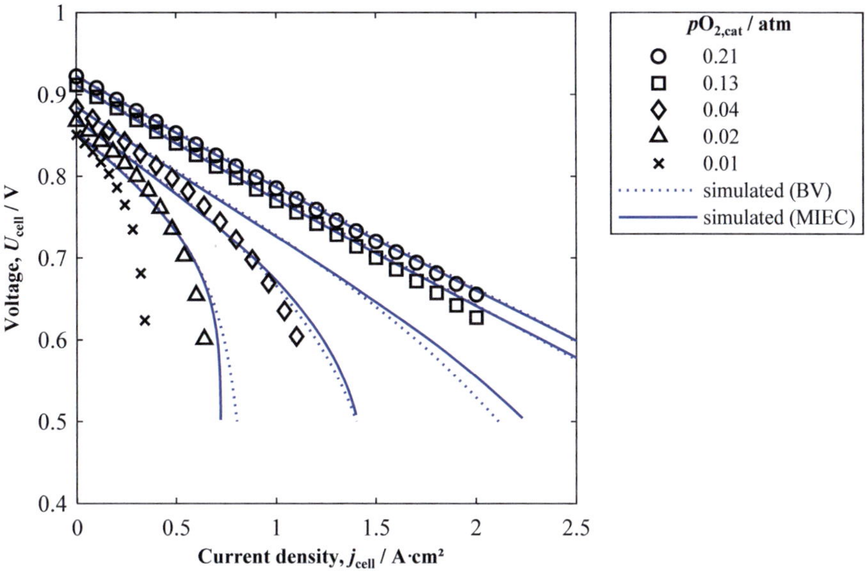

Figure 6.18.: Simulated (BVM - dotted line, MIEC - solid line) and measured (symbols) C/Vs, recorded at cell Type-B at varied $pO_{2,cat} = 0.21 \ldots 0.01$ atm and balance N_2. Constant $pH_{2,an} = 0.4$ atm, $pH_2O_{an} = 0.6$ atm, $T = 800$ °C and $\dot{v}_{an/cat} = 250$ sccm. [Cell Z1_188].

Hence, regardless of the applied electrochemical cathode model (BVM or MIEC), predicted OCVs are in precise agreement with the measured voltage (lines in Figure 6.18). Under load and for $pO_{2,cat} > 0.13$ atm, the simulated and measured results are also in good agreement and no further difference between the BVM and MIEC models can be observed. For $pO_{2,cat} < 0.13$ atm, however, the models cannot reproduce the drop in performance and overestimate the limiting current densities, without significant difference between each other. The results found in Section 6.4 indicate that the deviation between model prediction and measurement can be ascribed to the (until now) disregarded gas conversion, which is taking place in the cathode gas channel even at high gas-flow rates. As mentioned before, this matter is discussed further in Section 6.5. Nevertheless, the results in Figure 6.18) demonstrate that the 2D-RPU model is sufficient for predicting measured C/Vs at $pO_{2,cat} > 0.13$ atm. These are important results from an application-oriented view. There, high gas-flow rates of air ($pO_{2,cat} < 0.21$ atm) at the cathode are used to remove heat and thus to cool down the stack and it is therefore very unlikely to encounter $pO_{2,cat} < 0.1$ atm along the gas channel length.

With these results, the basic model validation is completed. An excellent agreement of measurement and simulation is found for the ideal contacted ASC, conducted under operating conditions relevant to application and far beyond. By variation of temperature and electrode gas compositions (separately and up to extreme conditions), it is shown in the previous sections, how the model reproduces occurring physical phenomena and their individual domination of the overall loss with high precision.

6.5. Modeling Considerations and Accuracy Aspects

The model validation presented in the previous sections is the result of a careful evaluation of multiple modeling aspects. The aim of this section is to supply the reader with valuable information, which ultimately leads to the final model framework setup. This is used in the next Chapter 7 to produce the results presented therein.

To begin with, characteristic microstructure parameters of anode and cathode may be determined experimentally by two different approaches: (i) Calculation of $\Psi_{\mathrm{EIS,i}}$ from impedance spectra (Section 5.1.7) or (ii) determination using 3D reconstructions of porous electrode samples (Sections 2.6 and 5.1). The impact on simulated anode gas diffusion loss ($R_{1\mathrm{A}}$) at small current stimuli $j_{\mathrm{EIS}} = 60$ mA/cm^2 and on simulated C/Vs at various fuel gas compositions is analyzed and discussed. Secondly, the impacts of various gas transport models commonly used in literature (Section 4.3.2.3) are analyzed by comparing the results at low and high current densities. Thereafter, it is discussed how the use of mixed unit systems in the gas transport model equation incorporating a coupling of convective and diffusive transport phenomena leads to an erroneous gas diffusion overpotential prediction and/or non-physical gas distribution within the anode. The impact of gas conversion on the modeling results is discussed by the use of different model geometry dimensions (Section 4.2). These results are assessed by weighing solution accuracy and model intent against required computational effort (due to the increase in numerical mesh elements) and the use of different cathode charge transfer model approaches (Section 4.3.3). A short analysis of contact mesh microstructure parameters is given at the end of this section.

Gas Conversion Implementation in the 2D-RPU Model

The results presented in Section 6.3 regarding the variation of anode fuel gas composition and in Section 6.4 regarding the variation of cathode gas composition revealed a certain deviation between predicted and measured C/Vs. This accounts for the $p\mathrm{O}_{2,\mathrm{cat}}$-variation at $p\mathrm{O}_{2,\mathrm{cat}} < 0.1$ atm. In the subsequent analysis, it was discussed how the deviation is related to occurring gas conversion in the fuel gas channel, especially at lower inlet gas composition of the consumed species. Due to the nature of the 2D-RPU model geometry, no gas channel and therefore no change in the gas concentration could be regarded along the gas channel due to increasing consumption under load. Hence, a workaround was implemented by calculating a current dependent inlet gas composition at $\mathrm{IF}_{\mathrm{GC,an/cat}}$ (Figure 4.3) with Equations 6.5 and 6.6. Its derivation is briefly explained in the following.

Gas conversion can be expressed as the ratio of current supplied by a cell to the current equivalent, j_{eq}, of its input gas volume. The current equivalent can be calculated by

$$j_{\mathrm{eq}} = (N_{\mathrm{A}}/V_0) \cdot n_{\mathrm{e}} \cdot e_0 \cdot (\dot{v}_{\mathrm{in}}/60) \tag{6.4}$$

where N_{A} denotes the Avogadro constant, V_0 the molar volume (ideal gas: $\mathrm{V}_0 = 22.41$ L/mol), n_{e} the fictional charge number, e_0 the elementary charge and $\dot{v}_{\mathrm{in}}$ the applied volumetric gas-flow in sccm. At the cathode n_{e} depends on the $p\mathrm{O}_{2,\mathrm{cat}}$ with $n_{\mathrm{e}} = 4 \cdot p\mathrm{O}_{2,\mathrm{cat}}/p_0$, while at the anode $n_{\mathrm{e}} = 2 \cdot p\mathrm{H}_{2,\mathrm{an}}/p_0$ accounts.

Table 6.3 gives an overview of how for example $pO_{2,\mathrm{cat}}$ determines the available current equivalent. Taking a look at current equivalents calculated for $pO_{2,\mathrm{cat}} < 0.1$ atm, where simulated and measured data deviate more strongly, it becomes clear that for a current density of 1 Acm^{-2}, already more than a third of the available current equivalent has been consumed. This means that for $pO_{2,\mathrm{cat}} < 0.1$ atm the oxygen concentration along the gas channel length cannot be assumed as constant, despite the high oxidant flow rates. In order to take the described effects of a decreased partial pressures in the anode and cathode gas channel into account the boundary condition at $\mathrm{IF}_{\mathrm{GC,an/cat}}$ is modified. Assuming linear gas utilization along the channel length, the total gas consumption in sccm for a certain current density j_{cell} can be calculated with the help of Equation 6.4 for anode or cathode to:

$$\Delta p_{\mathrm{i}} = 6 \cdot 10^4 \cdot \left(\frac{j_{\mathrm{cell}} \cdot V_0}{N_{\mathrm{A}} \cdot \dot{v}_{\mathrm{in}} \cdot n_{\mathrm{e}} \cdot e_0} \right) \quad \mathrm{sccm}. \tag{6.5}$$

In order to include gas conversion into the 2D-RPU model, the partial pressure set as boundary condition at the $\mathrm{IF}_{\mathrm{an/cat}}$ is now modified to:

$$p_i = \frac{p_{i,0}}{\dot{v}_{\mathrm{in}}} \cdot \left(\dot{v}_{\mathrm{in}} - \frac{\Delta p_{\mathrm{i}}}{2} \right) \quad \mathrm{atm}. \tag{6.6}$$

In Equations 6.5 and 6.6 the subscript i denotes H_2 or O_2, depending on the electrode. This is only a rough estimation due to the linear approximation of gas conversion. Nevertheless, no fitting of experimental data is needed. In reality gas utilization cannot be assumed linear and evolves differently for different gas compositions.

The inclusion of gas conversion leads to a better match of measured and simulated data. While the influence is almost non-noticeable at the anode (Figure 6.19) a considerable deviation is given at the cathode (Figure 6.20) at $pO_{2,\mathrm{cat}} < 0.21$ atm. The matter is further attended to in the next Section 6.5.

Table 6.3.: Calculated current equivalents for various cathode oxidant partial pressures $pO_{2,\mathrm{cat}}$.

$pO_{2,\mathrm{cat}}$ / atm	n_{e}	$\dot{v}_{\mathrm{in}}$ / sccm	j_{eq} / A
1	4	250	71.74
0.21	0.84	250	15.07
0.13	0.52	250	9.33
0.04	0.16	250	2.87
0.02	0.08	250	1.43
0.01	0.04	250	0.72

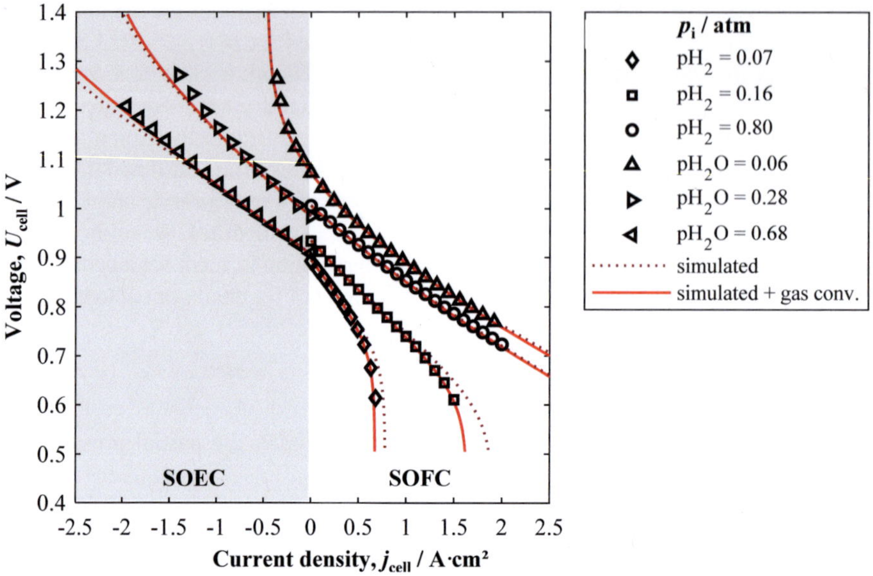

Figure 6.19.: Simulated (lines) and measured (symbols) C/Vs, recorded on cell Type-A at fuel gas compositions (SOFC: varied $pH_{2,an}$ with balance N_2 and constant $pH_2O_{an} = 0.2$ atm; SOEC: varied pH_2O_{an} with balance N_2 and constant $pH_{2,an} = 0.4$ atm). Constant $pO_{2,cat} = 0.21$ atm, $T = 800$ °C and $\dot{v}_{an/cat} = 250$ sccm. [Cells Z2_275, Z1_349].

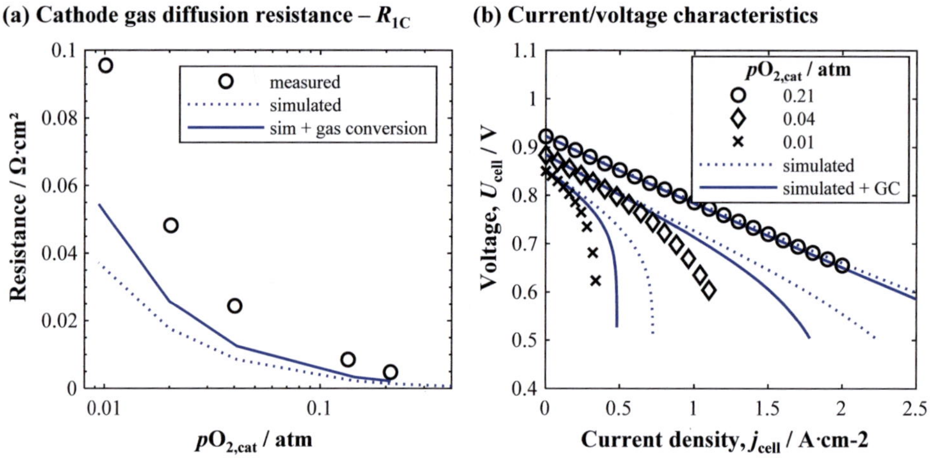

Figure 6.20.: Comparison of measured (symbols) and predicted (line) C/Vs, considering gas conversion at the cathode by Equation 6.6 in the 2D-RPU model approach. Oxidant: $pO_{2,cat} = 0.21 \ldots 0.01$ atm and balance N_2. Fuel: Constant $pH_{2,an} = 0.4$ atm, $pH_2O_{an} = 0.6$ atm, $T = 800$ °C, $\dot{v}_{an/cat} = 250$ sccm. [Cell Z1_188].

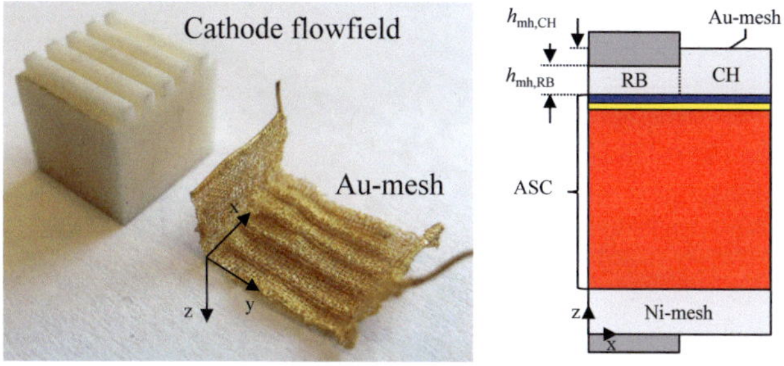

Parameter	Value
$h_{mh,CH}$	240 µm
$h_{mh,RB}$	120 µm
$\varepsilon_{mh,CH}$	0.64
$\varepsilon_{mh,CH}$	0.32
$\tau^{CH}_{mh-por,x}$	1.1
$\tau^{CH}_{mh-por,y}$	1.0
$\tau^{RP}_{mh-por,x}$	2.2
$\tau^{RP}_{mh-por,y}$	2.0

Figure 6.21.: (a) Image of an Au-mesh after several hundred hours of operation under various typical SOFC operating temperatures and current loads. (b) Adapted 2D-RPU model geometry to regard an adjusted mesh domain beneath the contact rib and (c) corresponding mesh parameters using Equation 6.7 with $\beta_{\mathrm{mh}} = 0.5$.

Contact Mesh Influence on Performance Prediction

A closer look at the Au contacting-mesh used on the cathode layer for in the measurements (Section 3.2.1) revealed a noticeable deformation of the mesh structure. Depressions are located where the flowfield ribs are positioned during the measurement. This is shown in Figure 6.21 on the left, showing the lower surface of an Au-mesh which has been operated for several hundred hours at various standard SOFC operating temperatures. The lower melting point of Au compared to Ni apparently affects the cathode mesh, while nothing similar is observed at the anode.

A lower thickness in the affected area means smaller volume space, but with the same amount of material. Consequently, less pore volume is available for O_2 gas transport. Furthermore, the impact on the mesh's semi-porous microstructure has to be considered as well. The 2D-RPU model geometry is adapted accordingly by implementing an extra mesh domain underneath the rib as displayed in Figure 6.21 in the middle by introducing an additional factor β_{mh}.

The mesh layer thickness $h_{\mathrm{mh,RP}}$ and pore fraction $\epsilon_{\mathrm{mh,RP}}$ are set by:

$$h_{\mathrm{mh,RB}} = h_{\mathrm{mh,CH}} \cdot \beta_{\mathrm{mh}} \qquad \mu\mathrm{m}, \tag{6.7}$$

$$\epsilon_{\mathrm{mh,RB}} = \epsilon_{\mathrm{mh,CH}} \cdot \beta_{\mathrm{mh}}, \tag{6.8}$$

while the mesh tortuosity is adjusted as

$$\tau^{\mathrm{RB}}_{\mathrm{mh-por,x/z}} = \tau^{\mathrm{CH}}_{\mathrm{mh-por,x/z}} / \beta_{\mathrm{mh}}. \tag{6.9}$$

In order to estimate the influence on performance predictions a numerical parameter study is carried out. β is varied and $R_{1\mathrm{C}}$ is compared to measured data at using the adapted 2D-RPU model with gas conversion consideration.

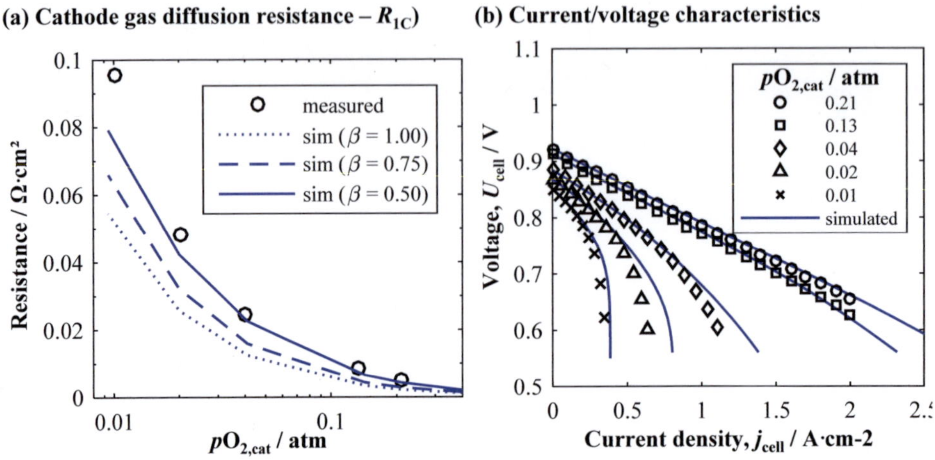

Figure 6.22.: Comparison of measured (symbols) and model predictions (lines), considering gas conversion at the cathode by Equation 6.6 in the 2D-RPU model approach. (a) Cathode gas diffusion resistance, $R_{1\mathrm{C}}$ and (b) C/Vs with adjusted mesh-properties beneath the flowfiel rib by Equation 6.7, using $\beta = 0.5$ in the C/V predictions. Oxidant: $pO_{2,\mathrm{cat}} = 0.21 \ldots 0.01$ atm and balance N_2. Fuel: Constant $pH_{2,\mathrm{an}} = 0.4$ atm, $pH_2O_{\mathrm{an}} = 0.6$ atm, $T = 800$ °C and $\dot{v}_{\mathrm{an/cat}} = 250$ sccm. [Cell Z1_188].

The results are displayed in Figure 6.22. Lowering β causes $R_{1\mathrm{C}}$ to increase incrementally as the O_2 transport is increasingly hindered. For $\beta_{\mathrm{mh}} = 0.5$ an acceptable agreement between measurement and simulation is given above $pO_{2,\mathrm{cat}} > 0.02$ atm. Below that, $R_{1\mathrm{C}}$ is still underestimated. A further decrease would however lead to an overestimation of $R_{1\mathrm{C}}$ at higher $pO_{2,\mathrm{cat}}$. Using $\beta_{\mathrm{mh}} = 0.5$ and the according calculated mesh-data as displayed in Figure 6.21 on the right, predicted C/Vs match the measurement at $pO_{2,\mathrm{cat}} = 0.21$ atm with high precision as the results in 6.22b demonstrate. With this and with a much improved agreement at lower $pO_{2,\mathrm{cat}}$, even at $pO_{2,\mathrm{cat}} = 0.01$ atm, an acceptable applicability of the procedure can be justified.

However, a small deviation at higher polarization still remains. The results presented in the next Section will show, how model predicted performance at lower $pO_{2,\mathrm{cat}}$ is influenced by adopting a geometry with gas channels.

Model Geometry Influence on Performance Prediction

The model validation for hydrogen operation has been carried out so far based on the 2D-RPU model geometry. It is shown that under standard, application-oriented operating conditions the two dimensional cross section is sufficient to reliably predict SOFC and SOEC single cell performance (Sections 6.2 and 6.3). Modeling of adjacent gas channels is only required to regard gas conversion via occurring reforming reactions for simulation of hydrocarbon fueled operation under higher polarization (Sections 6.2 and 6.3). The results presented in the previous Section 6.5 however indicate that under low $pO_{2,\mathrm{cat}} < 0.02$ atm the applied gas conversion approach is insufficient to regard gas conversion in the cathode gas compartment. Even though such low oxidant concentrations are far away from standard application conditions [79], it is important to verify the model framework's capability to reproduce measured performance even under such extreme conditions in order to fully trust in its reliability.

Hence, the 2D-GCh and 3D-RPU model geometries (Figure 4.3) are applied for simulations at varied $pO_{2,\text{cat}}$, thereby regarding adjusted mesh properties discussed in Section 6.5. Resulting $R_{1\text{C}}$-predictions (lines) are displayed in Figure 6.23a together with experimental results (symbols). It is shown therein that using the 2D-GCh model is insufficient because predicted $R_{1\text{C}}$ (dotted blue line) vastly underestimates the experimental data. In contrast to this the 3D-RPU model (red line) reproduces the measured values with even better precision than the 2D-RPU model (blue line). A very small deviation at $pO_{2,\text{cat}} = 0.01$ atm is still given, which could be explained by inhomogeneous mesh deformation along the gas channel length and its resulting impact (Section 6.5) with according high sensitivity at such low $pO_{2,\text{cat}}$.

Figure 6.23b depicts measured and predicted C/Vs using all three model geometries available in this work. Comparing experimental and numerical results in Figure 6.23b, it is clear that the 2D-GCh model (blue dotted line) is insufficient to reproduce SOFC performances recorded below $pO_{2,\text{cat}} < 0.21$ atm while the 3D-RPU model is fully able to reproduce it under all measured conditions with excellent precision, even at $pO_{2,\text{cat}} = 0.01$ atm. Hence, a better reproduction of occurring gas conversion along the gas channel length in the 3D model makes the difference compared to the 2D-RPU model.

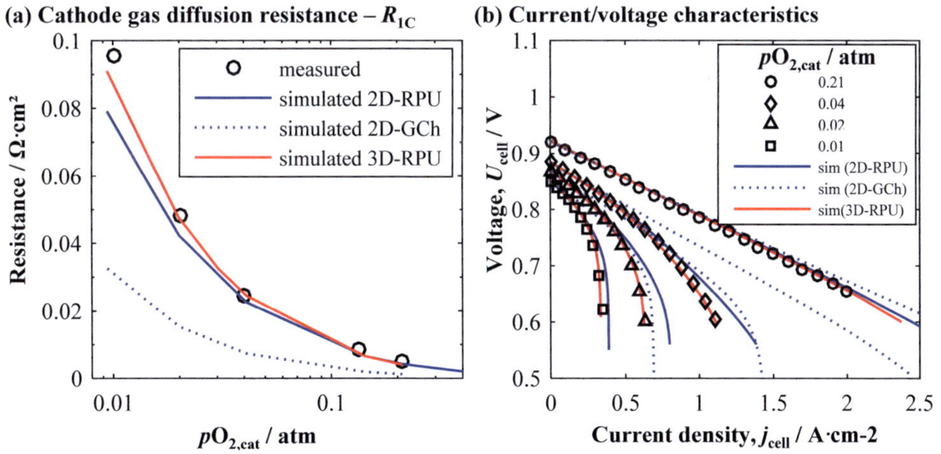

Figure 6.23.: Comparison of measured (symbols) and predicted (line) cathode gas diffusion resistance $R_{1\text{C}}$, considering gas conversion at the cathode by Equation 6.6 in the 2D-RPU model approach for adjusted mesh-properties beneath the flowfiel rib by 6.7. Measured and simulated C/Vs, using $\beta = 0.5$. Oxidant: $pO_{2,\text{cat}} = 0.21 \ldots 0.01$ atm and balance N_2. Fuel: Constant $pH_{2,\text{an}} = 0.4$ atm, $pH_2O_{\text{an}} = 0.6$ atm, $T = 800$ °C. Constant 250 sccm gas-flow rate. [Cell Z1_188].

The results demonstrate that it is necessary to regard gas transport perpendicular to the gas channel flow direction in order to predict SOFC performance reliably at lower $pO_{2,\mathrm{cat}}$. However, at standard operating conditions and intermediate polarization, the 2D-GCh geometry seems to be sufficient and the increased computational effort required for the 3D-RPU model can be saved (Table 6.5).

The last and probably most important point is that the 2D-RPU model geometry is able to regard the influence of O_2 transport limitations on SOFC performance predictions and a full 3D model with vastly increased computational effort (Table 6.5) is not required. Constant load $j = 0.5$ A/cm^{-2} $\lambda = 2.5$ is set for the cathode gas-flow rate of ambient air, which means that 40 % of the supplied O_2 is used up (standard operating conditions for SOFC stack operation at Jülich [252]). Hence, a minimum of $pO_{2,\mathrm{cat}} > 0.13$ atm is given at the gas channel end and is consequently higher in the middle. This is a very important fact to justify the use of the 2D-RPU model geometry to investigate cathode layer and contact parameters in Sections 7.2 and 7.3.

Influence of Microstructure Parameters on Predicted Species Transport Resistance and Overall SOFC Performance

In Section 5.1 various characteristic parameters are described, which are required to fully describe a porous electrode. Of interest in this section is $\Psi = \epsilon/\tau$ (Equation 5.6), in this work determined by two approaches: (i) Based on quantification of electrochemical impedance measurements (Section 5.1.7) and (ii) based on evaluation of porous electrode samples, reconstructed in 3D (Sections 2.6 and 5.1). Furthermore, the mean pore diameter $d_{\mathrm{por},el}$ is regarded in the following. It has a special role as it is one of the parameters gained from the 3D electrode reconstruction analysis, but it is also essential to calculate $\Psi_{\mathrm{EIS,an}}$ from impedance spectra evaluation (Section 5.1.7, Equation 5.14). To complicate the matter further, $d_{\mathrm{por},el}$ can be described by number, surface or volume averaging (Section 5.1.6), albeit in this work only the first and last method is applied. Usually no description is given in literature as to which approach was used in the determination procedure. The results presented in the following will show that the specification of the method-used is of vital importance for reliable model predictions.

Influence on the Anode Gas Diffusion Loss Prediction ($R_{1\mathrm{A}}$)

The results in Figure 6.24 show that it matters significantly. Compared are ASR values of the anode gas diffusion resistance, $R_{1\mathrm{A}}$, derived from impedance spectra (symbols) and corresponding model predictions (lines). Figure 6.24a features a $pH_{2,\mathrm{an}}$-variation with constant $pH_2O_{\mathrm{an}} = 0.2$ atm and balance N_2 and Figure 6.24b shows a fuel gas humidity variation ($pH_{2,\mathrm{an}}/pH_2O_{\mathrm{an}}$-variation) with $pN_{\mathrm{N,leak}} = 0.01$ atm. The operating temperature $T = 800$ °C and cathode O_2 partial pressure $pO_{2,\mathrm{cat}} = 0.21$ atm were kept constant.

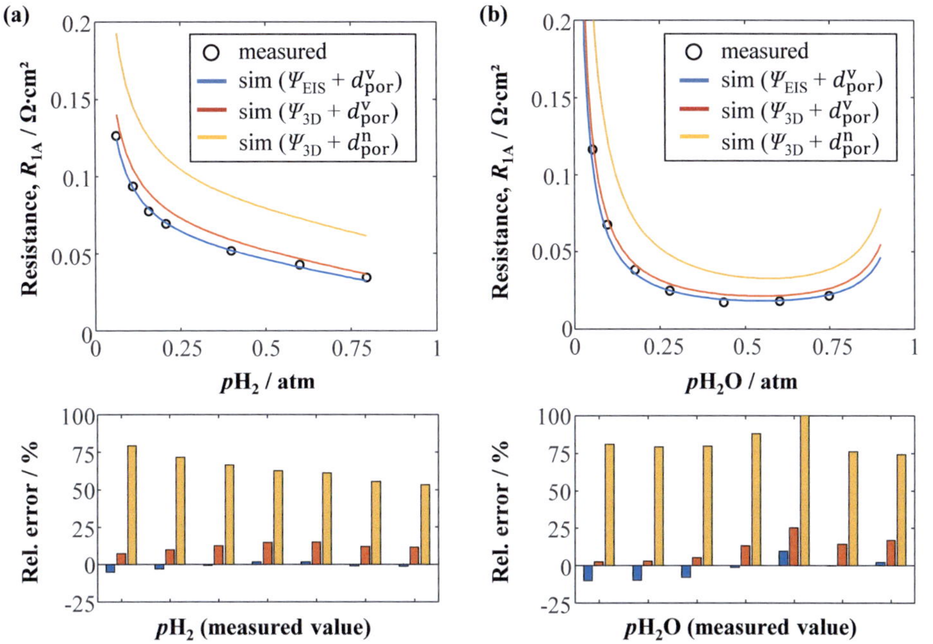

Figure 6.24.: Measured (symbols) and predicted (lines) anode gas diffusion resistances for ASC Type-A at varied fuel gas compositions and $T = 800$ °C, using different microstructure parameters in the simulations. Ψ_{EIS} determined from impedance measurements (Section 6.5), Ψ_{3D} determined form reconstructed electrode samples (Section 2.6), $d^{\mathrm{v}}_{\mathrm{por}}$ calculated by mean volume weight and $d^{\mathrm{n}}_{\mathrm{por}}$ by mean number count (Section 5.1.6). (a) Varied $p\mathrm{H}_{2,\mathrm{an}} = 0.8 \ldots 0.05$ atm and constant $p\mathrm{H_2O}_{\mathrm{an}} = 0.2$ atm, balance $\mathrm{N_2}$. (b) Varied $p\mathrm{H_2O}_{\mathrm{an}} = 0.05 \ldots 0.75$ atm with $p\mathrm{H}_{2,\mathrm{an}} = 1 - p\mathrm{H_2O}_{\mathrm{an}} - p\mathrm{N}_{2,\mathrm{leak}}$ atm. Bar graphs denote the calculated relative error between measured and predicted resistances. [Cell Z2_275].

Simulation results labeled with Ψ_{EIS} are produced using the characteristic microstructure parameter determined from impedance measurements (Section 5.1.7), while for the results labeled with Ψ_{3D} microstructure data gathered by the analysis of real anode reconstruction were used (ASC Type-A, Table 5.1). Furthermore, $d^{\mathrm{v}}_{\mathrm{por}}$ labels simulation results using the volume averaging method and $d^{\mathrm{n}}_{\mathrm{por}}$ model parameterization with a mean pore diameter gained by number averaging. While Ψ_{el} effects all transport processes in the porous electrode according to the homogenization approach (Section 5.1.5), d_{por} is only used in the calculation of the individual Knudsen diffusion coefficients $D_{\mathrm{Kn,i}}$ (Equation 4.31). As mentioned in the beginning of this section, $D_{\mathrm{Kn,i}}$ is also used in Equation 5.14 to calculate $\Psi_{\mathrm{EIS,an}}$, which inflicts an interdependence on the use of both parameters. The following results deliver a method of elimination based on logical consideration.

The analysis is restricted to the impact on $R_{1\mathrm{A}}$ where the longer diffusion pathways compared to those in the thin cathode layer are much more affected by variations in the microstructure modeling parameters. Furthermore, $\Psi_{\mathrm{EIS,cat}}$ is flawed due to the inseparable coupling of Au-mesh and cathode influence (Section 6.4).

It is in common for both comparisons illustrated in Figure 6.24 that all model predictions reproduce the measured trend with respect to each partial pressure variation. However, a significant overestimation of R_{1A} is observed, using the model parameterized with Ψ_{3D} and d^{n}_{por} (yellow line). The calculated relative error (displayed beneath the ASR-plots in Figure 6.24) ranges between $60...110\%$, independent from the varied fuel gas composition. However, while the relative error decreases from 75% down to 60% for increasing $pH_{2,an}$ (Figure 6.24a), resulting in a mean value of 64%, the relative error is more or less constant at 75% for the varied fuel humidity (Figure 6.24b), with two outliers for intermediate pH_2O_{an}. This may be ascribed to a higher error sensitivity in the CNLS-fitting method. Hence, using the parameter combination $\Psi_{3D} + d^{n}_{por}$ results in a significant error in the gas diffusion resistance prediction (R_{1A}). Using $d^{v}_{por} + \Psi_{3D}$ (red line) still results in an overestimation of R_{1A} with a mean relative error 11% in both fuel gas variations. In contrast to the first parameter set, this combination delivers an almost constant overestimation for the $pH_{2,an}$-variation (Figure 6.24a). At low pH_2O_{an} in the humidity variation (Figure 6.24b) the model predictions are actually in good agreement with the measured data and develop an increasing overestimation towards higher humidities. In total however, the mean relative error is small compared to the first parameter set and lies within reach of the measurable tolerance (2 mΩcm^2). Finally, the models parameterized with $\Psi_{EIS} + d^{v}_{por}$ (blue lines) deliver the best results with the lowest mean error of -1 to -2%, which is very precise considering the mentioned tolerances in the measurement and CNLS-fitting process.

The above described observations support the conclusion that using a mean pore radius averaged by volume (d^{v}_{por}) delivers far more accurate simulation results. This is attributed to a wider spread of detected pore sizes in the Ni/YSZ-anode, which is not as accurately represented by the number averaging method (Figure 5.2b). Furthermore, using $\Psi_{EIS,an}$ as microstructure parameter to calculate effective transport parameters delivers more accurate results. However, the results using $\Psi_{3D,an}$ are in close proximity. Based on the work of J. Joos, a high determination precision of $\epsilon_{an} = 0.415$ is secured for cell Type-A [120]. Together with Equation 5.14 a mean tortuosity of $\tau^{EIS}_{por,an} = 2.63$ is calculated, which still is ~15% lower compared to the value determined from the 3D-reconstruction of cell Type-A (Table 5.1). While this seems like a significant deviation, it lies within the possible margin of error using the finite volume method approach of M. Ender [121], where the staircase-effect may cause greater deviations due to the voxel-based resolution dependency [120]. In any case, the deviation is within the acceptable range for measured and predicted gas diffusion ASR at low current stimuli using microstructure data determined from 3D reconstructed electrodes. Hence, it is most interesting how the model predictions compare to measured data at larger current densities where an increased electrochemical gas conversion induces larger gas diffusion polarization.

Influence on SOFC Performance Prediction

Figure 6.25 illustrates measured (symbols) C/V characteristics at constant $T = 800$ °C and $pO_{2,cat} = 0.21$ atm, recorded on cell Type-A. The anode gas compositions are given in the caption. The operating conditions were set specifically at low H_2 (diamond, SOFC) and low H_2O (circle, SOEC) partial pressures to achieve operating conditions, where the regarding fuel species reaches a minimum and thereby causes a decreased limiting current density due to a domination of the overall losses by the anode diffusion overpotential $\eta_{\mathrm{cc,an}}$. This approach is ideal for analyzing the influence of $\Psi_{\mathrm{EIS/3D}}$ and $d^{\mathrm{v/n}}_{\mathrm{por,an}}$ on the diffusion overpotential predictions because the parameters directly influence the gas transport properties.

Parameterizing the model framework with $\Psi_{\mathrm{3D}} + d^{\mathrm{n}}_{\mathrm{por}}$ (yellow line) leads to overestimations of the gas transport diffusion polarization and causes the model to predict a lower limiting current density. Furthermore, the model predictions at low $p\mathrm{H_2O_{an}}$ (circles) are also lower with an increasing deviation compared to the measurement in SOFC operation. Comparing the model predictions with volume averaged pore diameter and microstructure data from 3D reconstructions, ($\Psi_{\mathrm{3D}} + d^{\mathrm{v}}_{\mathrm{por}}$, red line), a fairly good agreement with the measurement is given in SOEC/SOFC operation at low $p\mathrm{H_2O_{an}}$, while a small deviation is given at low $p\mathrm{H_{2,an}}$ in SOFC mode. The model predictions with $\Psi_{\mathrm{EIS}} + d^{\mathrm{n}}_{\mathrm{por}}$ (blue line) deliver precise predictions.

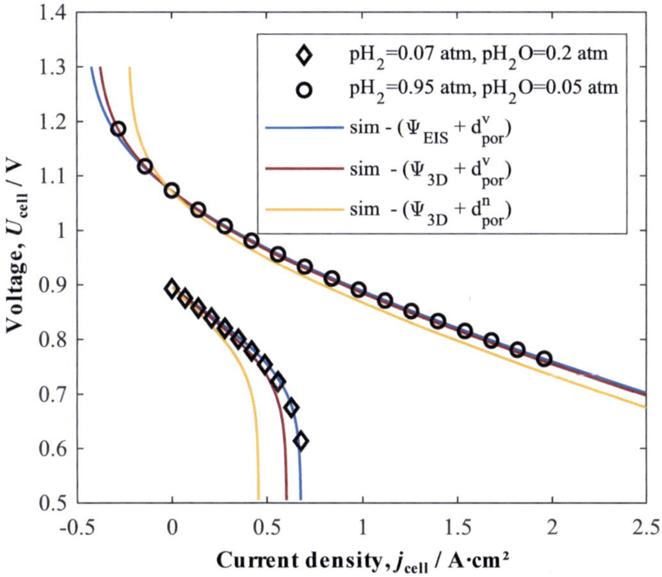

Figure 6.25.: Measured (symbols) and predicted (lines) C/Vs of varied fuel gas compositions at $T = 800\ ^{\circ}\mathrm{C}$ and ambient air as oxidant at ASC Type-A, using different microstructure parameters in the simulations. Ψ_{EIS} determined from impedance measurements (Section 6.5), Ψ_{3D} determined from reconstructed electrode samples (Section 2.6), $d^{\mathrm{v}}_{\mathrm{por}}$ calculated by volumetric weighing and $d^{\mathrm{n}}_{\mathrm{por}}$ by number count (Section 5.1.6). Fuel gas compositions are given in the figure, whereby $p\mathrm{N_{2,an}} = 1 - p\mathrm{H_{2,an}} - p\mathrm{H_2O_{an}}$ atm was set in the measurement. [Cell Z2_275].

The performance predictions in Figure 6.25 certainly support the results found for low current stimuli (Section 6.5). It is vital to have knowledge about the method of determination for the mean pore diameter d_{por} (number or volume averaging) and one should be sure to use parameters determined by a volume averaging algorithm. If the pore size distribution is as inhomogeneous as in Ni/YSZ-anodes (Figure 5.2) with a large number of small pore sizes, number averaging of d_{por} leads to an overestimation, and therefore to a smaller mean value for d_{por}. As a consequence, a lower effective Knudsen diffusion coefficient is calculated in the model (Equation 4.31) which decreases the effective gas transport properties. However, in the presence of few considerably large pores the species transport can proceed with less pore-wall contact probability, and therefore less impulse loss which ultimately implies less obstruction and larger species fluxes. Volume averaging of d_{por} seems to reflect this better: as the results presented in this section demonstrate.

It is also clearly demonstrated that using an effective microstructure parameter Ψ_{EIS} delivers the best results with respect to deviation from measured data. It is most encouraging however that only a small deviation is given using Ψ_{3D}. As mentioned in the previous section, the difference between Ψ_{EIS} and Ψ_{3D} might be caused by an overestimation of τ_{por} due to an insufficient resolution in the reconstructed volume. It is also possible that a larger electrode volume size may have an influence on τ_{por} and further investigations regarding a representative volume element (RVE) for such inhomogeneous structures as the Ni/YSZ anode need to be undertaken. Nevertheless, the results encourage the use of Ψ_{3D}, where no data for Ψ_{EIS} is available or can only be obtained with insufficient accuracy, as it is the case for the cathode. Based on these results and considerations: $\Psi_{\mathrm{3D,cat}}$ it is used with confidence for producing the numerical results in Chapter 7.

Comparison of Porous Media Gas Transport Models

This section is dedicated to analyzing the influence of various porous media gas transport models, which are frequently used in SOFC modeling literature. These are the mass-based Fick Model (*FM-mass*), the molar-based Fick Model (*FM-molar*), the molar-based Dusty Gas Model (*DGM-c*), the Dusty Gas Model with molar-based diffusion and mass-based convection flux coupling (*DGM-p*) and finally, the Mean Transport Pore Model (*MTPM*). The various modeling approaches are described by corresponding species transport expressions in Section 4.3.2.3. As in the previous sections, measured anode gas diffusion ASR values, $R_{1\mathrm{A}}$ (gained by impedance measurements under OCV), are compared with results predicted by the 2D-RPU model with Butler-Volmer charge transfer model and microstructural data from Type-A ASCs. Thereafter, the C/Vs recorded at selected operating conditions are compared with model predictions under load. In a final section, the results predicted by different Dusty Gas Model approaches are compared. In all experiments the temperature and cathode partial pressure were kept

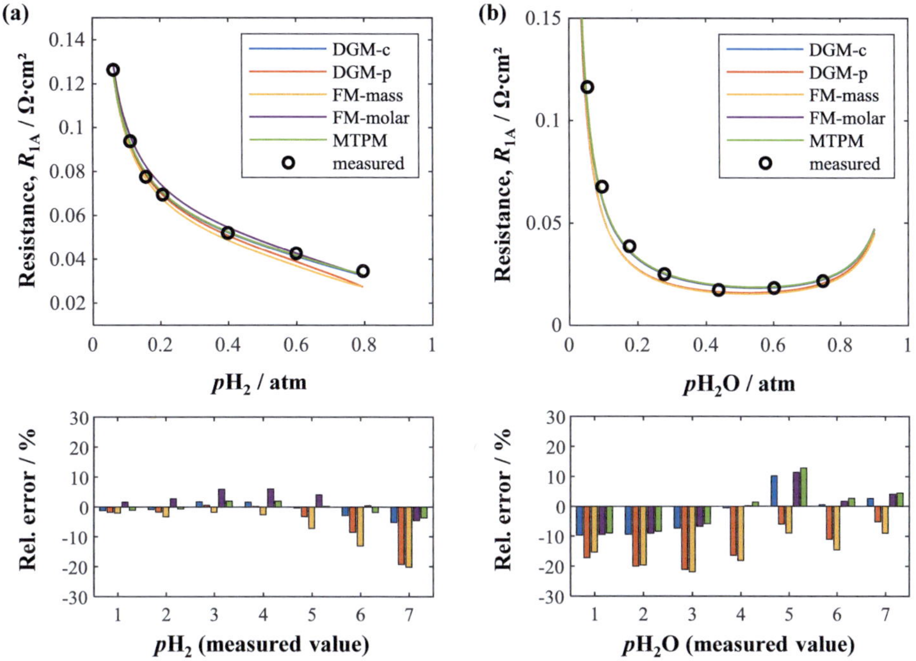

Figure 6.26.: Measured (symbols) and predicted (lines) ASR values of the anode gas diffusion resistances at varied fuel gas compositions at $T = 800$ °C, using different anode gas transport models. DGM-c: Pure Molar-based Dusty Gas Model (Section 4.3.2.3), DGM-p: Mixed unit system Dusty-Gas Model (Sections 4.3.2.1 and 4.3.2.3), FM-mass: Mass-based Fick Model (Section 4.3.2.3), FM-molar: Molar-based Fick Model, MTPM: Mean Transport Pore Model (Section 4.3.2.3). Anode microstructure properties based on cell Type-A (Table 5.1) with $\Psi_{\text{text}}EIS, an$ and $d^{\text{v}}_{\text{por}}$. (a) Constant $p\text{H}_2\text{O}_{\text{an}} = 0.2$ atm, $p\text{H}_{2,\text{an}} = 0.8 \ldots 0.05$ atm, balance N_2- (b) Varied fuel gas humidity with $p\text{H}_2\text{O}_{\text{an}} = 0.05 \ldots 0.75$ atm and $p\text{H}_{2,\text{an}} = 1 - p\text{H}_2\text{O}_{\text{an}} - p\text{N}_{2,\text{leak}}$.(c) and (d) Calculated relative error between measured and predicted resistance. [Cell Z2_275].

Influence of FM, DGM and MTPM on the Gas Diffusion Loss Prediction

In Figure 6.26, the measured $R_{1\text{A}}$ values are denoted by symbols, recorded under varied $p\text{H}_{2,\text{an}}$ (Figure 6.26a) and with constant $p\text{H}_2\text{O}_{\text{an}} = 0.2$ atm and balance N_2. Meanwhile, Figure 6.26b illustrates the results of $R_{1\text{A}}$ determined for a humidity variation with N_2 originating only from leakage. The ASR-results displayed below are the corresponding, calculated relative errors in % for each gas transport model at the various measurement points. Calculated average error values are listed in Table 6.4.

Analysis of Figure 6.26a reveals that all models generally reproduce the measured decreasing $R_{1\text{A}}$-trend towards higher $p\text{H}_{2,\text{an}}$. In detail, all model predictions are in excellent agreement with the measured data at lower $p\text{H}_{2,\text{an}}$, while an increasing underestimation of $R_{1\text{A}}$ by the FM-mass and DGM-p model is observed (red and yellow line) up to ~20% at $p\text{H}_{2,\text{an}} = 0.8$ atm and with an average relative error of -5 to -7%. The FM-molar model (violet line) overestimates $R_{1\text{A}}$ for intermediate $p\text{H}_{2,\text{an}}$ slightly with an average relative error of +2.3%.

The best agreement with measured data is delivered by the DGM-c and MTPM model over total $p\mathrm{H}_{2,\mathrm{an}}$-range with an average relative error of -0.43 to -1.02%.

The results in Figure 6.26a imply that models with molar-based convective flux calculation (FM-molar, DGM-c, MTPM) seem to cope better with an increasing amount of smaller H_2 molecules in the gas composition (or a more inhomogeneous gas composition) in terms of more accurate agreement with measured data. Before this matter is discussed further, the results in Figure 6.26b are analyzed, where the fuel gas humidity is varied. As displayed, all model predictions underestimate $R_{1\mathrm{A}}$. Here as well, all models generally reproduce the measured decreasing $R_{1\mathrm{A}}$-trend from low $p\mathrm{H_2O_{an}}$ towards intermediate fuel gas humidities and the following increase of $R_{1\mathrm{A}}$ towards higher $p\mathrm{H_2O_{an}}$.

The detailed analysis reveals a general underestimation by all model predictions at low $p\mathrm{H_2O_{an}}$, whereby the FM-mass and DGM-p show (with a deviation of 20%) twice as great a relative error compared to the predictions by FM-molar, DGM-c and MTPM. Towards increasing $p\mathrm{H_2O_{an}}$, the relative error decreases more strongly for the model predictions with molar-based convective flux calculation, ending up in a small overestimation of $R_{1\mathrm{A}}$. The larger overestimation at intermediate $p\mathrm{H_2O_{an}}$ is accounted to an increased error sensitivity, due to the small absolute value. The results predicted by the models with mass-based flux computation, however, exhibit a decreasing, but constant underestimation of $R_{1\mathrm{A}}$ over the total $p\mathrm{H_2O_{an}}$-range. constant at standard conditions.

It seems obvious that the greater deviation between mass-based model predictions (FM-mass, DGM-p) and measured $R_{1\mathrm{A}}$ is related to pressure induced convective fluxes. Hence, isobaric simulations were performed for both models. The results are displayed in Figure 6.27 and they demonstrate several important points (from a modeling point of view): First, while the isobaric FM-mass results (yellow line) still agree well with the measured data at high $p\mathrm{H}_{2,\mathrm{an}}$ (Figure 6.27a), they increasingly deviate towards high $p\mathrm{H}_{2,\mathrm{an}}$ (which is the opposite behavior when convective fluxes are included, Figure 6.26). Even more noteworthy are the results predicted for varied fuel humidities (Figure 6.27b), where a low overestimation is given at low $p\mathrm{H_2O_{an}}$, but an increasing overestimation of $R_{1\mathrm{A}}$ towards higher $R_{1\mathrm{A}}$ with an relative error up to 100%.

The isobaric DGM-p (red lines) simulation produces in both anode fuel gas variations the most accurate results with respect to the measured $R_{1\mathrm{A}}$. It should be pointed out that the isobaric DGM-c and DGM-p are actually equal as both models differ only in their convective flux calculation expression (Section 4.3.2.3). Furthermore, mathematical transformation shows that both models are equal to the FM-molar expression, therefore the isobaric DGM-c and FM-molar are not regarded in this comparison. However, these excellent modeling results predicted by the isobaric DGM-p are not surprising, since the finite length Warburg element, used in the ECM-CNLS-fitting process, is derived from Fick model based considerations [18]. Basically, the same gas transport model is applied in quantification of measured impedance spectra with the ECM and in the FEM model prediction of $R_{1\mathrm{A}}$. It is somewhat reassuring that the results match so well.

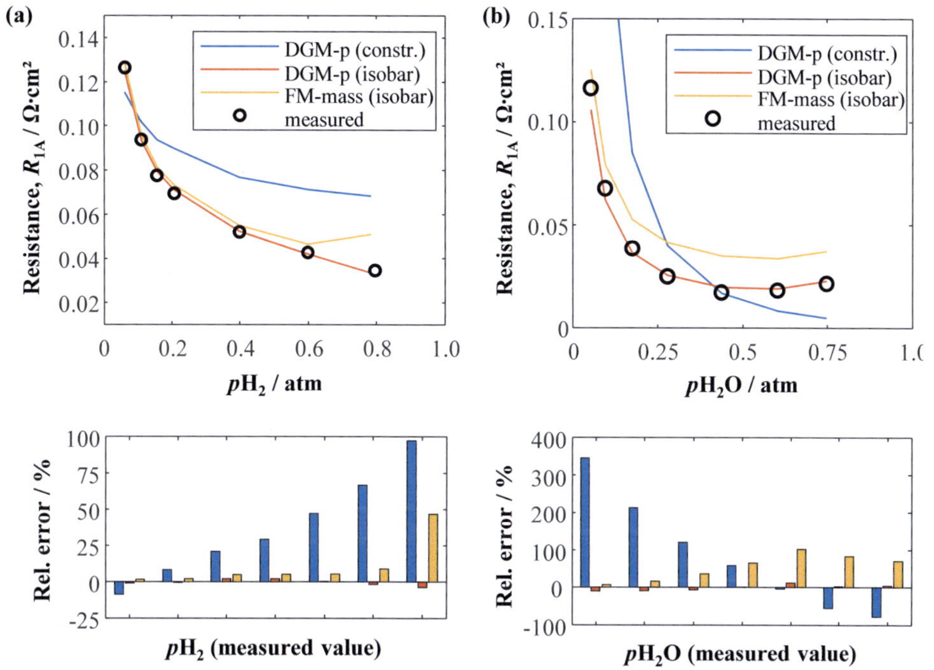

Figure 6.27.: Measured (symbols) and predicted (lines) ASR values of the anode gas diffusion resistance at varied fuel gas compositions at $T = 800$ °C, using the DGM-p and FM-mass under isobar and constraint conditions. (a) Constant $p\mathrm{H_2O_{an}} = 0.2$ atm, $p\mathrm{H_{2,an}} = 0.8 \ldots 0.05$ atm, balance N_2- (b) Varied fuel gas humidity with $p\mathrm{H_2O_{an}} = 0.05 \ldots 0.75$ atm and $p\mathrm{H_{2,an}} = 1 - p\mathrm{H_2O_{an}} - p\mathrm{N_{2,leak}}$.(c) and (d) Calculated relative error between measured and predicted resistance. [Cell Z2_275].

The last results included in Figure 6.27 are the $R_{1\mathrm{A}}$-predictions obtained by the DGM-p model (blue line) with constraint total concentration $\sum_{i=1}^{N} c_i = p/R/T$, which is, in fact, the ideal gas law. While the relative error of predicted $R_{1\mathrm{A}}$ is comparatively small at low $p\mathrm{H_{2,an}}$ (Figure 6.27a) it increases drastically up to 100% for higher $p\mathrm{H_{2,an}}$. The results are even more noteworthy for varied fuel gas humidity predictions in Figure 6.27, where at $p\mathrm{H_2O_{an}}$ an error of 345% is given. It decreases, however, towards intermediate $p\mathrm{H_2O_{an}}$, but underestimates $R_{1\mathrm{A}}$ thereafter at higher humidities up to the point where it actually predicts a negative $R_{1\mathrm{A}}$. This would mean that H_2 is produced in SOFC operation, which is maybe desirable but physically impossible. The reason to include this model in this comparison is to point out the risk of using a gas transport model built of modeling expressions from different unit systems and to still expect basic physical laws to hold (e.g. ideal gas law). If disregarded and therefore without enforced ideal gas law conditions, the result may be acceptable at a first glance: the resulting $R_{1\mathrm{A}}$ of DGM-p does not deviate tremendously from the measured data. However, in order to to equalize the species fluxes according to the different species molar weights M_i (Section 4.3.2.1), ideal gas conditions are no longer ensured in the regarded computational domain. This leads to incorrect calculations of the predicted gas distributions, which the calculated $R_{1\mathrm{A}}$ might not reflect as it only regards total species concentration gradients by Equation 4.87.

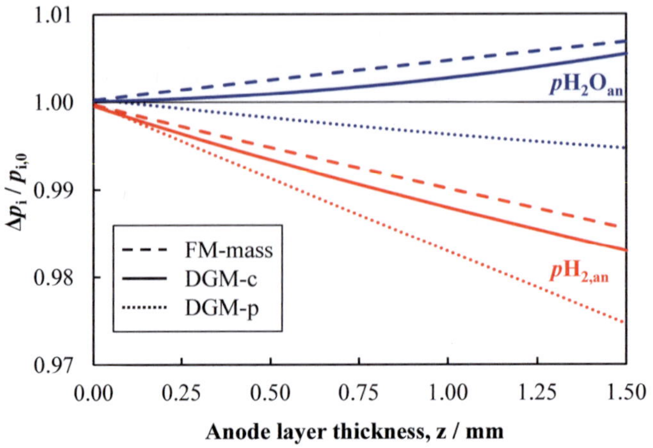

Figure 6.28.: Relative change in hydrogen ($pH_{2,an}$) and steam (pH_2O_{an}) partial pressures along a one dimensional profile from gas channel to anode TPB, as predicted by different gas transport models.

This is demonstrated in Figure 6.28, where the H_2/H_2O partial pressure distribution predictions are plotted in z-direction from gas channel to TPB. The fuel gas composition is 60% humidified hydrogen and N_2 from leakage at $T = 800$ °C and the applied current density $j = 60$ mA/cm^2. Displayed by solid lines are the results predicted by the DGM-c, dashed lines display predictions by the FM-mass and then dotted lines display the simulation results of the DGM-p. FM-mass and DGM-c deliver species distributions in accordance with the electrochemical reaction (Equation 2.1) for SOFC operation with increasing pH_2O_{an} and decreasing $pH_{2,an}$. DGM-p however, predicts a decreasing trend for pH_2O_{an}, which is clearly not physically correct.

At this point it should be noted that in thesis a comparison of all gas transport models with measured C/Vs at higher current densities has not been performed in this thesis. Even though minor deviations between the various model predictions are to be expected, based on what is presented above, no greater surprises occur. It has been shown in the previous sections that the DGM-c model parameterized with $\Psi_{EIS,an}$ and $d^V_{por,an}$ is fully able to predict all measured C/Vs with excellent precision, and none of the here discussed models are better suited. The MTPM delivers similar results to the DGM-c, as it differs only in the convective flux description, but the influence is negligible and the additional computational effort leads randomly to non-converging solutions. Furthermore, the FM-mass is a suitable candidate, if the DGM-c model implementation seems too elaborate. However, a small underestimation of gas diffusion losses has to be accepted. Finally, the DGM-p approach, as commonly used in literature, should not be implemented, because it mixes model expressions from different unit systems, ergo ideal gas conditions are violated and if enforced, a serious error in the gas diffusion loss prediction is the consequence.

Table 6.4.: List of relative deviation in % between measured anode gas diffusion resistance R_{1A} and predicted value using different gas transport model approaches.

Gas transport model	relative error / %	
	$pH_{2,an}$-variation	humidity-variation
FM-molar	2.31	-1.09
FM-mass	-7.18	-15.36
FM-mass (isobar)	10.80	54.67
DGM-p	-4.82	-13.83
DGM-p (constr.)	37.39	85.97
DGM-p & DGM-c (isobar)	-0.34	-0.27
DGM-c	-1.02	-1.92
Mean Transport Pore Model (MTPM)	-0.43	-0.24

Influence of Numerical Mesh Resolution

The FEM requires a specific numerical mesh resolution in geometrical model domains of interest in order to supply adequate solution accuracy. As explained in Section 2.7, species flux and equations of state are solved at every grid node and depend on the individual species gradients between the various node interconnections. However, simply increasing the total mesh resolution everywhere in the individual domains wastes computational power. It is more efficient, to only resolve areas with high species fluxes and gradients with increased resolution. Hence, a specific *a priori* knowledge is required. An optimized mesh resolution is found by customizing the mesh iteratively with respect to computational time and solution accuracy.

A good example is given by the MIEC-cathode model implementation (4.3.3.2) and its required mesh resolution in the regarding domain. Figure 6.29a shows predicted normalized ionic current densities in spatial dependence to the distance from the electrolyte ($\vec{j}_{O^{2-}}/\vec{j}_{O^{2-},max}$) [120], predicted by different cathode models. Solid lines depicts the results obtained by this work's MIEC-model approach with homogenized microstructure properties. Symbols display the results predicted by the 3D FEM model of J. Joos and T. Carraro [156, 120]. In their approach the cathodes actual microstructure is resolved with high detail in 3D by implementation of a reconstructed cathode volume element obtained from FIB/SEM analysis (Section 2.6). The depicted results in Figure 6.29a show that using the homogenized model with a course mesh (red line) the ionic flux is predicted with a discontinuous progression. While using a fine mesh (black line) a smooth course is predicted, much more similar to the real microstructure model prediction, which is assumed to supply the highest accuracy. However, the fine mesh prediction does not exactly reproduce the real microstructure model results, but mildly underestimates the ion flux prediction. This can be explained by the homogenization approach relying on global parameter validity, although the cathode microstructure becomes more denser the closer to the electrolyte [120].

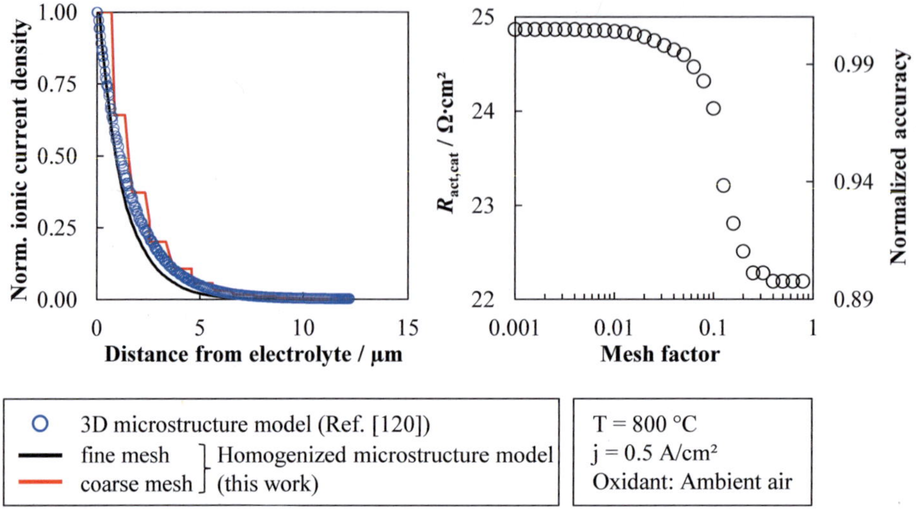

Figure 6.29.: (a) Normalized ionic current profiles predicted by a real microstructure 3D FEM model [120] and by this works homogenized MIEC-cathode model approach using different numerical mesh resolutions. Kinetic data of LSCF used as listed in Table 5.3 and given in Ref. [120]. (b) Predicted cathode activation polarization resistance, $R_{\mathrm{act,cat}}$, in dependence of a mesh resolution factor.

Nevertheless, using the fine mesh, the calculated cathode charge transfer resistance, $R_{\mathrm{act,cat}}$, reproduces the value determined by J. Joos with sufficient precision over a wide temperature range (Figure 6.2), this emphasizes the adequate trustworthiness of the homogenized MIEC-cathode model approach. The error due to using a coarse mesh is illustrated in Figure 6.29b, where predicted $R_{\mathrm{act,cat}}$ and the according deviation to the predicted maximum resistance are depicted depending on a mesh factor. The factor represents a mesh distribution step variable at $\mathrm{IF}_{\mathrm{cat-el}}$ subdividing the interface's mesh distribution into a corresponding number of nodes. Hence, smaller values result in a higher mesh resolution. The results displayed show that at a factor < 0.4 the solutions accuracy increases and converges at ≈ 0.02. Hence, this value is regarded as sufficient in the MIEC-cathode model to produce accurate results. It should be noted that the value is not universally applicable as it depends on a model width of $b_{\mathrm{RPU}} = 1$ mm, and needs to be adjusted to different model widths accordingly.

Increasing the numerical mesh resolution results in increased degrees of freedom (DFOs) and thus in an increased computational effort. Basically, it means higher memory demand and longer computational time. However, it is not reliable to translate this directly to a fixed value because the results strongly depend on the computational power available. The 2D-RPU and 2D-GCH models applied in this work are all manageable on a desktop workstation (Intel i7-2600 CPU and 16Gb RAM) with acceptable performance, while the coarse-meshed 3D-RPU models struggles to works on the desktop machine and the fine-meshed model requires much more memory (~250Gb).

Table 6.5 lists the corresponding results. The FM-mass model requires less DFOs as it only solvs $n-1$-species (Section 4.3.2.3), while the DGM-c requires more than twice the amount of DFOs in case of H_2 as fuel. Using reformate as fuel, the required DFOs increase by almost a factor of 5, due to the complex diffusion coefficient calculation in the DGM model (Section 4.3.2.3). While this does not affect the computational time for the 2D-RPU model, it is notable in the 2D-GCH for H_2 as fuel, and it also considerably (but still acceptably) notable for the reformate performance prediction. These results scale up according to the model size, however, with equal proportion and the comparison is restricted to models with small active electrode area. An even higher number of DFOs is produced by using the MIEC-cathode model instead of the Butler-Volmer approach as cathode charge transfer model, due to the increased mesh resolution. It is explained in the paragraph above that an increase of mesh grid nodes by more than an order of magnitude results from the demanded model accuracy. From this it is easy to comprehend that the MIEC-model is only applicable in the 2D-RPU model and not in 3D, as the number of DFOs is already enormous using the BVM approach. The choice becomes even more obvious when comparing the increased steps to convergence and resulting computational time for the stack contacting models. Hence, modeling the RPU in 3D is necessary to ensure the frameworks validity at low $pO_{2,\mathrm{cat}}$ (Figure 6.23), but becomes prohibitively expensive for computational costs. The results listed in Table 6.5 for the 3D-RPU model were achieved on a high-performance workstation (2x Intel Xeon E5-2699 CPUs with 384GB RAM), whereby the fine resolved model is required to achieve better accuracy.

Table 6.5.: Overview of numerical results in terms of computational cost with respect to model geometry and implemented sub-models. 2D results are achieved on a standard desktop workstation, while 3D models require a high-performance workstation.

Model geometry	Fuel	Contact Type	Gas Transp. Model	Cathode Model	DFOs	steps	time [s]
2D-RPU	H_2	ideal	FM-mass	BVM	12247	4	3
			DGM-c	BVM	26964	2	3
			DGM-c	MIEC	340505	4	59
		stack	DGM-c	BVM	19879	16	8
			DGM-c	MIEC	345836	17	134
2D-GCh	H_2	ideal	FM-mass	BVM	54009	3	6
			DGM-c		97353	3	14
	REF		FM-mass	BVM	57736	2	6
			DGM-c		294496	3	40
3D-RPU	H_2	ideal	FM-mass	BVM	590713	3	143
			DGM-c (coarse)		635561	5	253
			DGM-c (fine)		2986422	5	3152

7. Results of SOFC Stack Layer Performance Analysis

This chapter deals with the simulation of SOFC stack layer performance using the FEM model framework.

First, experimental results are presented to demonstrate how non-ideal contacting conditions reduce SOFC single cell performance. The experimental results are reproduced by an accordingly set up FEM model using the 2D-RPU geometry (Figure 4.3), which further strengthens the validity of the framework and the applicability of 2D-RPU geometry to predict SOFC stack layer performance. Occurring loss mechanisms are discussed by analysis of numerically predicted gaseous species and current density distributions.

Based on the conclusions drawn thereon, a detailed numerical parameter variation is presented as the main part of this chapter (Section 7.2). As reference interconnector (IC) geometry the state-of-the-art F-design for planar SOFC stacks of Forschungszentrum Jülich is applied. Characteristic cathode parameters are varied systematically and their impact on SOFC stack layer performance is analyzed at various operating temperatures and polarizations. In the next Section 7.4.1, the concept of an additional *cathode contact layer* (CCL) is discussed by different numerical parameter variations for the standard F-design and a state-of-the-art SOFC with LSCF-cathode. Consequently, the IC-design is subject of numerical variation in the next section (Section 7.3). In the last section of this chapter, a FEM model-based cell and IC design optimization is presented, based on the conclusions gathered from the previous sections.

7.1. SOFC Performance under Ideal and Stack Contacting Conditions

One of the motivations for this work is to develop a deeper understanding of how SOFC performance is influenced when a single cell is operated in a stack compared to testing in ideal laboratory conditions (Section 2.5).

M. Kornely has shown in his work [37], by a straight-forward electric FEM model and by a detailed experimental analysis that stack contacting conditions cause additional losses due to gas diffusion and electronic in-plane conduction limitations, foremost at the cathode. For an IC with thin contact ribs, Kornely's experimental and numerical results are in good agreement. It was possible to attribute the additionally-created-loss to increased ohmic, in-plane conduction occurring in the cathode layer, caused by elongated electron transport pathways from IC contact ribs into the cathode layer and to the electrochemically active zones beneath the gas channel (Figure 2.16). Using an Au-IC with wide ribs in an experimental analysis, it was possible to attribute the further increased loss to the mechanism of limited gas diffusion properties. However, limited to the description of ohmic loss and not regarding gas diffusion as physical process in the model, it was not possible with Kornely's model to reproduce the measured C/Vs recorded using an IC with wide contact ribs [37].

In this work, the experiment from Kornely (who used an ASC with LSM/YSZ composite cathode, Section 2.4.1) was repeated with an ASC Type-A equipped with a LSCF-cathode (Table 5.1) in order to use a comparable, state-of-the-art SOFC throughout the work. EIS spectra and C/Vs were recorded at various oxygen partial pressures ($pO_{2,\mathrm{cat}} = 1 \ldots 0.01$ atm), while all other operating conditions were kept constant ($pH_2O_{\mathrm{an}} = 0.6$ atm, $T = 800$ °C). The highly electronically conductive Ni contained in the cermet anode allows the use of a Ni-IC at the anode, equipped with thin contact ribs only at the outer rims of the flowfield geometry. Thereby, no additional ohmic loss is created [37, 68] and a homogeneously distributed fuel gas atmosphere is assured. This setup is comparable to stack contacting, where no distinct gas channel flowfield exists, instead a course Ni-mesh is placed between the anode and the next stack layer IC (Figure 2.9). In the first experiment, an ideal contact setup at the cathode is realized by the combination of permeable Au-mesh (Section 5.4) and Al_2O_3-housing, hosting the required planar flowfield (Figure 3.4). In the second measurement, the cathode contact setup was exchanged for an Au-flowfield with only two, yet wide, contact ribs. Geometric details are given elsewhere in Figure 3.5g and in Figure 7.6, where the applied model geometry RPUs used for the numerical analysis are displayed.

A comparison of both cell contact methods is presented in the following: at first by means of analyzing DRT spectra, calculated from EIS spectra recorded at various $pO_{2,\mathrm{cat}}$ under *OCV*. Thereafter, the measured C/Vs and the C/Vs predicted by accordingly set up FEM models are subjected to a performance comparison at higher current densities (and additionally serve as model validation). The FEM approach thereby supplies a spatial species resolution and furthermore a loss process quantification, thus leading to a comprehensive understanding of individual loss process contributions to the overall loss of SOFCs under stack contacting conditions.

7.1.1. DRT Analysis

Figure 7.1 displays DRTs of two $pO_{2,\mathrm{cat}}$-variations, calculated from impedance spectra recorded at ideal and stack contacting conditions. Analyzing the DRTs in Figure 7.1a (ideal contacting conditions) from high to low $pO_{2,\mathrm{cat}}$, a prominent peak is observable in the low frequency range (0.9 Hz) at $pO_{2,\mathrm{cat}} = 0.01$ atm, denoted with P_{1C} (Section 2.3.2. The loss process P_{1C} is associated with the gas diffusion polarization in the cathode layer and becomes noticeable only at $pO_{2,\mathrm{cat}} < 0.1$ atm, as shown by A. Leonide [9]. At higher $pO_{2,\mathrm{cat}}$ P_{1C} decreases because the permeable Au-mesh procures a sufficient O_2 supply to the thin cathode layer. The characteristic frequency of P_{1C} increases and the process becomes inseparable from the anode gas diffusion polarization (P_{1A}). P_{2C} denotes in Figure 7.1a the process associated with the electrochemical activation loss at the cathode, exhibiting its sensitivity in the middle frequency range (Section 2.3.2, [9]). At lower $pO_{2,\mathrm{cat}}$ the electrochemical reaction is hindered, which results in an increase of P_{2C} (Section 6.4). P_{2A} and P_{3A} denote the anodes charge transfer polarization in the DRT spectrum at higher frequencies and are independent from changes in $pO_{2,\mathrm{cat}}$.

Figure 7.1b displays the DRTs from recorded impedance spectra under stack contacting conditions. P_{1C} exhibits an obviously different sensitivity compared to ideal contacting conditions (Figure 7.1a). It is well developed at standard conditions ($pO_{2,\mathrm{cat}} = 0.21$ atm), dramatically increased at $pO_{2,\mathrm{cat}} = 0.01$ atm and vanishes beneath P_{1A} only at $pO_{2,\mathrm{cat}} = 1.00$ atm. P_{2C}, however, is less affected and shows a comparable sensitivity to ideal contacting conditions.

These findings agree with the work of M. Kornely and underline the statement that (depending on the stack contacting conditions) gas diffusion losses in the cathode layer may contribute a significant loss to the overall cell resistance, even at standard conditions. Consequently, gas diffusion has to be to regarded in a modeling approach in order to deliver accurate SOFC performance predictions. The model framework presented in this work incorporates porous media gas diffusion by the DGM model (Section 4.3.2.3). It is therefore suitably equipped, as demonstrated in the previous chapter in Sections 6.3 and 6.4.

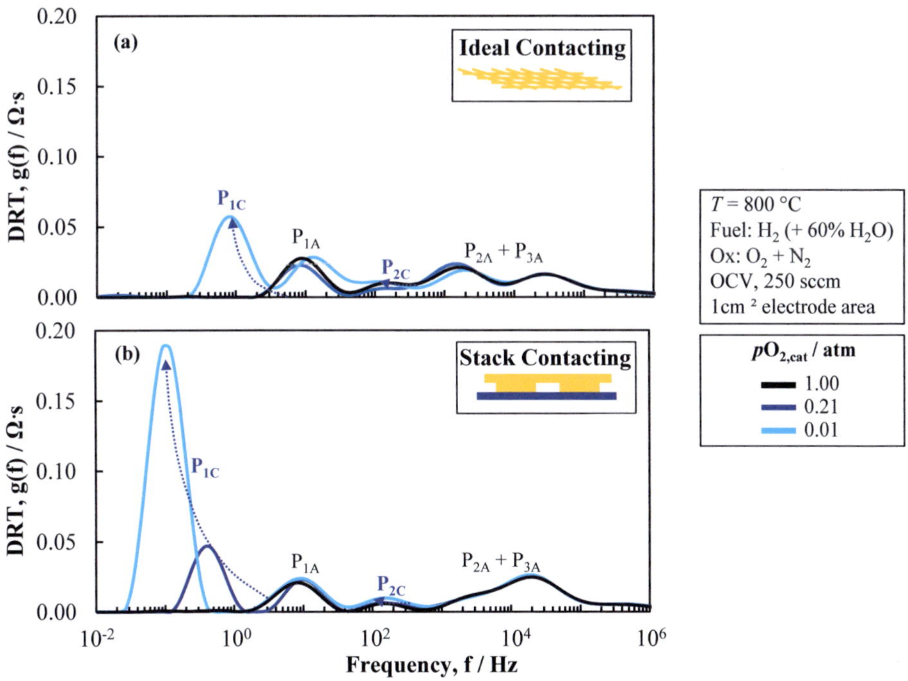

Figure 7.1.: From EIS spectra calculated DRTs, recorded at ASC Type-A at varied $pO_{2,\mathrm{cat}} = 1 \ldots 0.01$ atm and contacting conditions. (a) Ideal contact using a Au-mesh and (b) stack contact, using an Au-IC as displayed in Figure 3.5f. All other operating conditions are shown in the figure. [Cells Z1_275, Z1_276].

The experimental results in this subsection are a product of electrochemical impedance spectroscopy under *OCV*, wherein only a very small current stimuli ($j_{\mathrm{EIS}} = 60$ mA/cm^2) is used. In the next subsection, the SOFC performances of ideal and stack contacting at higher, application-oriented current densities are analyzed and occurring loss behavior resolved and discussed.

7.1.2. Performance and Loss Process Analysis

Figure 7.3 displays the measured and simulated C/Vs, recorded for cell Type-A and predicted by adequately set-up 2D-RPU FEM models under equal operating conditions, but with the above-mentioned differing contact setup at the cathode. Applied model geometries are displayed in Figure 7.2 and resemble the cell assembly at ideal and stack contact as used in the measurement. The Au-IC used to emulate stack contacting conditions at the cathode features only two, yet wide, contact ribs (Figure 3.5g). Hence, the smallest possible RPU is represented by half of the actual cell assembly, displayed in Figure 7.2b on the right. The ideal contacting model geometry is accordingly adapted to the same model width for reasons of better comparison and is displayed in Figure 7.2a on the left.

The results in Figure 7.3 show a decrease in performance under stack contacting conditions, where the deviation already becomes significant at low current density $j_{\mathrm{cell}} = 0.5$ A/cm^2 and increases with increasing load. The measured C/Vs (symbols) are thereby reproduced with high precision by each model prediction (lines). In Chapter 6, the agreement between measurement and model prediction under ideal conditions was already discussed. The results displayed in Figure 7.3 are further proof for the model validity under both ideal, but also under stack contacting conditions. Hence, placing confidence in the correct physical reproduction of occurring loss mechanisms, it is the feature of numerical FEM models to resolve distributions of regarded species spatially and allow an individual loss process analysis and quantification.

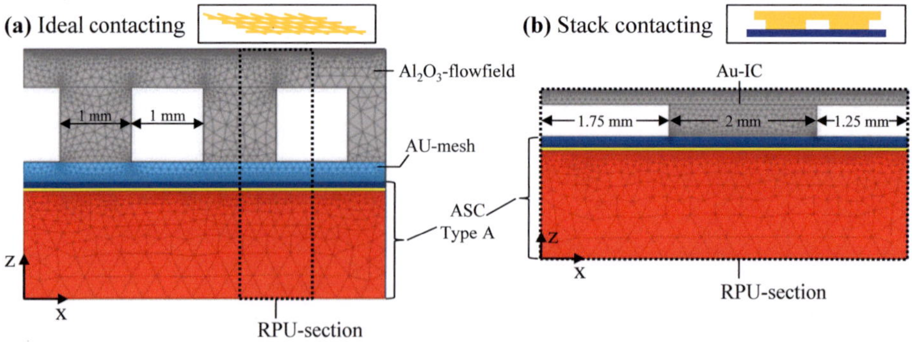

Figure 7.2.: Display of applied model geometries for comparison of ideal and stack contacting (inhouse measurement with AU-interconnector, Figure 3.5g).

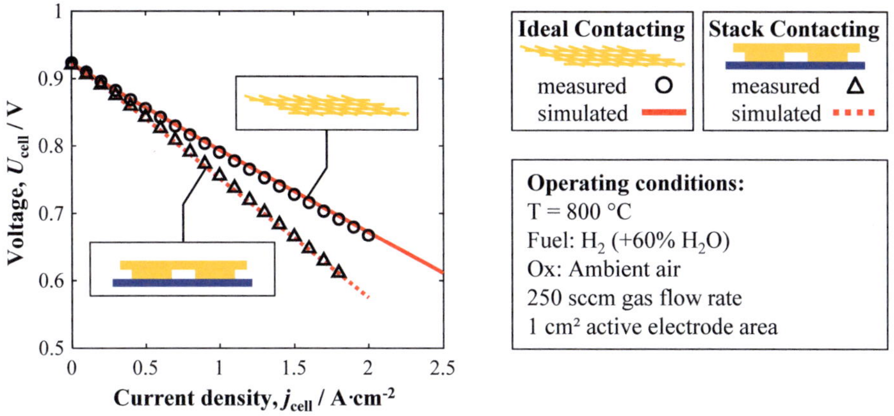

Figure 7.3.: Comparison of measured (symbosl) and predicted (lines) C/Vs, recorded at ASC Type-A at ideal (solid line) and stack (dotted line) contacting conditions. Operating conditions given in the figure above. [Cells Z1_275, Z1_276].

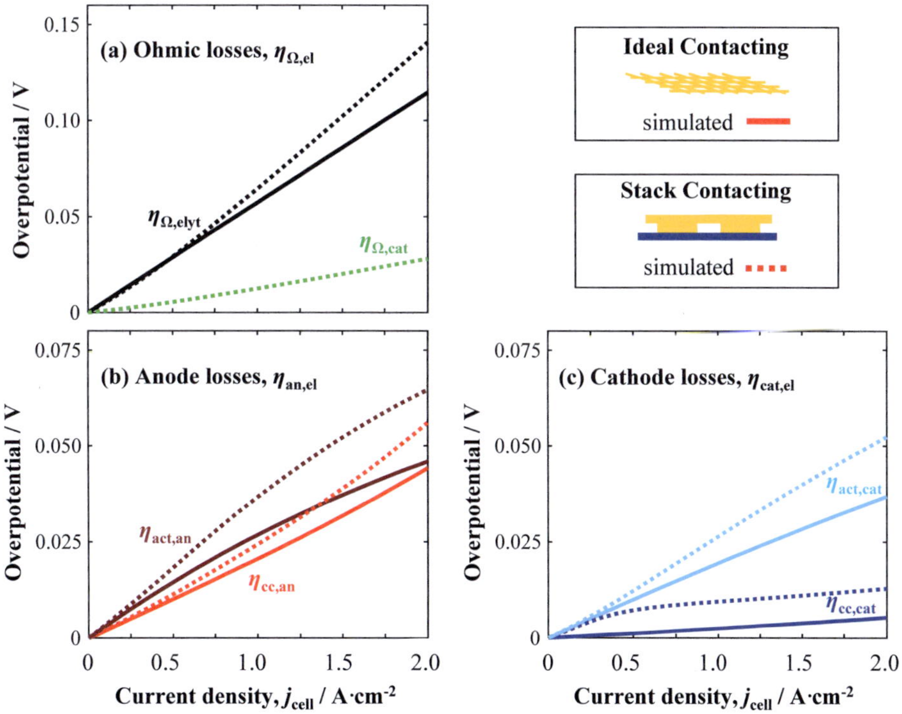

Figure 7.4.: Comparison of measured (symbosl) and predicted (lines) C/Vs, recorded at ASC Type-A at ideal (solid line) and stack (dotted line) contacting conditions. Operating conditions given in the figure above. [Cells Z1_275, Z1_276].

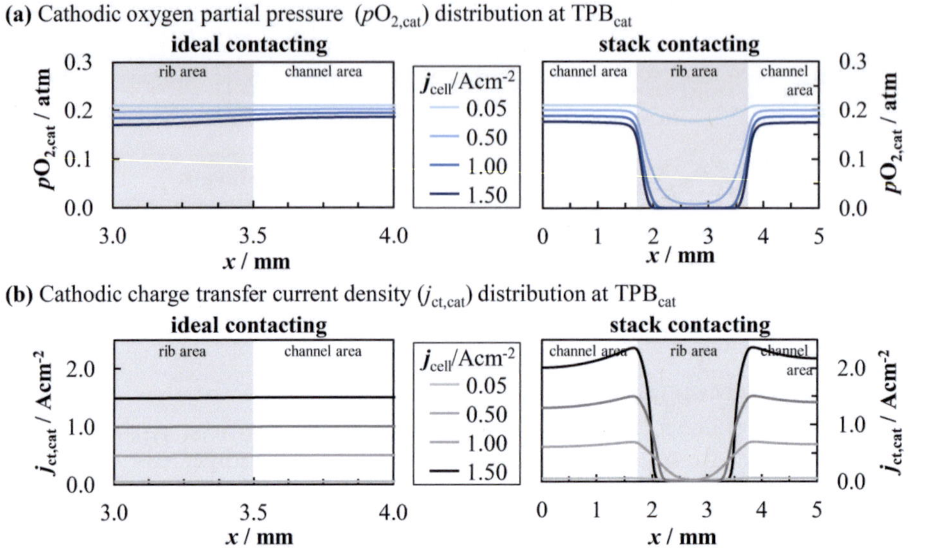

Figure 7.5.: Display of applied model geometries (a) and simulation results (b-e) for comparison of ideal and stack contacting (inhouse measurement with AU-interconnector, Figure 3.5g). Operating conditions: $T = 800\ °C$.

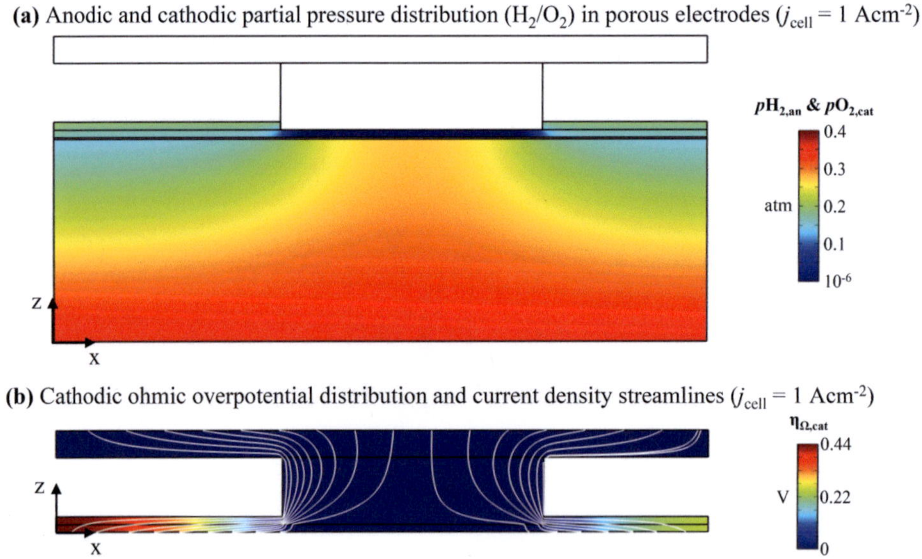

Figure 7.6.: Display of applied model geometries (a) and simulation results (b-e) for comparison of ideal and stack contacting (inhouse measurement with AU-interconnector, Figure 3.5g). Operating conditions: $T = 800\ °C$.

Figures 7.4a-c depict the overpotentials η_i in V, occurring at a cell current density j_{cell} under ideal (solid line) and stack contacting conditions (dotted line), and subdivided into thematic groups. However, before these results are discussed, it is important to point out that in order to calculate occurring losses in non-ideal (stack) contacting conditions, the local current density, $\vec{j}$, needs to be taken into account. This is achieved by calculation of the local power dissipation, $P_{i,\text{loss}}$, caused by the individual loss process i and correlating this with j_{cell}. Detailed equations with regarding explanations are given in Section 4.4.3.

Figures 7.4b/c display in V calculated gas diffusion ($\eta_{\text{cc,an/cat}}$) and electrochemical charge transfer activation loss ($\eta_{\text{cc,an/cat}}$) occurring at the TPBs of each porous electrode, while Figure 7.4a illustrates the calculated ohmic loss, induced in the electrolyte/interdiffusion layer ($\eta_{\Omega,\text{elyt}}$ and in the cathode layer ($\eta_{\Omega,\text{cat}}$). Ohmic loss in the anode layer, the contact meshes and Au-IC are all disregarded due to the corresponding material's high electronic conductivity.

It is clear from Section 4.4.1 that the cell voltage, U_{cell}, at a specific current load, j_{cell}, is the result of operating condition dependent OCV (Equation 4.80) subtracted by the sum of occurring overpotentials η_i (Equation 4.84). Under stack conditions the created loss is in sum greater compared to ideal conditions and the corresponding U_{cell} is lower. This is clearly confirmed by the predicted results in Figs.7.4a-c. All calculated overpotentials for stack conditions are greater compared to those predicted for ideal conditions. While $\eta_{\Omega,\text{elyt}}$ starts to diverge at slightly increased $j_{\text{cell}} = 0.5$ A/cm^2, a diverging course can be observed for the remaining overpotentials from the beginning. The slope of $\eta_{\text{cc,cat}}$ is steep at $j_{\text{cell}} < 0.5$ A/cm^2 and it decreased thereafter, while $\eta_{\Omega,\text{cat}}$ stands out from the rest as it only occurs under stack contacting conditions. The underlying reason for the increased performance loss in stack conditions have been introduced in Section 2.5 and will be explained in detail in the following with the help of local species distributions predicted by the FEM approach.

Figure 7.5 illustrates the predicted oxygen partial pressure ($pO_{2,\text{cat}}$) distribution at the cathodes TPB or various $j_{\text{cell}} = 0.05...1.5$ A/cm^2. In ideal conditions (Figure 7.5a, left) a homogeneous $pO_{2,\text{cat}}$ distribution exists at all current densities with a low decrease from the initial O_2-concentration ($pO_{2,\text{cat}} = 0.21$ atm) due to gas conversion. In the area above the TPBs occluded by the flowfield rib of the Al_2O_3 (grey area) a small drop in $pO_{2,\text{cat}}$ can be observed due to longer O_2 transport pathways compared to the non-occluded area directly beneath the gas channels (Section 2.5.2). The resulting charge transfer current density, $j_{\text{ct,cat}}$, is displayed in Figure 7.5b on the left. A homogeneous distribution of $j_{\text{ct,cat}}$ is predicted as sufficient O_2 molecules are available to secure a continuous reaction rate along the cathode/electrolyte interface and current is transported only in z-direction.

The situation is dramatically different at stack contacting conditions. The predicted $pO_{2,\text{cat}}$-distribution at the TPBs (Figure 7.5a, right) already shows a drop in the area at 0.05 A/cm$_2$, where the Au-IC rib is placed directly on top of the cathode layer. The ongoing charge transfer reaction consumes O_2 faster as resupplied via the electrodes pore network, where the gas needs to be transported not only in z-direction, but mainly in x-direction to reach the electrochemically active zones beneath the middle of the contact rib. The consequence is a developing O_2 starvation, which is already at $j_{\text{cell}} = 0.5$ A/cm^2 quite pronounced and fully developed at higher j_{cell}. The $pO_{2,\text{cat}}$/$pH_{2,\text{an}}$ display (Figure 7.6) at $j_{\text{cell}} = 1.0$ A/cm^2 reveals a complete O_2 starvation beneath the Au-IC rib and an inhomogeneous H_2 distribution in the anode as well. In the area, where no O_2 is present, the electrochemical reaction is hindered and slowed down, consequently less H_2 is consumed. This statement is confirmed by the inhomogeneous $j_{\text{ct,cat}}$ distribution in (Figure 7.5b) on the right. So $j_{\text{ct,cat}}$ drops to zero in the area where no oxygen is available and increases in the area beneath the gas channels, where the O_2 supply is still sufficient. However, with increasing j_{cell}, electrons are transported in x-direction within the cathode layer in order to supply the ongoing electrochemical reaction (displayed in Figure 7.6b by the current density streamline course). This induces an additional ohmic overpotential in the cathode layer, which explains the increasing ohmic overpotential created in the cathode (Figure 7.4c). Furthermore, the inhomogeneous electrochemical activity at the cathode causes increased anodic gas diffusion and activation loss, as well as an elongated charge carrier transport in the electrolyte, which starts to take effect with the O_2 starvation. For the chosen cell and contact design, this point is reached at $j_{\text{cell}} \approx 0.5$ A/cm^2 and consequently $\eta_{\Omega,\text{elyt}}$ begins to diverge in Figure 7.4a.

Figure 7.7a illustrates measured and predicted performance values at $U_{\text{cell}} = 0.8$ V, while Figure 7.7b shows a graphical comparison between the individual occurring loss processes by displaying R_i. For the given operating conditions and geometric cell and contact parameters, a performance decrease of 24.4% occurs as well as an overall cell resistance increase of 32.3%. $R_{\Omega,\text{cat}}$ increases by far the most due to the very low value under ideal conditions. A vast increase (425%) was predicted for $R_{\text{cc,cat}}$, while $R_{\text{act,cat/an}}$ increase by a third. $R_{\Omega,\text{elyt}}$ increases 14.8% and the smallest growth is seen for $R_{\text{cc,an}}$ with 8.9%. The relative increase has to be put into perspective. Correlating the individual process resistance increases with the total cell resistance increase delivers a different view, where it becomes clear that besides the anode gas diffusion loss all other loss mechanisms contribute almost equally to the overall loss. These values, however, should not be taken as general results. It will be shown in the following sections that various material properties and cell/IC design parameters change the loss process contribution significantly and the overall performance of SOFC single cells in stack contacting conditions may reach the performance of single cells operated under ideal conditions.

In summary, the experimental and modeling results presented in this section demonstrate how power output of high performance, anode supported SOFCs with an LSCF cathode is reduced in stack operation. The planar interconnector design with alternating gas channel and contact rib array allows only a non-ideal O_2 and electron supply to the cathode and its electrochemically active zones. As a consequence, additional ohmic and polarization losses of all occurring cell processes are created due to limited porous media gas transport and electronic in-plane conduction properties in the cathode compared to ideal cell contacting in laboratory operation. The FEM model framework presented in this work is able to reproduce this accurately, demonstrated by a precise reproduction of measured performance. Furthermore, the spatial species distribution within the numerical domains allows a detailed analysis of the complex coupled loss process.

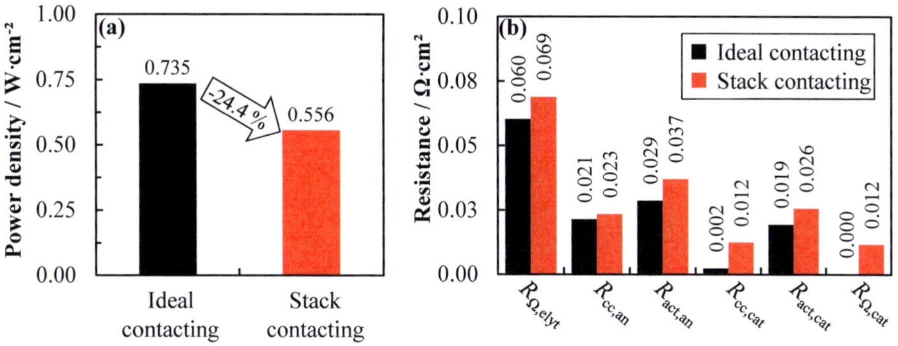

Figure 7.7.: Comparison of (a) predicted power density and (b) individual loss process resistances using ideal and stack contacting conditions. $U_{\text{cell}} = 0.8$ V, $T = 800$ °C, $p\text{H}_{2,\text{an}} = 0.4$ atm, $p\text{H}_2\text{O}_{\text{an}} = 0.6$ atm, $p\text{O}_{2,\text{cat}} = 0.21$ atm, cell Type-A, no gas conversion.

7.1.2.1. Gas Transport Modeling Aspect

It is found that the FM-mass model (Section 4.3.2.3) combined with Darcy's law (Equation 4.12) as gas transport model is not applicable in the stack contact geometry. With increasing O_2-consumption an increased velocity gradient is imposed by the boundary condition set for the Darcy equation at $\text{IF}_{\text{cat}-\text{elyt}}$. While mass is actually diminished by the electrochemical reaction as oxygen is incorporated into the bulk material and the gas mixtures density is affected accordingly, it is not physically correct to assume alongside an interface velocity gradient.

Numerically, it results in a total pressure drop within the gas mixture according to the Darcy equation (Equation 4.12), as the results displayed in Figure 7.8a show. Even if O_2 is completely consumed at some point, N_2 still remains in the mixture as it cannot simply vanish and $p_{\text{total}} = 0.79$ atm should prevail. Consequently, the simulation encounters numerical errors as negative and discontinuous p_{total} distribution at the interface are predicted and the calculation no longer continues (in this case $j = 0.67$ A/cm^{-2}). The problem is inherent in the Darcy equation, which was originally proposed to describe fluid transport in porous solids where species actually bypass an interface, whereas in fuel cells species are converted from thermodynamically different states.

The Darcy equation in molar-based description combined with the DGM-c model delivers physically comprehensive results, as displayed in Figure 7.8a. A decreasing p_{total} is also predicted in the area with O_2 depletion, however p_{total} remains above $p_{\text{total}} > 0.9$ atm. Nevertheless, with $p_{\text{total}} = p\text{O}_{2,\text{cat}} + p\text{N}_{2,\text{cat}}$ and an almost complete depletion of O_2 one should expect $p_{\text{total}} = 0.79$ atm. Hence, more N_2 molecules remain in the area beneath the rib compared to the open gas channel volume or $\text{pN}_{2,\text{cat}}^{\text{TPB}} > \text{pN}_{2,\text{cat}}^{\text{GC}}$. The sum of convective flux imposed on the total mixture combined with a possible small amount of diffusion drag of O_2 molecules imposed on the N_2 molecules is greater than the N_2 diffusion flux towards the open gas channel.

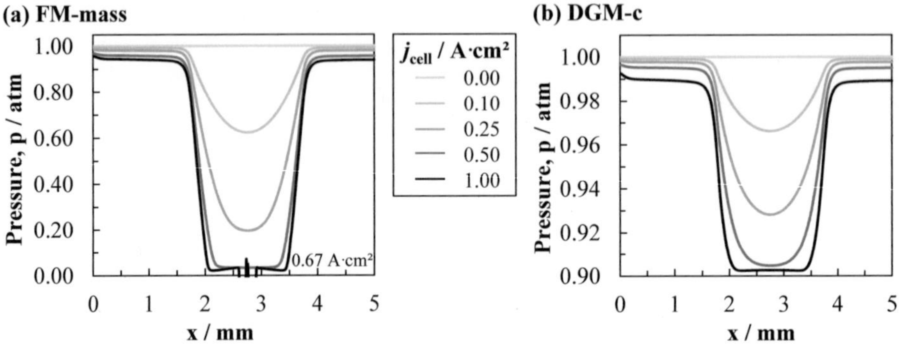

Figure 7.8.: Simulated total pressure p_{total} distribution at the cathode/electrolyte interface using (a) FM-mass+Darcy and (b) DGM-c as gas transport models.

Hence, using the DGM-c combined with the molar-based Darcy approach delivers comprehensive results for the cathode gas transport description, whereas it should be refrained from using the FM-model with the standard Darcy equation. At the anode, a different discussion has to be made as steam enters and hydrogen leaves the model domain. The unequal molecular mass causes under certain conditions an increase of p_{total}, which should also be carefully regarded. Depending on the implemented permeability value, very high p_{total} can cause an increase of convective fluxes which create a numerical result where hydrogen is actually transported away from the TPBs. It is therefore very important to revise a numerical solution for its physical meaning.

7.1.3. Comparison of Model Predictions with Stack Measurements

The results presented in the previous section demonstrate the model frameworks ability to reproduce the complex coupled loss mechanisms occurring under non-ideal contacting conditions at the cathode. However, the measurements were carried out in test bench A_1 (Section 3.2.1) on ASCs with an active electrode area of 1 cm^2 and high gas flow rates, thus ensuring homogeneous distribution of relevant operating parameters. In real stack operation, these ideal conditions are not guaranteed, an increased active electrode area and high fuel utilization rather cause inhomogeneous partial pressure and temperature distribution across the cell area. Hence, it is interesting to compare performance predictions made by the isothermal model framework using the 2D-RPU model geometry with what is actually measured in a real stack.

Forschungszentrum Jülich (IEK-3) kindly provided corresponding experimental stack-measurements data. The ASCs active cell area in each stack layer were 80 cm^2, using interconnectors with planar F-design flowfield (Figure 2.10) at the cathode and a coarse Ni-mesh at the anode between each layer. The applied ASC resembles cell Type-B (Table 5.1), however with a CGO interdiffusion layer applied between the 8YSZ electrolyte and LSCF cathode using PVD-coating instead of tape casting (cell Type-B).

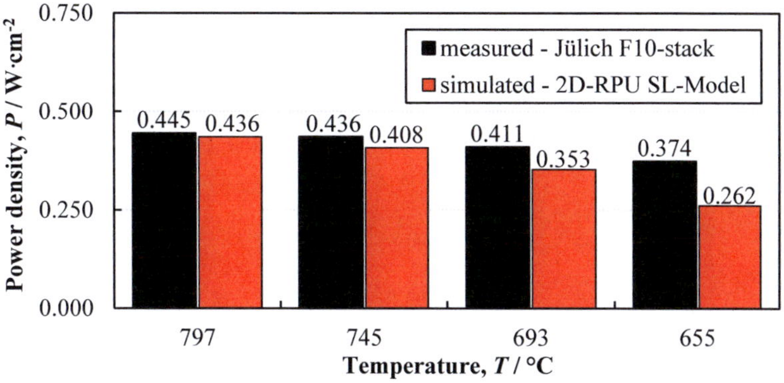

Figure 7.9.: Comparison of measured (black) and predicted (red) stack layer power density at $j = 0.5$ A/cm^2 and various operating temperatures. The experimental data was kindly supplied by Forschungszentrum Jülich (IEK-3). Operating conditions: 20% humidified H_2 as fuel with fuel utilization 40% and ambient air with $\lambda = 2$ as oxidant.

A current-collector layer (CCL) is applied on top of the IC flowfield by atmospheric plasma spraying (APS), which results in a very dense layer. An accordingly parametrized 2D-RPU stack layer model was set up and the performance at $j = 0.5$ A/cm^2 calculated at various temperatures. The operating conditions were set according to the stack measurements: 20% humidified H_2 as fuel with a flow rate accordingly set to 40% fuel utilization at the given current load and ambient air with $\lambda = 2$ as oxidant. A contact resistance is assumed in the model, based on the measurements carried out in Jülich (Section 5.6).

Figure 7.9 displays the at Jülich experimentally measured power density (black) together with numerically predicted performances using the 2D-RPU model (red) at various temperatures. At high temperatures ($T = 797$ °C), numerical and experimental results are in good agreement, however with a slightly smaller (2.2%) power density predicted by the model. At lower temperatures, an increasing deviation is observed, whereby the model predictions underestimate the measured performance. While at high temperatures the deviation lies within the range of measurable tolerance, it then increases up to 30% at $T = 655$ °C. The mismatch can be explained by the varying GDC interdiffusion manufacturing. J. Szasz showed in Refs. [72, 232] that using tape casting (ASC Type-B) instead of PVD-coating (FZJ stack) to apply the GDC layer leaves small pores, which cause (in combination with intermediate sinter temperature) the formation of secondary phases at the electrolyte/interdiffusion layer interface. While the impact on the performance is lower at high temperatures, a significant decrease is measured by J. Szasz at lower operating temperatures. Combined with an FEM modeling approach, the performance decrease is correlated with poor oxygen ion transport properties in the secondary phases. PVD-coating produces an absolutely dense GDC layer, wherefore no secondary phases disturb the O^{2-} transport. Further reasons for the mismatch (especially at low temperatures) may be found in the isothermal modeling approach. Nevertheless, the good agreement at higher temperatures strengthens the model's validity and applicability to reproduce occurring SOFC loss mechanisms under stack operation. Hence, a numerical parameter variation is presented with confidence in the next sections to investigate possible optimization potentials.

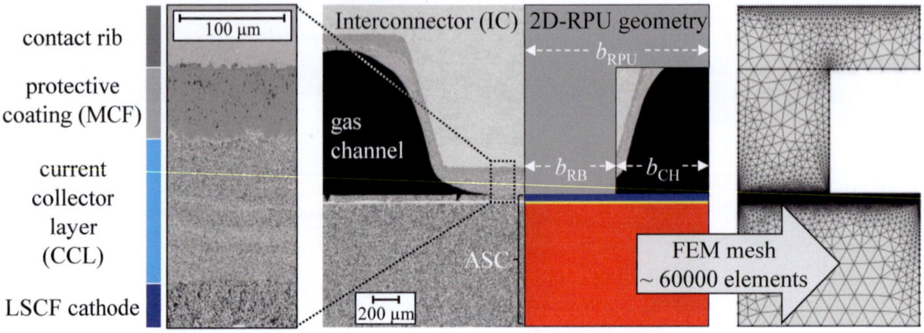

Anode supported SOFC (ASC Type-A)		**Planer Interconnector (IC) F-Design**
Cathode: $La_{0.58}Sr_{0.4}Co_{0.2}Fe_{0.8}O_{3-\delta}$	h_{cat} = 45 µm	Contact rip: b_{Rp} = 0.75 mm
Electrolyte: 8YSZ + CGO	h_{elyt} = 14 µm	Gas channel: b_{Ch} = 0.75 mm
Anode: Ni/8YSZ-cermet	h_{an} = 1500 µm	Crofer22APU-IC + LCC10-CCL

Figure 7.10.: SEM image of planar F-design interconnector (IC) developed at Forschungszentrum Jülich (middle) with magnified section of cathode and corresponding IC (left) [79]. Thereof developed 2D-RPU model geometry and numerical mesh (right). Technical details are displayed underneath.

7.2. Influence of Cathode Parameters

From the results earlier presented in this chapter, it has become clear that the cathode and its associated contact geometry under stack contacting conditions have significant influences on the utilizable performance of SOFC stacks. In order to investigate the matter rigorously and find out potential measures for optimization, intrinsic cathode design and material parameters are the subjects of a detailed numerical parameter study, at first presented in this section, followed by the IC geometry influence study in the next Section 7.3.1.

The numerical model employed in the following is based on the FEM model framework described in Chapter 4. In order to use an application-oriented approach, the planar SOFC stack design used at Forschungszentrum Jülich (F-design, Section 2.4.2) is used as a reference case with an anode supported SOFC Type-A (Table 5.1) featuring a LSCF cathode. Figure 7.10 illustrates the 2D-RPU geometry development, based on a F-design stack layer SEM cross section image. In Jülich, a current collector layer (CCL) is applied on top of the IC flowfield using atmospheric plasma spraying (APS), which results in a very dense microstructure. This is qualitatively deduced from the magnification in Figure 7.10 on the left. At low porosity a drastic increase of the pore fraction's tortuosity (Figure 5.1a) is predicted by the stochastic LSCF microstructure generator [120], therefore notable effective gaseous species transport fluxes within the CCL are prevented. A high material fraction however enable excellent electric conduction, wherefore the CCL is not resolved as an individual numerical domain in the FEM geometry, but attributed to the IC contact rib domain. The benefit of various CCL forms is analyzed individually in Section 7.4.1. Furthermore, the protective coating layer, shown in Figure 7.10 on the left, is not resolved either, but regarded by an interface contact impedance between IC rib and cathode layer (IF_{RP-cat}, Figure 4.3). An ongoing degradation due to corrosion is not regarded and a constant contact resistance $ASR_{contact}$ assumed, based on measured data from Jülich (Section 5.6). All simulation results are produced assuming 60% humidified hydrogen as fuel and ambient air as oxidant at constant stack operating temperature $T = 800$ °C (unless noted otherwise). Potentiostatic operation is applied in order to gain comparable results at different temperatures. Gas conversion is neglected at both electrodes, because the related processes occurring at the anode are not of interest at this point and high gas flow rates at the cathode side (intended to cool the stack under standard operating conditions) prevent a drop in the oxygen partial pressure along the gas channel length. With this setup, the following results are produced.

7.2.1. Layer Thickness

The results described in the previous section reveal that gas diffusion limitation within the porous cathode layer is the primary cause for the overall loss increase under stack contacting conditions. One of the first counter measures is to increase the cathode layer thickness in order to enhance the horizontal (x-direction) gas transport to prevent a premature O_2 depletion in the area farthest away from the gas channels. In this way, the cathodes pore network is extended and more space is added to increase the gas transport flux. On the other hand, above a certain layer thickness the total cathode gas diffusion loss $\eta_{cc,cat}$ will increase, compared to the loss created in the thick anode substrate. Furthermore, from an economic point of view, higher material usage and possible additional manufacturing steps (repeated screen printing) cause an increase in production cost, therefore it is desirable to keep the cathode layer thickness to a minimum. Hence, a compromise between additional loss creation and economic considerations must be found. Instead of elaborate and expensive experimental work, the task is perfectly suited for a numerical approach, the results of which are presented later.

Figure 7.11 displays predicted $pO_{2,cat}$ (Figure 7.11a) and $j_{ct,cat}$ distribution (Figure 7.11b) at the cathode/electrolyte interface ($IF_{cat-TPB}$, Figure 4.3) for varied cathode layer thickness $h_{cat} = 10 \ldots 300$ µm at $U_{cell} = 0.8$ V. The species distributions displayed in Figure 7.11 are in accordance with the results presented in Section 7.1. At lower h_{cat} a depletion of $pO_{2,cat}$ in the area beneath the ICs contact rib due to limited gas transport properties results in an electrochemical reaction slowdown, recognizable by the $j_{ct,cat}$ drop in the same area. Furthermore, the limited, electronic in-plane conduction capability causes a drop of $j_{ct,cat}$ in the area farthest away from the contact rib in the middle of the gas channel. With increasing h_{cat} however, $pO_{2,cat}$ increases at the TPB beneath the rib, which results in an increased electrochemical activity and consequently of $j_{ct,cat}$. At the same time, $pO_{2,cat}$ drops slightly at the TPB beneath the gas channel due to the elongated transport pathways within the porous cathode microstructure. However, the increased gas diffusion polarization does not impact on the electrochemical activity, therefore at $h_{cat} = 300$ µm the distribution of $j_{ct,cat}$ is almost homogeneous at the TPBs.

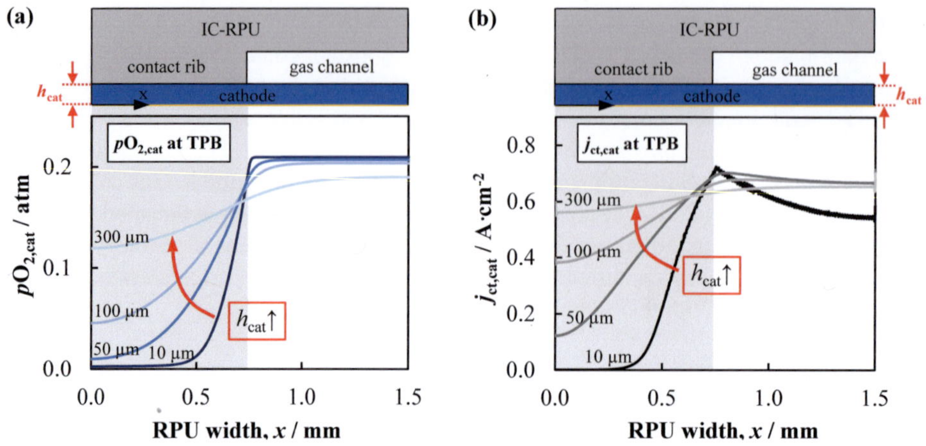

Figure 7.11.: Calculated species distributions at cathode/electrolyte interface (TPB) for various cathode layer thickness $h_{\mathrm{cat}} = 10\ldots300$ µm. (a) Oxygen partial pressure $pO_{2,\mathrm{cat}}$ and (b) charge transfer current density $j_{\mathrm{ct,cat}}$ distribution. Operating conditions: $U_{\mathrm{cell}} = 0.8$ V, $pH_{2,\mathrm{an}} = 0.4$ atm, $pH_2O_{\mathrm{an}} = 0.6$ atm, $pO_{2,\mathrm{cat}} = 0.21$ atm, $T = 800$ °C. ASC Type A, FZJ-IC, no gas conversion.

The numerical approach allows us to quantify the contribution of each occurring loss process (R_i) to the overall stack layer resistance (R_{SL}). Figure 7.12a displays the course of all R_i calculated by the equations stated in Section 4.4.3 for varied cathode layer thickness h_{cat}. Comparing the individual loss contribution's course, it is observed in Figure 7.12a that besides the constant loss created by the non-ideal, electronic contact between cathode layer and IC-surface ($R_{\Omega,\mathrm{contact}}$), all other loss contributions show an increased resistance at low h_{cat} and decrease with increasing h_{cat}. Above $h_{\mathrm{cat}} > 150$ µm the predicted change of all resistances becomes insignificant. The ohmic loss in the electrolyte ($R_{\Omega,\mathrm{elyt}}$) and cathode layer ($R_{\Omega,\mathrm{cat}}$) as well as the electrodes electrochemical activation loss ($R_{\mathrm{act,an/cat}}$) decrease with increasing h_{cat}. This is explained by the enhanced O_2 supply enabling an increasingly homogeneous electrochemical activity at the TBPs (Figure 7.11) and thus less loss of associated processes. $R_{\Omega,\mathrm{cat}}$ exhibits the highest sensitivity against a change in h_{cat} followed by $R_{\Omega,\mathrm{elyt}}$ while $R_{\mathrm{act,an/cat}}$ are less severely affected. In contrast to these dependencies, the gas diffusion loss at the cathode ($R_{\mathrm{cc,cat}}$) increases slightly in the range of 10 µm $< h_{\mathrm{cat}} <$ 50 µm and only afterwards decreases (as the rest) with increasing h_{cat}. This divergent trend is due to the increase in total current density j_{cell} at constant operating voltage $U_{\mathrm{cell}} = 0.8$ V as a consequence of an enhanced O_2 supply at the TPBs and therefore an increased O_2 consumption. $R_{\Omega,\mathrm{cat}}$ converges to zero at higher layer thicknesses whereas the other loss contributions converge at a certain level. The gas diffusion loss in the anode ($R_{\mathrm{cc,an}}$) is the only process that is unaffected by changes of h_{cat}. Figure 7.12b displays the percentage distribution of R_i for $h_{\mathrm{cat}} = 45$ µm, which is the standard cathode layer thickness of Type-A ASCs. It is revealed that $R_{\Omega,\mathrm{elyt}}$ and $R_{\Omega,\mathrm{contact}}$ account for more than half of the overall loss, while $R_{\mathrm{act,an/cat}}$ account for a quarter, followed by $R_{\mathrm{cc,an/cat}}$ and with $R_{\Omega,\mathrm{cat}}$ as the smallest contribution. However, electric in-plane conduction loss ($R_{\Omega,\mathrm{cat}}$) is apparently a factor to have in mind, even for the higly electronically conductive conduction LSCF. Its influence is further examined in the next Section 7.2.2.

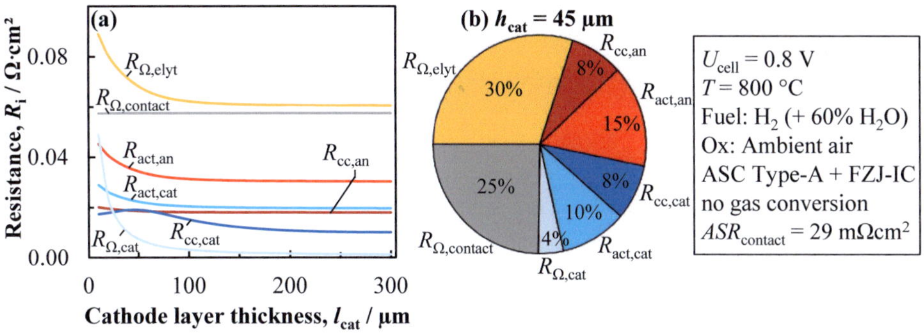

Figure 7.12.: (a) Model predicted evolution of individual loss process contributions (R_i) to overall stack layer resistance (R_{SL}) at various cathode layer thickness $h_{cat} = 10 \ldots 300$ µm. (b) Loss composition at $h_{cat} = 45$ µm. Operating conditions, cell and contact type stated in figure.

Figure 7.13a displays predicted stack layer power density P_{SL} in W/cm^{-2} calculated by the 2D-RPU model for a varied cathode layer thickness $h_{cat} = 10 \ldots 300$ µm at various operating voltages $U_{cell} = 0.9 \ldots 0.7$ V at $T = 800$ °C. Under low polarization at $U_{cell} = 0.9$ V, the influence of h_{cat} on P_{SL} is small, only at $h_{cat} < 30$ µm is a decreased performance predicted by the model. With decreasing cell voltage (or higher polarization), more current is drawn from the cell; consequently, P_{SL} increases. However, this also means that more O_2 is consumed and the affects of limited gas diffusion transport properties have an increased impact on the performance. The results in Figure 7.13a show how the predicted maximum power density shifts to higher h_{cat} with a more pronounced drop towards lower h_{cat}. While at $U_{cell} = 0.8$ V the maximum of P_{SL} is reached at $h_{cat} \approx 150 - 200$ µm, a further increase of P_{SL} is observed at $h_{cat} = 300$ µm for $U_{cell} = 0.7$ V.

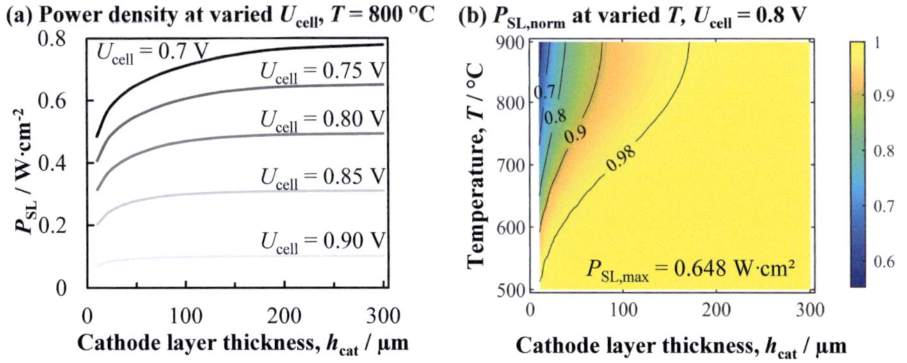

Figure 7.13.: (a) Model predicted evolution of stack layer power density (P_{SL}) at various cathode layer thickness $h_{cat} = 10 \ldots 300$ µm and cell voltage $U_{cell} = 0.9 \ldots 0.7$ V. (b) Display of model predicted power density, $P_{SL,norm.}$, at various temperatures $T = 900 \ldots 500$ °C, normalized to the respective maximum power density. Operating conditions, cell and contacting type stated in Figure 7.12.

Table 7.1.: Model predicted performance loss in % at given operating temperature T and cell voltage U_{cell} using an ASC Type-A with planar F-design interconnector geometry from Forschungszentrum Jülich.

	Temperature T / °C	Cell voltage, U_{cell} / V 0.90	0.85	0.80	0.75	0.70
	800	-8.8	-11.6	-14.8	-16.9	-18.0
Rel. power density loss / %	700	-3.9	-4.9	-6.5	-8.8	-11.4
	600	-1.0	-1.2	-1.3	-1.6	-1.8

Figure 7.13b depicts predicted power densities $P_{\text{SL,norm}}$ at $U_{\text{cell}} = 0.8$ V for varied h_{cat} at various operating temperatures $T = 900 \ldots 500$ °C. The values have been normalized to the individual maximum calculated at $h_{\text{cat}} = 300$ µm for reasons of better comparison. The power enhancing influence of h_{cat} is high at elevated operating temperatures and decreases at low T. This behavior is explained by the strong temperature dependence of the ohmic and electrochemical activation losses (Section 6.2) compared to the almost temperature independent gas species transport. At equal operating voltage, U_{cell}, ohmic and activation losses dominate the overall loss at lower operating temperature, causing a decreased electrochemical gas conversion compared to what is converted at higher temperatures. Consequently, the influence of elongated gas species transport pathways decreases with decreasing temperature. Table 7.1 lists the calculated power loss in % for the standard cathode layer thickness of ASC Type-A ($h_{\text{cat}} = 45$ µm) at varied cell voltage and operating temperature.

To sum up this section, the above presented numerical results show the influence of cathode layer thickness on SOFC stack performance. It is demonstrated that the power density generated by each layer in a stack can be improved significantly, depending on the chosen operating conditions. Adding more pore space improves the gaseous species transport fluxes and thereby the O_2 supply in the electrochemical active areas beneath the IC contact ribs and thus ensures a homogeneous charge transfer reaction across the active cell area. The simulation results further demonstrate that an insufficient gas supply causes an increase in occurring loss processes and even creates additional loss due to in-plane ohmic loss in the cathode layer. This effect increases with decreasing cathode layer thickness, even for the highly electronically conductive cathode material LSCF. The performance enhancing effect of a thicker cathode decreases with decreasing operating temperature as temperature activated loss processes prevent a stronger influence of gas depletion in affected areas. Nevertheless, the results have shown that electronic conduction is a sensible factor, therefore its influence on stack performance is examined in the following section.

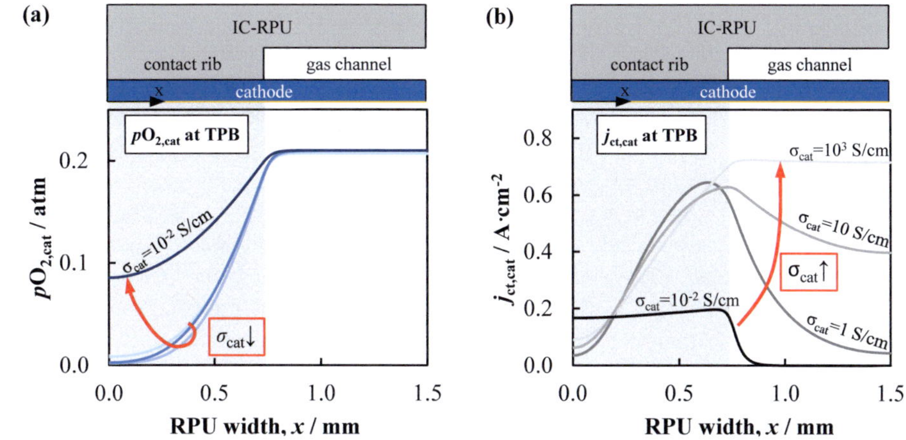

Figure 7.14.: Calculated species distributions at cathode/electrolyte interface (TPB) for various electronic, cathode layer conductivities σ_{cat}. (a) Oxygen partial pressure $pO_{2,\mathrm{cat}}$ and (b) charge transfer current density j_{ct} distribution. Operating conditions: $U_{\mathrm{cell}} = 0.8$ V, $pH_{2,\mathrm{an}} = 0.4$ atm, $pH_2O_{\mathrm{an}} = 0.6$ atm, $pO_{2,\mathrm{cat}} = 0.21$ atm, $T = 800$ °C. ASC Type A, FZJ-IC, no gas conversion.

7.2.2. Electronic Conductivity

The results presented in Section 7.2.1 have demonstrated that additional ohmic loss is created in the cathodes material phase under stack contacting conditions. Therefore, the electronic conductivity of the applied cathode material σ_{cat} is certainly a parameter of interest. The state-of-the-art cathode material LSCF exhibits a high capability to transport electrons compared to other potential cathode materials, as Table 5.4 shows. The listed materials are considered to be state-of-the-art SOFC cathode materials (LSM, LSCF) or are mentioned in literature as potential candidates (PCO, STF, BSCF, LSC). The σ_{cat}-range of these materials extends over several orders of magnitude, therefore it is of interest how σ_{cat} alone influences SOFC stack performance, regardless of the electrochemical reactivity. Hence, σ_{cat} is varied in the following, using the 2D-RPU SL Model according to the IC geometry and ASC Type-A (as displayed in Figure 7.10) with cathode charge transfer kinetics of LSCF (Section 5.2) and the standard cathode layer thickness $h_{\mathrm{cat}} = 45$ µm.

Figure 7.14 displays the model predicted species distributions at the cathode/electrolyte interface ($\mathrm{IF}_{\mathrm{cat-TPB}}$) for varied electronic conductivity of the cathode material phase (σ_{cat}). For reference value the electronic conductivity of LSCF (σ_{LSCF}) is chosen and varied over several orders of magnitude to capture the great variety of potential cathode materials (Tab.5.4). The simulations were carried out at standard operating condition as stated in the caption of Figure 7.14. The predicted oxygen partial pressure ($pO_{2,\mathrm{cat}}$) distribution at $\mathrm{IF}_{\mathrm{cat-TPB}}$ (Figure 7.14a) shows only a very small sensitivity to changes of σ_{cat}, except at very low $\sigma_{\mathrm{cat}} < 10^{-1}$ S/cm. As the cathode layer thickness is not varied (constant with $h_{\mathrm{cat}} = 45$ µm) the limited gas transport properties cause the already familiar drop in $pO_{2,\mathrm{cat}}$ in the area beneath the rib (Section 7.2.1) with a small increase at very high

σ_{cat}. It decreases slightly at lower σ_{cat}, but increases again for $\sigma_{\text{cat}} < 1$ S/cm. In the area beneath the gas channel no significant influence is observable. Figure 7.14b shows the corresponding predicted charge transfer current density ($j_{\text{ct,cat}}$) distribution at $\text{IF}_{\text{cat-TPB}}$. In contrast, therein a significant influence of σ_{cat} is observable. First of all, as discussed in Section 7.2.1, $j_{\text{ct,cat}}$ drops significantly in the area beneath the contact rib due the hindered electrochemical reactivity as a consequence of the O_2 undersupply. For $\sigma_{\text{cat}} > 10^{-1}$ S/cm the $j_{\text{ct,cat}}$ generation is low in accordance to the predicted low $pO_{2,\text{cat}}$ distribution in that area (Figure 7.14a), but increases slightly at lower σ_{cat}. The mentioned significant impact of σ_{cat} is revealed at the TPB beneath the gas channel. $j_{\text{ct,cat}}$ remains constant at a high level from the rib edge to the gas channel middle for high σ_{cat}, but it decreases further with decreasing σ_{cat} the further away the reaction zone is from the contact rib. A slowdown up to a total breakdown of electrochemical reactivity in the middle of the gas channel is predicted for materials with $\sigma_{\text{cat}} >= 10^{-1}$ S/cm. As a consequence, the $j_{\text{ct,cat}}$ distribution develops a peak in the area at the rib edge where O_2 and electron supply are in a certain equilibrium. The electrochemical reaction slowdown in the gas channel middle favors a further increase of electrochemical activity beneath the rib, which explains the drop in $pO_{2,\text{cat}}$ at $\sigma_{\text{cat}} \approx 10^{-1}$ S/cm. At $\sigma_{\text{cat}} < 10^{-1}$ S/cm the ohmic loss created in the cathode exceeds a certain level and limits the electrochemical activity to an extent, where it occurs only in the area beneath the ribs at a decreased rate. Hence, the oxygen consumption is reduced to an amount where the gas transport properties are less limiting, thus explaining the increase in $pO_{2,\text{cat}}$ beneath the rib area at $\sigma_{\text{cat}} < 10^{-1}$ S/cm. The results in Figure 7.14 clearly demonstrate the importance of a highly electronically conducting cathode material in order to avoid the creation of additional ohmic loss/es. The next section will discuss how the various occurring loss processes and the resulting stack layer power density are influenced by σ_{cat}.

Figure 7.15a displays the model-based loss distribution, R_i, of occurring loss processes for varied σ_{cat}. The results are plotted on a logarithmic scale at the x-axis due to the great range of σ_{cat}. Analyzing R_i starting from high σ_{cat} towards lower values in the range of $10^3 > \sigma_{\text{cat}} > 10^2$ S/cm, no influence is observed besides the increase of $R_{\Omega,\text{cat}}$ and that the individual amount of loss distribution is the same as displayed in Figure 7.12 in the middle. The increase of $R_{\Omega,\text{cat}}$ is obviously triggered by the decrease in σ_{cat}. However, below $\sigma_{\text{cat}} \geq 10^2$ S/cm, the cathode diffusion loss $R_{\text{cc,cat}}$ starts to increase as a result of the overall decreasing j_{cell} and the constant gas consumption with limited supply in the area beneath the ribs. In the range of $10^1 \geq \sigma_{\text{cat}} \geq 1$ S/cm, $R_{\Omega,\text{cat}}$ remains at a certain level while the remaining loss processes start to increase their contribution to the overall loss. Looking at Figure 7.11b, it is the range where $j_{\text{ct,cat}}$ drops significantly in the area beneath the channel, thus increasing the inhomogeneous electrochemical activity distribution further, which consequently affects the related loss mechanisms. In this range $R_{\text{cc,cat}}$ contributes already as much to the overall loss as $R_{\Omega,\text{ely/contact}}$. Below $\sigma_{\text{cat}} \geq 1$ S/cm $R_{\text{cc,cat}}$ starts to increase again, while the remaining loss contributions seem to level off at their individual amount. The behavior is concurrent to the decreasing peak of $j_{\text{ct,cat}}$ in Figure 7.11b. Then σ_{cat} reaches a level where it is as low as the ionic conductivity in the electrolyte layer σ_{elyt}. Decreasing σ_{cat} further elevates $R_{\Omega,\text{cat}}$ to the dominating loss with no significant power production.

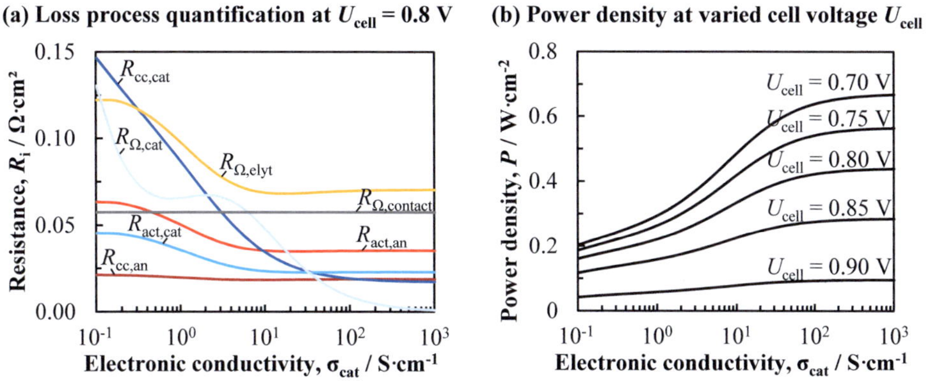

T = 800 °C	pH_{2an} = 0.4 atm	ASC Type-A	no gas conversion
$pO_{2,cat}$ = 0.21 atm	pH_2O_{an} = 0.6 atm	FZJ-IC	$ASR_{contact}$(FZJ)

Figure 7.15.: (a) 2D-RPU SL model predicted evolution of individual loss process contributions (R_i) to overall stack layer resistance (R_{SL}) at various cathode layer electronic conductivities σ_{cat}. (b) 2D-RPU SL model predicted evolution of stack layer power density (P_{SL}) at various cathode layer thickness σ_{cat} and cell voltage U_{cell} = 0.9 ... 0.7 V. Operating conditions, cell and contact type stated in figure.

Naturally, the power density increases with increasing cell voltage U_{cell}. Figure 7.15b illustrates this fact by displaying the model predicted stack layer power density P_{SL} in W/cm^2 for the σ_{cat}-variation. While at $U_{cell} = 0.9$ V the polarization is low, the influence of σ_{cat} is insignificant. With decreasing U_{cell} more current is generated and the limiting in-plane conduction influence (and the associated chain of loss mechanisms as discussed above) becomes more relevant. Consequently, at $U_{cell} = 0.7$ V the performance loss is greater, ending up $\sigma_{cat} = 10^{-1}$ S/cm with a power density only slightly higher than predicted for $U_{cell} = 0.75$ V.

The influence of σ_{cat} under F-design stack contacting conditions at various operating temperatures $T = 900 \ldots 500$ °C at constant $U_{cell} = 0.8$ V and standard gas flow conditions is displayed in Figure 7.16. The calculated power density is normalized to the individual maximum at $\sigma_{cat} = 10^3$ S/cm for better comparison, again with a logarithmic scale of the x-axis. The color code reveals a larger influence of σ_{cat} at elevated operating temperatures. While at $T = 800$ °C the performance drops from $95 - 80\%$ in the conductivity range of $10^2 \geq \sigma_{cat} \geq 10^1$ S/cm, $P_{SL,norm}$ is only mildly decreased at $T = 600$ °C in the equal conductivity range. The trend is similar to the temperature influence at varied cathode layer thickness h_{cat} (Figure 7.13). At elevated operating temperature increased ohmic loss in the electrolyte and contact resistance as well as activation loss of the electrodes charge transfer reactions prevent a significant current density generation, therefore fewer electrons need to be conducted in-plane through the cathode and the low conductivity loses its influence compared to at high temperatures. Nevertheless, at $T = 600$ °C a performance loss of $\approx 10\%$ is predicted by the model, using a cathode material with an inherent electronic conductivity of $\sigma_{cat} = 1$ S/cm.

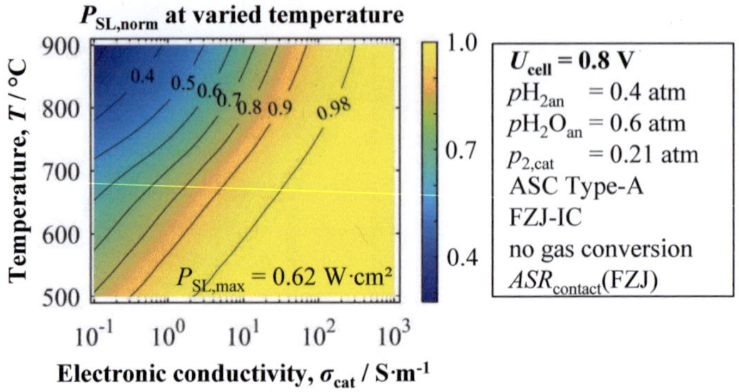

Figure 7.16.: Display of model predicted power density $P_{\mathrm{SL,norm.}}$ at various temperatures $T = 900 \ldots 500$ °C, normalized to the respective maximum power density for varied cathode layer electronic conductivity σ_{cat}. Operating conditions, cell and contact type stated in figure.

The results presented above demonstrate the influence of the cathodes electronic conductivity σ_{cat} at various cell voltages and operating temperatures. It shows how the performance of the SOFC stack layer decreases when using the Jülich F-design with an ASC Type-A due to additional ohmic losses caused by electronic in-plane conduction limitations in the cathode layer. The limited charge carrier transport in the standard LSCF cathode layer with the thickness $h_{\mathrm{cat}} = 45$ µm causes a decreased charge transfer reaction activity in the area beneath the gas channels. At higher polarization, the influence of σ_{cat} is increased as more electrons are required for the electrochemical reaction, and electron transport is governed by σ_{cat}. Furthermore, the power decreasing influence of σ_{cat} decreases at lower operating temperatures as increased ohmic and activation loss prevent an equal current density generation compared to what is generated at higher temperatures, thus fewer electrons need to be transported in-plane through the cathode layer. The results predict that at $U_{\mathrm{cell}} = 0.8$ V and $T = 800$ °C insignificant additional ohmic loss generation using a cathode material such as LSC with a high electronic conductivity, small loss of ~4 % for LSCF and up to ~18 % using LSM as cathode material. The performance loss is thereby independent of the material electrochemical kinetics or the layer's microstructural properties. Other promising cathode materials proposed in the literature (Table 5.4) exhibit an even lower σ_{cat}, wherefore their application is questionable regardless of their potential high electrochemical activity. This concludes the numerical analysis how the cathode layer's electronic conductivity influences SOFC stack performance. The next section investigates the cathode layer microstructure influence on stack performance by numerical evaluation.

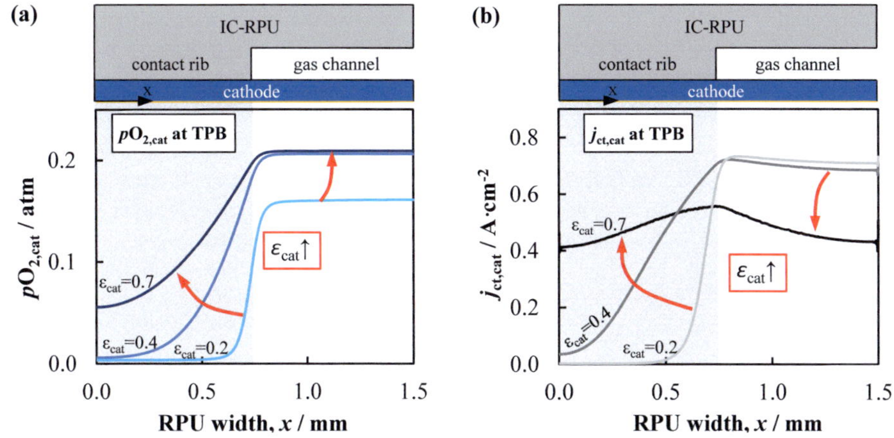

Figure 7.17.: Calculated species distributions at cathode/electrolyte interface (TPB) for varied cathode layer pore fractions ε_{cat}. (a) Oxygen partial pressure $pO_{2,\text{cat}}$ and (b) charge transfer current density j_{ct} distribution. Operating conditions: $U_{\text{cell}} = 0.8$ V, $pH_{2,\text{an}} = 0.4$ atm, $pH_2O_{\text{an}} = 0.6$ atm, $pO_{2,\text{cat}} = 0.21$ atm, $T = 800$ °C. ASC Type A, FZJ-IC, no gas conversion.

7.2.3. Microstructure Properties

Section 5.1 describes in detail how the cathode pore fraction ε_{cat} determines material and pore phase tortuosity $\tau_{\text{por/mat,cat}}$ (Figure 5.1a), the electrochemical active surface area a_{miec} (Figure 5.1b) and the permeability κ_{cat} (Figure 5.7). These interrelations accordingly affect species transport fluxes and reaction kinetics. For example, a low ε_{cat} generates high $\tau_{\text{por,cat}}$ and both low $\tau_{\text{mat,cat}}$ and κ_{cat}. Hence, effective gas species transport fluxes are decreased, while electronic conduction is increased due to the opposing microstructure interrelation. Both transport fluxes are coupled with the electrochemical charge transfer reaction, which takes place at the decreased a_{miec} (at low porosity). It is therefore of particular relevance to regard these complex coupled mechanisms in a modeling approach, in order to generate meaningful results. In contrast to similar works available in literature (Ref.[145, 154]), this work presents a (so far) unique approach by combining effective transport and reaction mechanisms modeling with microstructure properties determined from 3D reconstructions from real LSCF cathode samples (Section 2.6). Furthermore, inherent microstructure parameter interdependencies are regarded by implementation of results produced by a stochastic LSCF microstructure generator (Ref. [120]), which itself has been calibrated by several 3D-reconstructions of different, real LSCF cathode samples.

Figure 7.17 displays numerical results for varied cathode layer pore fraction using the F-design IC geometry and an ASC Type-A at standard operating conditions ($T = 800$ °C, $H_2 + 60\%H_2O$ as fuel and ambient air as oxidant). Figure 7.17a illustrates the resulting $pO_{2,\text{cat}}$ distribution prediction, prevailing at the cathode/electrolyte interface ($\text{IF}_{\text{cat-TPB}}$) at $U_{\text{cell}} = 0.8$ V at the given ε_{cat}. In the area beneath the ICs contact rib, $pO_{2,\text{cat}}$ increases continuously with increasing ε_{cat}, whereby only above $\varepsilon_{\text{cat}} > 0.5$ can O_2-depletion due to limited transport properties be avoided. In the area beneath the gas channel, a notable microstructure influence on $pO_{2,\mathrm{cat}}$ is predicted only for $\varepsilon_{\mathrm{cat}} < 0.3$. Hence, an increased pore fraction is beneficial to avoid gas depletion, but the in-plane electron transport in the cathode layer has to be regarded as well. Figure 7.17b displays the corresponding, predicted $j_{\mathrm{ct,cat}}$ distribution. The illustrated numerical results show that for $\varepsilon_{\mathrm{cat}} > 0.5$ the $j_{\mathrm{ct,cat}}$ actually decreases in the area beneath the channel while is increases up to the value beneath the rib area. The latter is clearly caused by the increasing $pO_{2,\mathrm{cat}}$ while at higher porosity an increasing $\tau_{\mathrm{mat,cat}}$ limits the charge carrier transport flux and thereby the electrochemical reaction rate.

The individual loss process contribution displayed in Figure 7.18a reflects the trends given by the species distribution predictions discussed above. $R_{\Omega,\mathrm{elyt}}$ and $R_{\mathrm{act,an}}$ decrease with increasing $\varepsilon_{\mathrm{cat}}$, because at low $\varepsilon_{\mathrm{cat}}$ an electrochemical reaction slow-down is caused by O_2 transport limitation; electronic in-plane conduction limitation at higher $\varepsilon_{\mathrm{cat}}$ does not reach a limiting level due to the highly conductive LSCF. In opposition to this, $R_{\Omega,\mathrm{cat}}$ remains constant at a low level and increases from $\varepsilon_{\mathrm{cat}} > 0.5$. $R_{\mathrm{act,cat}}$ exhibits the highest sensitivity to varied $\varepsilon_{\mathrm{cat}}$ as it increases at low and high $\varepsilon_{\mathrm{cat}}$ due to its dependency on a_{miec} (Figure 5.1b) with a minimum at $\varepsilon_{\mathrm{cat}} \approx 0.4$. $R_{\mathrm{cc,cat}}$ contributes only mildly to the overall loss and only increases from $\varepsilon_{\mathrm{cat}} < 0.3$ at the applied polarization. Similar to the results shown in the previous section, $R_{\mathrm{cc,an}}$ only exhibits a low sensitivity to changes of the cathodes microstructure, and $R_{\Omega,\mathrm{contact}}$ shows no sensitivity at all. These individual loss contributions determine a maximum power density prediction at $\varepsilon_{\mathrm{cat}} > 0.6$ for low polarization with a trend towards lower $\varepsilon_{\mathrm{cat}}$ at higher polarization, as the results in Figure 7.18b show. The maximum power density is predicted for $\varepsilon_{\mathrm{cat}} \approx 0.5$.

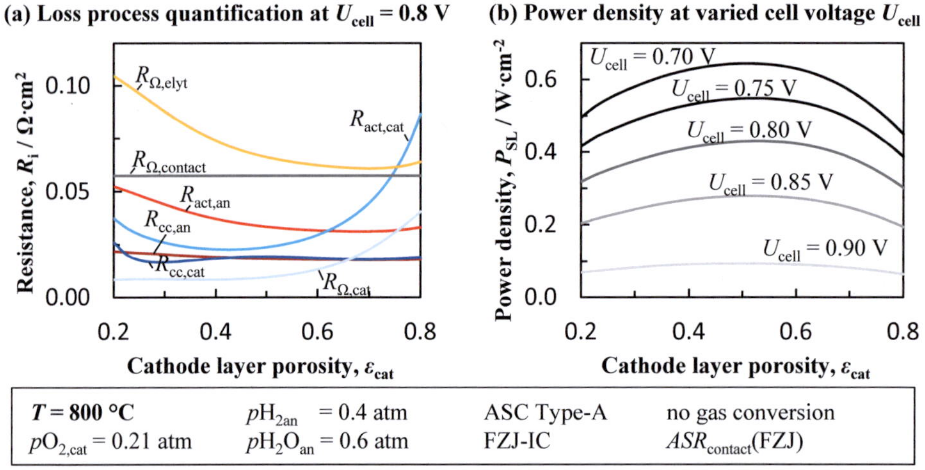

Figure 7.18.: (a) Predicted evolution of individual loss process contributions (R_i) to overall stack layer resistance (R_{SL}) at various cathode layer pore fraction $\varepsilon_{\mathrm{cat}}$. (b) By 2D-RPU SL model predicted evolution of stack layer power density (P_{SL}) at various cathode layer pore fraction $\varepsilon_{\mathrm{cat}}$ and cell voltage $U_{\mathrm{cell}} = 0.9 \ldots 0.7$ V. Operating conditions, cell and contact type stated in figure.

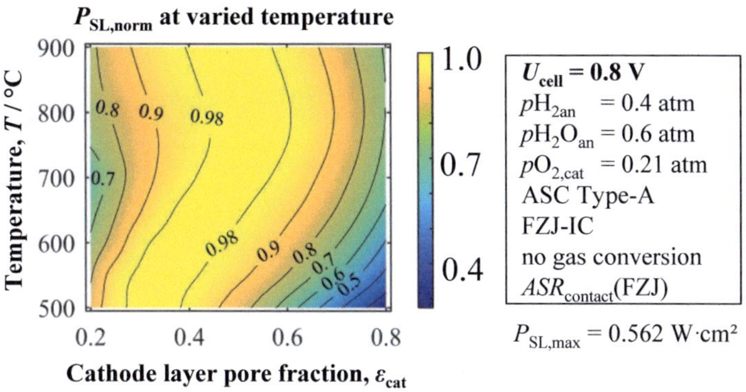

Figure 7.19.: Display of model predicted power density $P_{\mathrm{SL,norm}}$. at various temperatures $T = 900 \ldots 500$ °C, normalized to the respective maximum power density for varied cathode layer pore fraction $\varepsilon_{\mathrm{cat}}$. Operating conditions, cell and contact type stated in figure above.

Figure 7.19 illustrates the predicted power density $P_{\mathrm{SL,norm}}$ at varied cathode layer pore fraction $\varepsilon_{\mathrm{cat}}$ and operating temperature T. $P_{\mathrm{SL,norm}}$ is normalized to the respective maximum predicted for each temperature while the remaining operating conditions (Figure 7.19) were kept constant for each numerical parameter variation. While at higher temperatures the best performance is generated by a cathode layer with equal pore and material fraction ($\varepsilon_{\mathrm{cat}} \approx 0.5$), a shift towards lower pore fractions $\varepsilon_{\mathrm{cat}} < 0.4$ at lower temperatures is predicted by the model. Furthermore, Figure 7.19 shows that an increased bandwidth of potential power loss is given at lower T. From the individual loss contribution deconvolution displayed in Figure 7.18, it can be concluded that $R_{\mathrm{act,cat}}$ and $R_{\Omega,\mathrm{cat}}$ are responsible for both phenomena at lower T as their temperature dependence is stronger compared to gas transport loss, and that it is consequently beneficial to supply less pore and more material fraction with less tortuous transport pathways for effective electronic conduction.

The results presented in this section show the influence of the cathode layer's microstructure properties on SOFC stack layer performance (with an interconnector design from Jülich at various operating conditions). It is shown that at low $\varepsilon_{\mathrm{cat}}$ the performance is generally decreased due to gas transport limitations. Furthermore, low $\varepsilon_{\mathrm{cat}}$ result in a decreased active surface area (Figure 5.1b) and causes diminished power generation because it directly scales the electrochemical reaction rate. However, a turning point is reached at a certain $\varepsilon_{\mathrm{cat}}$ and ohmic loss increases in the cathode layer due to limited electronic conduction properties because less material phase and elongated electronic transport pathways result at higher $\varepsilon_{\mathrm{cat}}$. Hence an optimum strongly depends on the polarization level and operating temperature because the overall loss determines the gaseous and electronic species demand. These highly complex coupled transport and reaction mechanism can only be taken into account in a modeling approach by accurately regarding the microstructure parameter interdependencies and combining these with the electrochemical MIEC-cathode model. In contrast to other modeling approaches in literature, the model framework presented here is able to do just that.

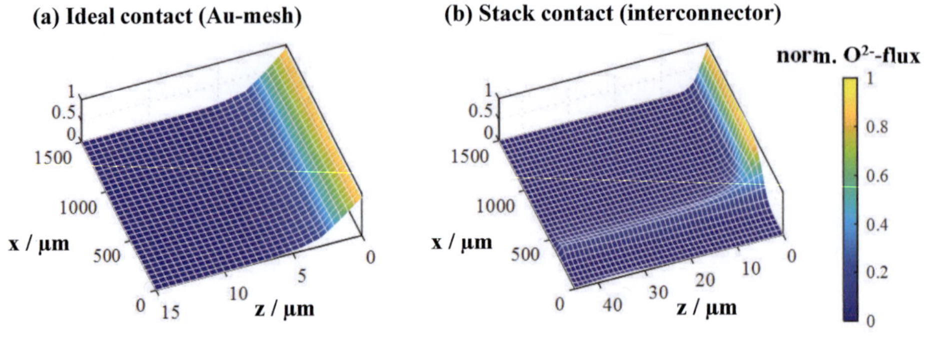

Figure 7.20.: Model predicted O^{2-}-flux, normalized to the individual maximum using (a) ideal an (b) stack contacting conditions (Section 7.1). LSCF surface exchange coefficient, k^{δ}, and bulk diffusion coefficient, D^{δ}, taken from Table 5.3. Operating conditions: $U_{\text{cell}} = 0.8$ V, $p\text{H}_{2,\text{an}} = 0.4$ atm, $p\text{H}_2\text{O}_{\text{an}} = 0.6$ atm, $p\text{O}_{2,\text{cat}} = 0.21$ atm, $T = 800$ °C.

7.2.4. Charge Transfer Kinetics (MIEC-Material)

Mixed-ionic-electronic conductive (MIEC) cathode layers are state of the art for high performance, anode supported SOFCs [63] and it is naturally of great interest how the use of different cathode materials affect SOFC stack performance. With the homogenized MIEC-cathode approach (Section 4.3.3.2) this can be numerically simulated using the implementation of *surface exchange* (k^{δ}) and *bulk diffusion coefficients* (D^{δ}), called $k\&D$ in the following. In this way, virtually any desired MIEC material can be tested for its performance as cathode material in SOFC/SOEC application. While the Butler-Volmer approach (Section 4.3.3.1) is more robust and much less computationally expensive (Section 6.5), it is in this work validated only for LSCF and therefore limited to this material. Furthermore, the BV-approach does not regard true MIEC-properties as it is limited to electronic transport in the cathodes material phase.

The MIEC-approach attends to both electronic and ionic charge transport, thus regarding the microstructure dependency on the charge transfer reaction polarization and accounts for variable reaction-penetration depth into the cathode material. This is visualized in Figure 7.20, where the penetration depth is depicted by the normalized ionic current flux density in the MIEC phase. Figure 7.20a displays a cathode section with ideal contacting and Figure 7.20b the same with stack contacting, as used in Section 7.1. It is noticeable how the inhomogeneously distributed penetration depth reflects the IC geometry (placed on top of the cathode layer) and how a complete section of the cathode layer is electrochemically inactive while under ideal conditions a homogeneous reaction distribution would occur. With the MIEC-cathode approach, other materials qualifying as potential cathode materials such as LSC or BSCF can be numerically tested for their performance with a physically realistic reproduction of occurring transport processes with respect to microstructure properties.

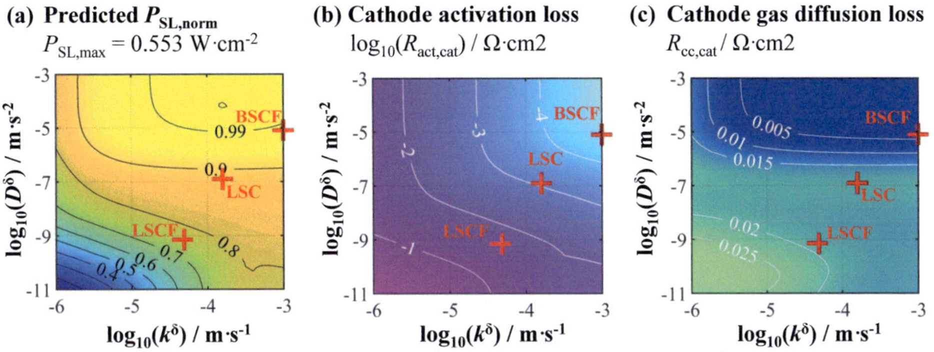

Figure 7.21.: Stack contact model predictions for varied MIEC-cathode kinetic parameters, surface exchange coefficient k^δ and bulk diffusion coefficient D^δ with constant $\sigma_{\mathrm{cat}} = \sigma_{\mathrm{LSCF}}$. (a) Predicted and maximum-normalized power density $P_{\mathrm{SL,norm}}$, (b) cathode activation loss $R_{\mathrm{act,cat}}$, displayed in decimal logarithmic color scale and with according exponents as contour lines and (c) cathode diffusion loss $R_{\mathrm{cc,cat}}$ Operating conditions: $U_{\mathrm{cell}} = 0.8$ V, $p\mathrm{H}_{2,\mathrm{an}} = 0.4$ atm, $p\mathrm{H_2O}_{\mathrm{an}} = 0.6$ atm, $p\mathrm{O}_{2,\mathrm{cat}} = 0.21$ atm, $T = 800$ °C. ASC Type A, FZJ-IC, no gas conversion.

The work of B. Rüger [65] (called *cube model* in the following) provides similar capabilities and laid the foundation to this works MIEC-cathode model approach. Nevertheless the cube model proposed by B. Rüger is limited to cathode loss prediction under ideal contacting conditions. Furthermore, it is numerically much more expensive and therefore limited to small spatial geometry widths in the micrometer scale. Hence, it is valuable progress that with this work the influence of MIEC-cathode electrochemical properties can be tested for stack contacting conditions as applied in actual applications in the form of the 2D-RPU model.

Figure 7.21 displays the 2D-RPU model predicted results relevant for varied $k\&D$-values, using the Jülich stack design (Figure 7.10). The results show how stack layer power density increases with increasing $k\&D$-values (Figure 7.10a), which is directly linked to the decreasing cathode activation (Figure 7.10b) and diffusion loss (Figure 7.10c). Red crosses mark the corresponding $k\&D$-values of LSCF, LSC and BSCF at $T = 800$ °C (Section 5.2.2). According to the displayed results, BSCF promises the highest performance due to its high kinetic properties. However, the materials individual electronic conductivity is not regarded in this calculation and BSCF exhibits the lowest conductivity of mentioned materials (Section 5.3). Assuming ideal contacting (e.g. cube model by B. Rüger [65]), the conductivity is negligible, but the results in the following will demonstrate that under stack contacting conditions it does have an impact and has to be considered in order to give a realistic material performance prediction for application.

Similar to the previous sections, predicted species distributions ($p\mathrm{O}_{2,\mathrm{cat}}$ and $j_{\mathrm{ct,cat}}$) at the cathode/electrolyte interface are presented for LSCF, LSC and BSCF. The applied 2D-RPU model features the F-Design IC geometry (Figure 7.10) with a Type-A ASC cell design (Table 5.1). The kinetic parameters (k^δ and D^δ) are experimentally determined for LSCF in this work at cell Type-A and adopted from literature for LSC and BSCF (Section 5.2.2). Additionally, the effective electronic conductivities $\sigma_{\mathrm{cat}}^{\mathrm{eff}}$ are calculated from bulk conductivity data adopted from literature (Section 5.3). As before, standard operating conditions were chosen for this first result, displayed in Figure ??. Starting the analysis with the $pO_{2,\mathrm{cat}}$-distribution in Figure ??a, LSCF (blue line) shows the familiar O_2 starvation in the area beneath the contact rib due to the O_2 transport properties in gas and bulk phase. Similar results are predicted for LSC (green line), however with a slightly higher $pO_{2,\mathrm{cat}}$ in the critical area beneath the rib. Using BSCF in the model, the results (orange line) show almost no $pO_{2,\mathrm{cat}}$ decrease. In the area beneath the gas channel, the $pO_{2,\mathrm{cat}}$ prediction is equally high for all materials. The results demonstrate that, when using BSCF as cathode material, no gas starvation has to be feared, while the situation with LSC is only mildly improved compared to the reference material LSCF. The resulting charge transfer current density predictions are displayed in Figure ??b. Unsurprisingly, the predicted $j_{\mathrm{ct,cat}}$ distributions for LSC and LSCF resemble each other, however, with a small increase over the total RPU width for LSC. Especially, no drop of $j_{\mathrm{ct,cat}}$ in the area beneath the gas channel is observed due to the much higher electronic conductivity of LSC. The $j_{\mathrm{ct,cat}}$ prediction for BSCF as cathode material shows a completely different distribution compared to the other two materials. Beneath the rib, $j_{\mathrm{ct,cat}}$ is high and decreases towards the channel area. This opposing behavior compared to LSC and LSCF is caused by the extremely high k^{δ} and D^{δ} and the low electronic conductivity of BSCF. Oxygen is transported well in both the gas and bulk phase, therefore the electrochemical reaction is not hindered in the area beneath the rib due to transport limitations but in the area beneath the channel due to in-plane conduction limitations. This result dismisses the previously made high expectations imposed on BSCF, at least in comparison to LSC and with the chosen cell and contact design applied in this numerical investigation.

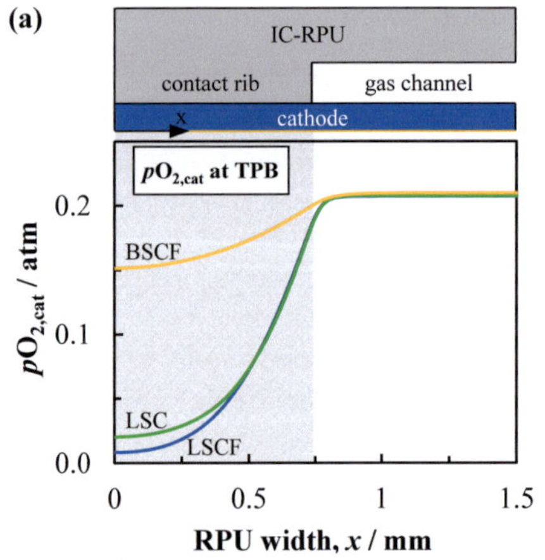

Figure 7.22.: Calculated species distributions at cathode/electrolyte interface (TPB) for various cathode materials or electrochemical properties. (a) Oxygen partial pressure $pO_{2,\mathrm{cat}}$ and (b) charge transfer current density j_{ct} distribution. Operating conditions: $U_{\mathrm{cell}} = 0.8$ V, $pH_{2,\mathrm{an}} = 0.4$ atm, $pH_2O_{\mathrm{an}} = 0.6$ atm, $pO_{2,\mathrm{cat}} = 0.21$ atm, $T = 800$ °C. ASC Type A, FZJ-IC, no gas conversion.

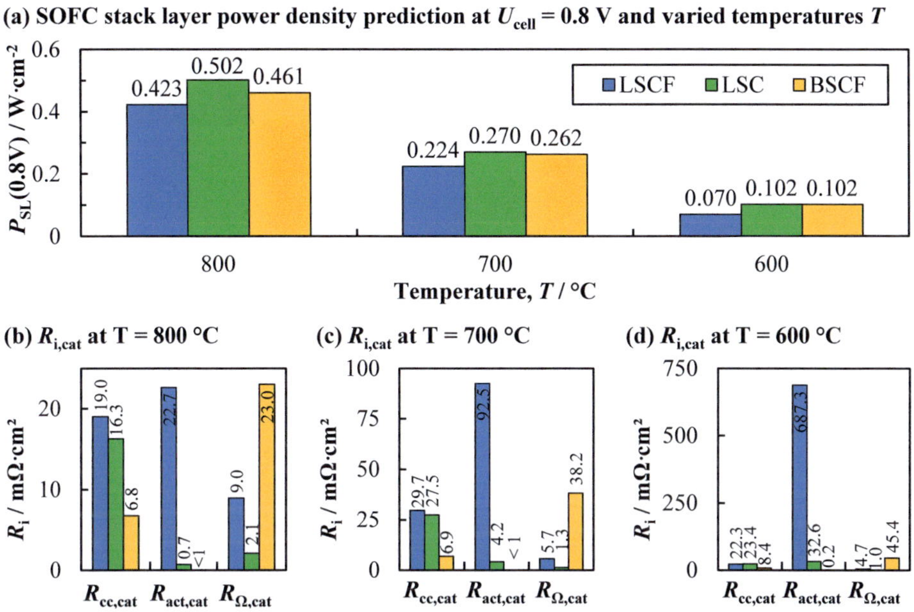

Figure 7.23.: Model predictions for various cathode materials (or electrochemical properties) at varied operating temperature, T. (a) Stack layer power density, P_{SL} and (b) cathode losses, R_i. Operating conditions: U_{cell} = 0.8 V, $pH_{2,an}$ = 0.4 atm, pH_2O_{an} = 0.6 atm, $pO_{2,cat}$ = 0.21 atm. ASC Type-A, FZJ-IC+$ASR_{contact}$, no gas conversion.

Figure 7.23 displays the calculated stack layer performance and resulting cathode loss predictions at U_{cell} = 0.8 V and various temperatures T = 800 . . . 600 °C. The results show that at higher temperatures LSC as cathode material delivers the highest performance, while BSCF produces -8.2% and LSCF -15.7% less power (Figure 7.23a). Towards lower temperatures, the performance loss using LSCF increases up to -31.4% compared to LSC, while the gap between BSCF and LSC decreases down to equal performance at T = 600 °C. The explanation is given in Figure 7.23b by the depicted cathode losses. BSCF creates the highest ohmic loss ($R_{\Omega,cat}$) and depletes thereby its advantage gained by the low diffusion ($R_{cc,cat}$) and activation loss ($R_{act,cat}$). LSC produces almost the same gas diffusion losses as LSCF, but only very low activation and ohmic loss. Similar observations account for lower temperatures, whereby the increasing performance loss of LSCF is accounted to the strong increase in activation loss and the diminishing performance loss using BSCF are explained by a stronger increase of activation loss using LSC and the decreasing impact of ohmic loss.

It can be concluded that not only high electrochemical kinetics, but also electronic conduction properties are of importance in the search of the next generation of cathode materials. Furthermore, these findings can only be made using a modeling approach such as presented in this work.

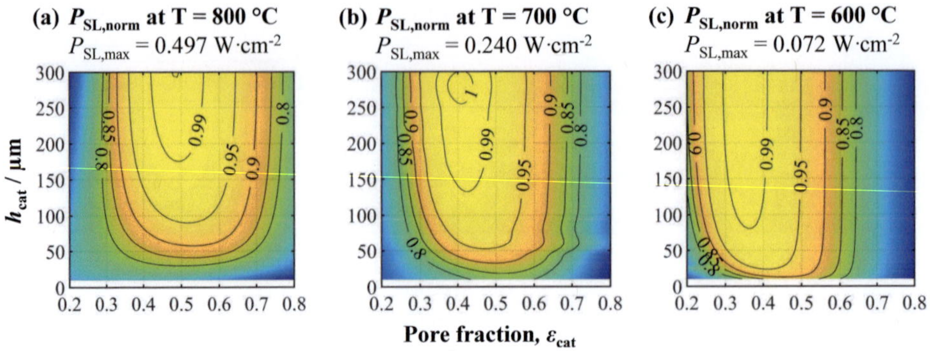

Figure 7.24.: Display of model predicted power density $P_{\mathrm{SL,norm}}$ at various temperatures $T = 800/700/600$ °C, normalized to the respective maximum, $P_{\mathrm{SL,max}}$, for varied cathode layer pore fraction, $\varepsilon_{\mathrm{cat}}$, and layer thickness, h_{cat}. Standard operating conditions at $U_{\mathrm{cell}} = 0.8$ V.

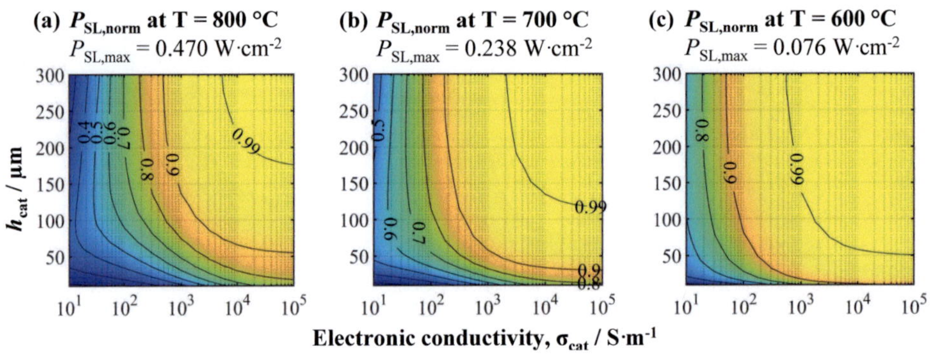

Figure 7.25.: Display of model predicted power density, $P_{\mathrm{SL,nom}}$, at various temperatures $T = 800/700/600$ °C, normalized to the respective maximum, $P_{\mathrm{SL,max}}$, for varied cathode layer thickness, h_{cat}, and electronic conductivity, σ_{cat}. Standard operating conditions at $U_{\mathrm{cell}} = 0.8$ V.

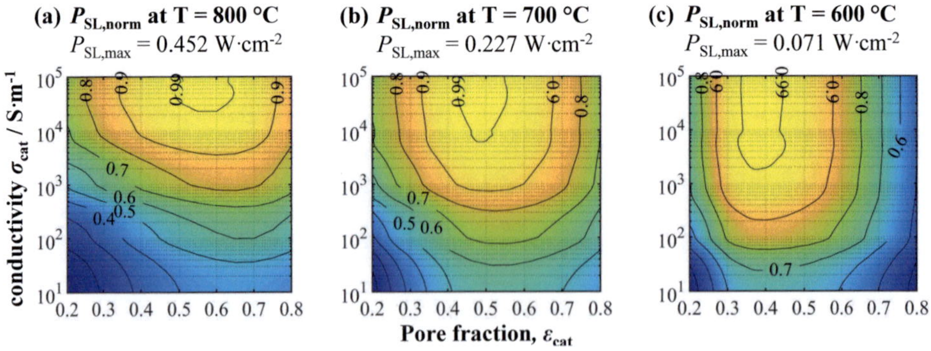

Figure 7.26.: Display of model predicted power density, $P_{\mathrm{SL,nom}}$, at various temperatures $T = 800/700/600$ °C, normalized to the respective maximum, $P_{\mathrm{SL,max}}$, for varied cathode layer pore fraction, $\varepsilon_{\mathrm{cat}}$, and electronic conductivity, σ_{cat}. Standard operating conditions at $U_{\mathrm{cell}} = 0.8$ V.

7.2.5. Parameter Interdependencies

In the previous subsections, the performance influences of individual cathode material and microstructure parameters were analyzed. However, due to the complex loss process coupling it is worthwhile to investigate parameter interdependencies. In the following, numerical results are presented where the layer thickness, h_{cat}, is varied with the pore fraction, ε_{cat}, and the electronic conductivity, σ_{cat}. Furthermore, ε_{cat} is varied with σ_{cat}. The simulations are performed at $T = 800/700/600$ °C, standard gas settings and $U_{\text{cell}} = 0.8$ V. The 2D-RPU model geometry of the Jülich F-design was used (Figure 7.10). The results are presented by displaying the stack layer power density, normalized to the maximum predicted at the regarding temperature.

Layer Thickness / Pore Fraction

Simulations results carried out on real cathode microstructure geometries in the work of J. Joos [120] revealed a minimum of cathode activation loss at a pore fraction of $\varepsilon_{\text{cat}} \approx 0.35$. The maximum penetration depth of the electrochemical reaction was thereby < 5 µm. However, these results regarded ideal contacting conditions. The results in Figure 7.24 show that, under stack contacting, a different setting is required, depending on the operating temperature. At $T = 800$ °C (Figure 7.24a), a minimum layer thickness of $h_{\text{cat}} > 175$ µm is required, combined with a pore fraction of $\varepsilon_{\text{cat}} \approx 0.5$, in order to achieve the maximum power density. Towards lower operating temperatures at $T = 700$ °C (Figure 7.24b), the optimum shifts to thinner and less porous layer at $h_{\text{cat}} > 130$ µm and $\varepsilon_{\text{cat}} \approx 0.43$. At $T = 600$ °C (Figure 7.24c), a thickness of $h_{\text{cat}} > 80$ µm and porosity of $\varepsilon_{\text{cat}} \approx 0.36$ are required to achieve maximum power output. The increasing ohmic and electrode activation loss limit the producible current density at lower temperatures, therefore lower gas concentration and current density gradients occur in-plane within the cathode layer. Even though the contact is still non-ideal, ideal conditions with decreasing temperature are approached regarding the optimum pore fraction value. In summary, the predicted power loss is $5 - 8\%$ if the cathode layer is optimized for high temperatures and operated at low temperature or *vice versa*.

Layer thickness / Electronic Conductivity

The electron conduction properties play an important role as the results presented in Sections 7.2.2 and 7.2.4 show. A slowdown of electrochemical activity due to gas starvation in the rib area of the cathode causes an increased electrochemical conversion rate in the better supplied channel areas. So, more electrons have to be conducted. Otherwise the reaction rate will be hindered and the overall power density decreases further. It is therefore easy to comprehend that the performance increases with increasing layer thickness, $h_{\mathbf{cat}}$, and electronic conductivity, $\sigma_{\mathbf{cat}}$. However, this only accounts for $T = 800$ °C as the results in Figure 7.25a show. The maximum power density is achieved only at $\sigma_{\mathbf{cat}} > 5e4$ S/m and $h_{\text{cat}} > 250$ µm or $\sigma_{\mathbf{cat}} > 1e4$ S/m and $h_{\text{cat}} > 175$ µm. The decreased electrochemical activity (due to increased ohmic and electron activation loss at lower temperatures) are the cause for lower current production and therefore the decreased influence of electronic conduction properties. While at $T = 700$ °C (Figure 7.25b) the influence is still notable with up to $5 - 10$ % power loss for any cathode material with a $\sigma_{\mathbf{cat}} > 1e3$ S/m and $h_{\text{cat}} > 50$ µm, the influence becomes insignificant at $T = 600$ °C (Figure 7.25c). It should be kept in mind however that these results are achieved using the electrochemical kinetics of LSCF, and it has been demonstrated in the previous Section 7.2.4 that the influence of $\sigma_{\mathbf{cat}}$ becomes relevant at lower temperatures for materials like BSCF, which feature high electrochemical kinetics but low electron conduction properties.

Pore fraction / Electronic Conductivity

This last section analyzes how electronic conductivity and pore fraction interdependently affect SOFC stack performance, while the layer thickness is kept constant. A smaller pore fraction corresponds naturally with a higher material fraction, therefore better in-plane conduction is assured. However, less pore volume increases the gas transport limitation and more electrons have to be transported in-plane. Hence, the numerical results at $T = 700$ °C (Figure 7.26a) predict at conductivities above $\sigma_{\mathbf{cat}} > 2e4$ S/m the best performance for pore fractions between $\varepsilon_{\text{cat}} = 0.5 - 0.6$. As in the previous comparisons, the influence decreases towards lower temperatures, but, the optimal pore fraction does not shift to the extent seen in Figure 7.24. At $T = 700$ °C (Figure 7.26b) it remains between $\varepsilon_{\text{cat}} = 0.45 - 0.55$ and at $T = 600$ °C (Figure 7.26c) between $\varepsilon_{\text{cat}} = 0.35 - 0.45$. In general, the influence of σ_{cat} is relevant, especially at higher operating temperatures. At $T = 800$ °C the performance loss can amount up to $20 - 30$ % for $\sigma_{\mathbf{cat}} = 1e3$ S/m, while the loss decreases down to $10 - 20$ % at lower temperatures, always depending on the pore fraction. However, the results are achieved (as in the previous section) with the electrochemical kinetic of LSCF, therefore the same considerations as above also account here.

Thus, the performance influence investigation of relevant cathode materials and microstructural parameters is completed. In the next section, numerical results cover the influence of cell contact properties on SOFC stack layer performance.

7.3. Influence of Interconnector Parameters

In order to increase the overall output, single cells are connected in series into stack units. With the help of interconnectors (ICs, also called bipolar-plates), the electrical interconnection between the individual stack layers is ensured. Simultaneously, the ICs feature gas channels in order to supply fuel and oxidant to the porous electrodes. Section 2.5 gives further details regarding design and additional loss process derivation induced by the use of ICs. The following results regard the planar IC design, based on the F-design employed as standard at Forschungszentrum Jülich (Figure 7.10). Consequently, an alternating flowfield design with ribs and channels exists only at the cathode. Meanwhile, at the anode, a course Ni-mesh functions as simultaneously electrical contacting and as gas channel. Hence, this work first analyzes the performance influences of IC parameters with according cathode rib (b_{RB}) and channel width (b_{CH}). Please note, due to the RPU model's inherent symmetry feature, both values denote half of the actual width. Because the contact between IC rib and cell surface is not ideal [37, 85], a certain contact resistance ($ASR_{contact}$, Section 4.4.2) has to be taken into account. Additional coatings against chromium evaporation and oxidation of the metallic IC structure further increase $ASR_{contact}$, wherefore its influence is analyzed as well in this section.

The following operating conditions were set as default: $U_{cell} = 0.8$ V, 60 % humidified H_2 as fuel and ambient air as oxidant (both without gas conversion), and the temperature is varied between $T = 800 \ldots 600$ °C.

7.3.1. Flowfield Design

The results presented in the previous sections revealed that gaseous O_2 transport and electronic in-plane conduction are coupled and, at the same time, competing loss processes. Optimizing a planar IC contact design with its parameters rib and channel width (b_{RB} and b_{CH}) against the minimization of one process will increase the other and vice versa. Hence, a stepwise variation of b_{RB} and b_{CH} is performed, with the 2D-RPU model based on the standard, planar, Jülich F-design as cathode IC, combined with a Type-A ASC (Figure 7.10) as reference case. b_{RB} and b_{CH} were varied independently with the constrain $b_{RPU} = b_{RB} + b_{CH}$. The employed parameter matrix contains 324 elements for each simulation result in order to achieve a meaningful resolution. All simulations are performed at $U_{cell} = 0.8$ V in order to maintain a meaningful comparison of the results from different operating temperatures.

Figure 7.27 shows how oxygen partial pressure $pO_{2,cat}$ and cathode charge transfer current density $j_{ct,cat}$ distributions at the cathode/electrolyte interface are affected by varying $b_{CH} = 0.05 \ldots 2$ mm while keeping $b_{RB} = 0.75$ mm constant. $pO_{2,cat}$ (Figure 7.27a) exhibits the familiar drop due to gas transport limitations in the rib area, with a very mild increase for increasing b_{CH} and a constant distribution beneath the gas channel area. In contrast, Figure 7.27b reveals a strong influence on $j_{ct,cat}$. In the rib area $j_{ct,cat}$ mildly increases due to better O_2 supply, while the increasing b_{CH} causes $j_{ct,cat}$ to drop. In-plane conduction limitations certainly effect the electron supply in the zones most distant from the contact rib. Furthermore, an overall decrease of the stack layer current density can be observed. Keeping $b_{CH} = 0.75$ mm constant and varying $b_{RB} = 0.05 \ldots 2$ mm in the same manner, a different effect on the predicted species distributions is revealed. As the results in Figure 7.28 show, $pO_{2,cat}$ remains constant beneath the channel and drops with increasing b_{RB} due to limited gas transport properties (Figure 7.28a). However, $j_{ct,cat}$ drops in the area beneath the rib for increasing b_{RB}, but also for smaller b_{RB} in the channel area (Figure 7.28b). The results displayed in Figures 7.27 and 7.28 imply that neither small contact ribs nor wide gas channels are beneficial for a high total power output, but a certain ratio between both parameters is required.

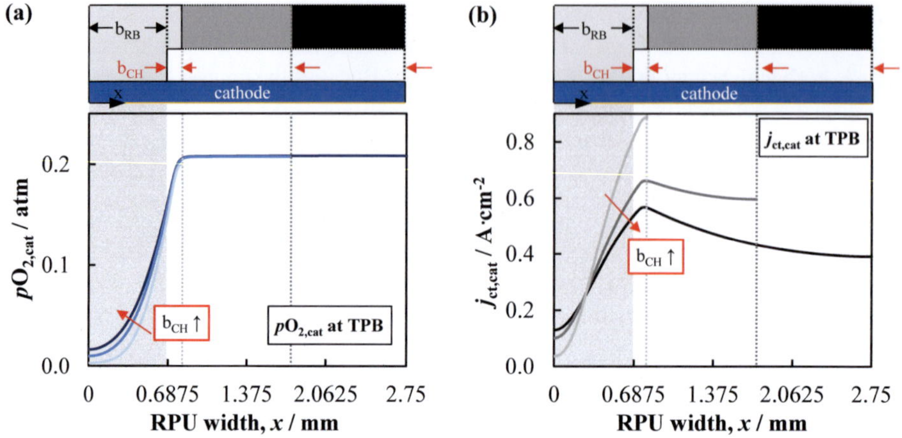

Figure 7.27.: Calculated species distributions at cathode/electrolyte interface (TPB) for varied IC gas channel width $b_{\mathrm{CH}} = 0.05 \ldots 2$ mm and constant contact rib width $b_{\mathrm{RB}} = 0.75$ mm. For each color, b_{CH} is up to the arrow indicated above. (a) Oxygen partial pressure $p\mathrm{O}_{2,\mathrm{cat}}$ and (b) charge transfer current density j_{ct} distribution. Operating conditions: $U_{\mathrm{cell}} = 0.8$ V, $p\mathrm{H}_{2,\mathrm{an}} = 0.4$ atm, $p\mathrm{H}_2\mathrm{O}_{\mathrm{an}} = 0.6$ atm, $p\mathrm{O}_{2,\mathrm{cat}} = 0.21$ atm, $T = 800$ °C. ASC Type A, FZJ-IC, no gas conversion.

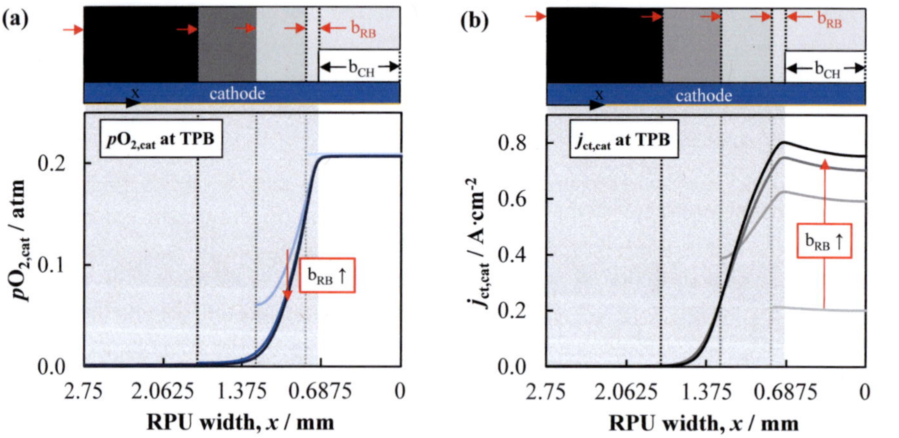

Figure 7.28.: Calculated species distributions at cathode/electrolyte interface (TPB) for varied IC contact rib width $b_{\mathrm{RB}} = 0.05 \ldots 2$ mm and constant gas channel width $b_{\mathrm{CH}} = 0.75$ mm. For each color, b_{RB} is up to the arrow indicated above. (a) Oxygen partial pressure $p\mathrm{O}_{2,\mathrm{cat}}$ and (b) charge transfer current density j_{ct} distribution. Operating conditions: $U_{\mathrm{cell}} = 0.8$ V, $p\mathrm{H}_{2,\mathrm{an}} = 0.4$ atm, $p\mathrm{H}_2\mathrm{O}_{\mathrm{an}} = 0.6$ atm, $p\mathrm{O}_{2,\mathrm{cat}} = 0.21$ atm, $T = 800$ °C. ASC Type A, FZJ-IC, no gas conversion.

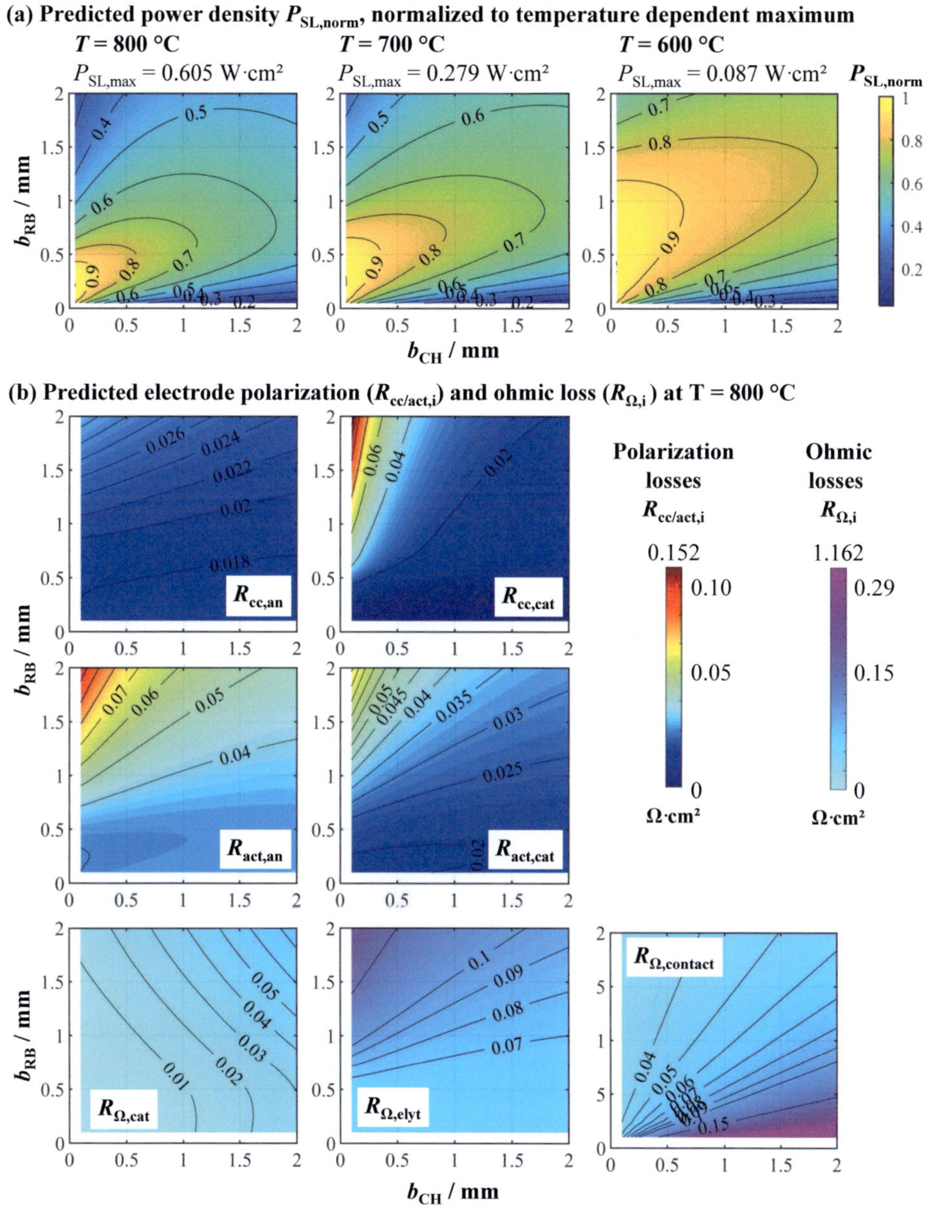

Figure 7.29.: Predicted (normalized) stack layer power density $P_{\mathrm{SL,norm}}$ at varied IC flowfield contact rib b_{RB} and channel width b_{CH}) at $T = 800 \ldots 600$ °C. (b) From simulation result at $T = 800$ °C calculated $R_{\mathrm{cc},i}$ (gas diffusion) and $R_{\mathrm{act},i}$ (electrode activation) loss distribution maps and (c) $R_{\Omega,i}$ (ohmic loss) distribution maps.

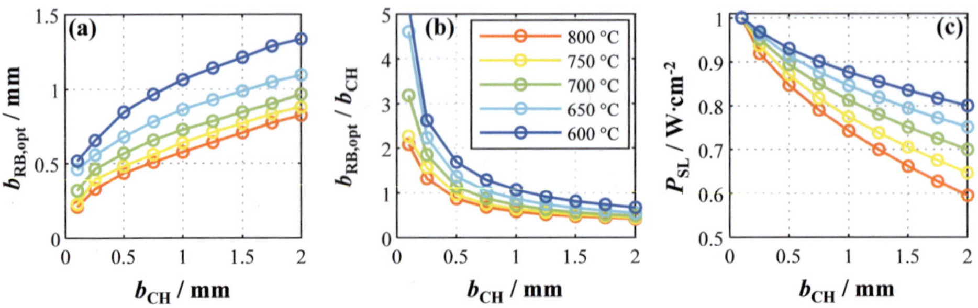

Figure 7.30.: Model determined optimum values for (a) interconnector flowfield gas channel width b_{CH} in dependence to contact rib width b_{CH}, (b) b_{CH} in dependence to channel/rib ratio and (c) corresponding predicted power densities using the IC parameters displayed in (a). Varied operating temperature, H_2 as fuel with 60 % steam and ambient air as oxidant. ASC Type-A (Table 5.1) with MCF as protective coating (Section 5.6).

Figure 7.29a displays the model predicted power density $P_{\mathrm{SL,norm}}$ for varied IC flowfield geometry at $T = 800/700/600$ °C, normalized at each operating temperature to the regarding power density maximum, $P_{\mathrm{SL,max}}$. The displayed results imply that $P_{\mathrm{SL,max}}$ is achieved at very small $b_{\mathrm{RPU}} = b_{\mathrm{RB}} + b_{\mathrm{CH}} < 0.25$ mm with a tendency to wider ribs as channels. This accounts for high and lower operating temperatures, while at $T = 800$ °C the IC geometry impact on P_{SL} is certainly larger due to the wider range in predicted $P_{\mathrm{SL,norm}}$ in both axis directions. Hence, a more detailed analysis is presented for $T = 800$ °C in order to gain further insights and identify performance limiting loss mechanisms.

Figure 7.29b displays calculated polarization loss maps while Figure 7.29c displays ohmic loss distribution maps. Although all processes occur simultaneously they are plotted by different colormaps in order to appropriately regard the varying data range. Within Figure 7.29b and Figure 7.29c, the data range marked by the colorbar on the right is kept equal. Thus it is easy to identify that gas diffusion loss occurring at the anode ($R_{\mathrm{cc,an}}$) is not highly affected by changes of the IC geometry and only plays a minor role in the overall polarization, whereas anode activation loss ($R_{\mathrm{act,an}}$) increases drastically at small channels and wide ribs. The same behavior is deduced for cathode gas diffusion ($R_{\mathrm{cc,cat}}$) and cathode activation loss ($R_{\mathrm{act,cat}}$). This behavior is caused by the gas transport limitations and thereof decreased electrochemical activity beneath the rib (Section 7.1.2). In the group of polarization losses, $R_{\mathrm{act,an}}$ has the largest effect, and surprisingly $R_{\mathrm{cat,an}}$ has only a minor impact, however the elevated operating temperature should be kept in mind (Section 7.2.4). Only under a very thin channel and wide rib IC design does $R_{\mathrm{cc,cat}}$ exhibit the largest loss. Solely based on the polarization loss distribution map analysis, it would make sense to keep $b_{\mathrm{RB}} < 1.0$ mm and choose an arbitrarily value for the channel width below $b_{\mathrm{CH}} < 1.5$ mm. Nevertheless, as the power density prediction distributions in Figure 7.29a already imply, ohmic loss has to be regarded as well. In fact, the maximum loss of $R_{\Omega,i}$ exceeds $R_{\mathrm{cc/act,i}}$ by almost a factor of 8. Analyzing the individual ohmic loss contributions, electronic in-plane conduction limitations induced loss, $R_{\Omega\mathrm{cat}}$, increases with a linear dependency towards symmetrically increasing b_{RPU} and for $b_{\mathrm{RB}} > 0.5$ mm. Wider channels cause longer electronic transport pathways while wider ribs increase the electrochemical activity in the channel area due to the limited gas transport beneath the ribs (Section 7.2.2). The highly electronically conducting LSCF as cathode material causes $R_{\Omega \text{cat}}$ to contribute the smallest amount to the total ohmic loss in this analysis.

Ohmic loss created by non-ideal ionic transport in the electrolyte increases independently from the chosen b_{CH} above $b_{\text{RB}} > 0.5$ mm. The same chain of arguments apply here as for the increased polarization loss: thicker ribs cause inhomogeneous electrochemical reaction distribution due to gas transport limitations beneath the rib (Section 7.1.2). Ohmic contact loss, $R_{\Omega,\text{contact}}$, however is the process with potentially the highest contribution to the overall loss, especially in an IC flowfield design configuration with wide channels and thin ribs. The area-related increase of current transported within thinner ribs causes high $R_{\Omega \text{contact}}$ due the low electronic conductivity of applied protective coatings and oxide scale formation (Section 5.6). Wider ribs would be beneficial, but are on the other side disadvantageous due to the the increasing polarization loss.

Hence, a theoretical optimum exists for different flowfield configurations. With the help of a coded MATLAB routine, the numerical results displayed in Figure 7.29a are analyzed and for each b_{CH} an optimum $b_{\text{RB,opt}}$ can be determined, which prevails at the individual operating temperature. Figure 7.30a displays the according results. The plotted results show that $b_{\text{RB,opt}}$ increases with increasing b_{CH}, independently of the temperature T. Furthermore, at lower temperatures thicker ribs (increased $b_{\text{RB,opt}}$) are optimal according to the model predictions. Figure 7.30b depicts the rib/channel width ratio between the determined optimum values. It is shown that $b_{\text{RB,opt}} > b_{\text{CH}} \approx 0.5$ mm at high temperatures and that the optimum moves towards wider channels for lower temperatures. At $b_{\text{CH}} = 0.1$ mm, the ratio can account between 2 and 5. Finally, Figure 7.30c denotes the normalized power density predictions for the individual optimum IC design with the corresponding $b_{\text{RB,opt}}$-values displayed in Figure 7.30a. The highest performance is predicted for thin channels.

Figure 7.31.: Model based determined (a) optimal IC flowfield contact rib width $b_{\text{RB,opt}}$ in dependence to repeat unit width b_{RPU} and (b) corresponding predicted power densities using the IC parameters displayed in (a). Varied operating temperature, H_2 as fuel with 60 % steam and ambient air as oxidant. ASC Type-A (Table 5.1) with MCF as protective coating (Section 5.6).

From a stack developer point of view the information given in Figure 7.30 may not be sufficient. Optimizing a planar IC flowfield design requires optimized parameters with regard to b_{RPU} in order to account for the symmetrical design with alternating ribs and channels. Figure 7.31 displays the according results for varying RPU-width b_{RPU} with $b_{\mathrm{CH}} = b_{\mathrm{RPU}} - b_{\mathrm{RB,opt}}$. The results show that regardless of the operating temperature, T, the corresponding maximum power output is predicted for the smallest possible value of b_{RPU}, with ribs almost as wide as the RPU itself. With increasing b_{RPU} the optimum value $b_{\mathrm{RB,opt}}$ increases equally, but at a smaller rate and with a different gradient at each temperature. At higher temperatures b_{RB} increases with an almost linear trend and non-linearly, but with steeper gradient at lower T. The impact of an optimal IC geometry is certainly higher at elevated operating temperature as the increased drop in $P_{\mathrm{SL,norm}}$ shows in Figure 7.30b.

The potential performance gain for a standard Jülich F-design amounts up to 30% at $T = 800$ °C while at 700 °C the potential power increase is lowered to 20%. However, from a practical point of view it is rather challenging to produce a flowfield geometry with $b_{\mathrm{RPU}} = 0.1$ mm. Furthermore, a certain gas channel width is also necessary to ensure the required gas flow rate under acceptable pressure conditions. Nonetheless, these results revealed vast optimization potentials. Table 7.2 lists calculated $b_{\mathrm{RB,opt}}$ values and corresponding power densities, $P_{\mathrm{SL,max}}$, at selected RPU-width, b_{RPU}.

Table 7.2.: List of model based determinedoptimal IC flowfield parameters for varied operating temperature, H_2 as fuel with 60 % steam and ambient air as oxidant. ASC Type-A (Table 5.1) with MCF as protective coating (Section 5.6).

Value/Parameter	T / °C	RPU-width, b_{RPU} / µm					Unit
		250	500	1000	1500	2000	
$P_{\mathrm{SL,max}}$	800	590	535	475	429	392	mW·cm^{-2}
	700	280	265	240	222	208	
	600	87	86	81	77	74	
$b_{\mathrm{RB,opt}}$	800	227	347	502	635	784	µm
	700	240	452	616	744	856	
	600	240	490	858	1030	1162	
$b_{\mathrm{RB,opt}}/b_{\mathrm{RPU}}$	800	0.91	0.69	0.50	0.42	0.39	-
	700	0.97	0.90	0.62	0.50	0.43	
	600	0.98	0.98	0.86	0.69	0.58	

IC Design in Dependence to Cathode Layer Thickness

Sections 7.2.1 and 7.2.2 discuss how cathode layer thickness and electronic conductivity affects the power output of SOFC stack layers. Consequently, IC flowfield design parameter variation simulations were performed at varied h_{cat} and $\sigma_{\mathrm{cat}}^{\mathrm{eff}}$ in order to find an optimal rib design at respective parameter settings and various operating temperatures.

To anticipate the results for $\sigma_{\mathrm{cat}}^{\mathrm{eff}}$, there was no significant influence on the IC design identifiable. In the optimal parameter range, the influence of $R_{\Omega\mathrm{cat}}$ (which is the main loss connected to $\sigma_{\mathrm{cat}}^{\mathrm{eff}}$) on the overall resistance is too low to have an impact (Figure 7.29c). The situation might change for a different cathode material with higher electrochemical reaction kinetics (*e.g.* BSCF, Section 7.2.4), which will be discussed shortly in Section 7.4.

Figure 7.32 displays the results for varied cathode layer thickness $h_{\mathrm{cat}} = 10 \ldots 300$ µm and IC flowfield design with $b_{\mathrm{RB/CH}} = 0.25 \ldots 2$ mm at varied operating temperature $T = 800/700/600$ °C. The previously mentioned MATLAB algorithm was applied to identify the optimal flowfield contact rib width value, $b_{\mathrm{RB,opt}}$, at selected RPU-widths, b_{RPU}. The results are displayed in Figure 7.32 on the left. The stack layer performance, P_{SL}, at $U_{\mathrm{cell}} = 0.8$ V predicted for the identified $b_{\mathrm{RB,opt}}$ values are displayed in Figure 7.32 on the right. The results demonstrate again the importance of applying a cathode layer with a certain thickness in order to increase the power output. In Section 7.2.1 the underlying mechanism is discussed in detail while here it is demonstrated how an optimization of the IC flowfield design can improve SOFC stack layer performance further. The results show how for thin cathode layers the predicted $b_{\mathrm{RB,opt}}$ is small in order to decrease the gaseous transport limitations impact and how $b_{\mathrm{RB,opt}}$ increases for thicker cathodes in order to decrease the generated $R_{\Omega\mathrm{contact}}$. Furthermore, the range of $b_{\mathrm{RB,opt}}$ increases with increasing b_{RPU} and in general at decreasing operating temperatures as well. Most certainly, the results demonstrate the importance of an optimized IC flowfield design. For example, P_{SL} can be pushed almost up to 250 mW·cm^{-2} or further at $T = 700$ °C using $h_{\mathrm{cat}} = 200$ µm.

It should be noted that a certain gas channel width b_{CH} is required to ensure sufficient gas flow along the gas channel range. To solve this problem a different kind of modeling approach is required, covering the total stack layer flowfield with inlet and outlet pressure boundary conditions. Having the required minimum value for b_{CH} at hand, the here presented results can be used to determine $b_{\mathrm{RB,opt}}$ in dependence to h_{cat} at a desired operation temperature to maximize the power output.

Nevertheless, it is evident that $R_{\Omega,\mathrm{contact}}$ has a very significant influence on SOFC stack performance, especially at lower temperatures. Moreover, ASR_{contact} is very sensitive towards the applied materials and method of fabrication and one of the parameters responsible for long-term degradation rate [247]. Hence, the influence of ASR_{contact} on an optimized IC flowfield design is analyzed in the next section.

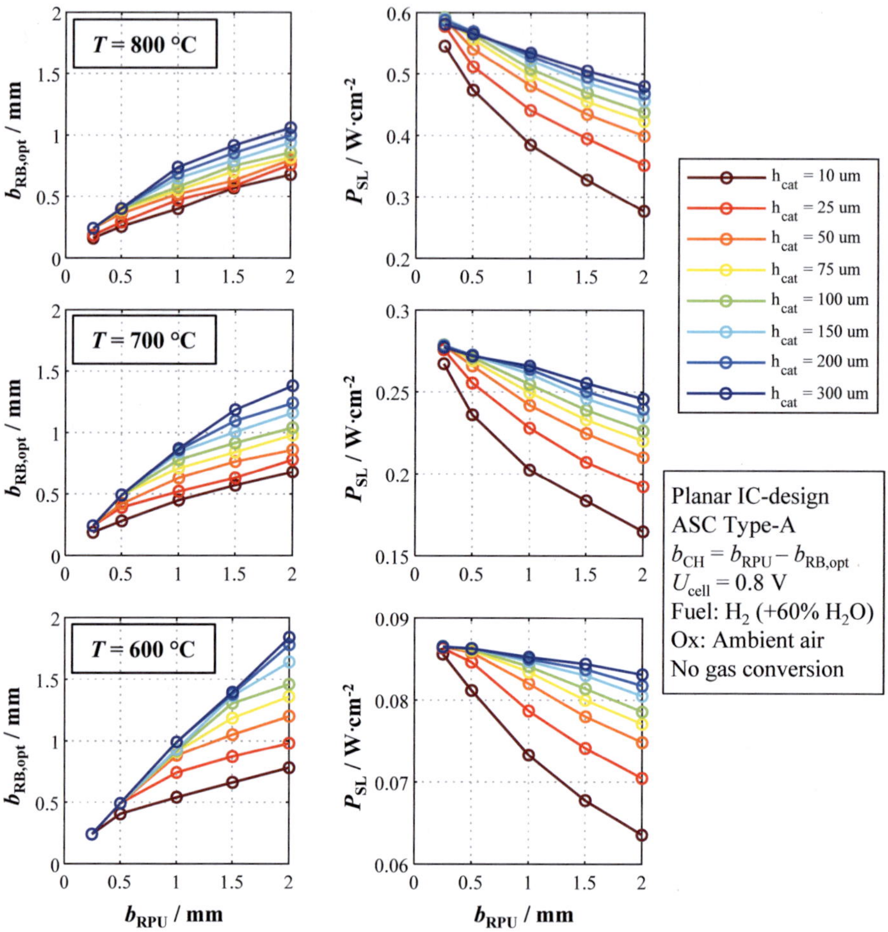

Figure 7.32.: Model based determined optimal IC flowfield contact rib width $b_{\mathrm{RB,opt}}$ in dependence to repeat unit width b_{RPU} and cathode layer thickness h_{cat} with corresponding predicted power densities. Varied operating temperature, H_2 as fuel with 60 % steam and ambient air as oxidant. ASC Type-A (Table 5.1) with MCF as protective coating (5.6).

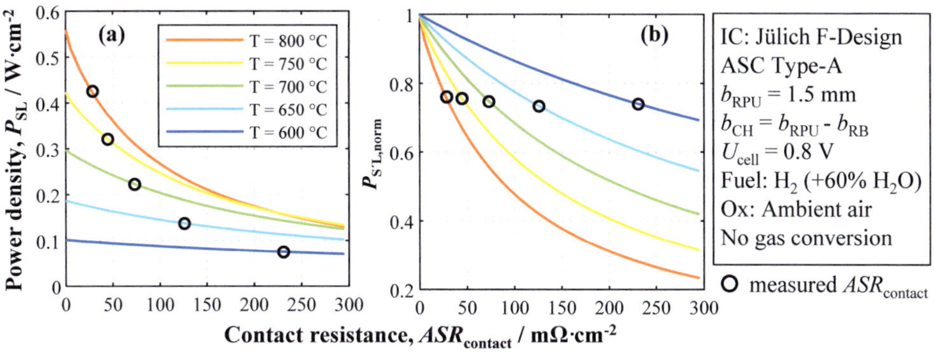

Figure 7.33.: SOFC stack layer performance at varied contact resistances, ASR_{contact}, predicted for various operating temperatures. The applied 2D-RPU model geometry is based on the Jülich F-design and an ASC Type-A. Operating conditions are stated in the figure above. (a) Power density P_{SL} and (b) temperature dependent normalized power density $P_{\text{SL,norm}}$. Black symbols highlight the ASR_{contact} values, measured at the corresponding temperature using MCF as protective coating [247].

7.3.2. Contact Resistance

The contact resistance, ASR_{contact}, arising from the poorly electronically conducting layers between the IC flowfield rib and cathode layer (Section 5.6), contributes significantly to the overall loss. The numerical results depicted in Figure 7.12 show that $R_{\Omega,contact}$ accounts at $T = 800$ °C and $U_{\text{cell}} = 0.8$ V for 25% of the total loss. Even though applied protective coatings prevent a fast corrosion rate, a complete prevention of oxide layer growth is not possible, therefore ASR_{contact} will increase over time [247]. The numerical results in Figure 7.33 demonstrate how SOFC stack performance decreases with increasing ASR_{contact}. The results are obtained using the 2D-RPU model parametrized with the standard Jülich F-design IC geometry and ASC Type-A at operating conditions stated in Figure 7.33. Marked by black circle symbols are the predicted values for P_{SL} (Figure 7.33a) and $P_{\text{SL,norm}}$ (Figure 7.33b) at the ASR_{contact}, measured in Jülich at the corresponding temperature. At higher operating temperatures the influence of ASR_{contact} is far more drastic, which results from the higher current density generated at equal cell voltage. Nevertheless, the displayed results show that even using the high performance protective coating from Jülich only $74 - 76\%$ of the maximum power density can be used. Alternatively, recently developed metallic coatings offer extremely low ASR_{contact} values [248].

Hence, bearing in mind the results presented in Section 7.3.1, it is of interested to examine further optimization potential regarding the IC flowfield design in dependence to ASR_{contact}. For this task the model framework using the 2D-RPU geometry is perfectly suited.

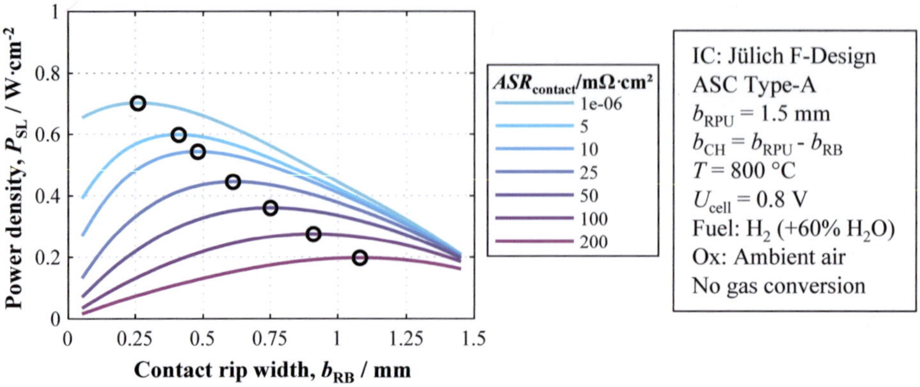

Figure 7.34.: Model predicted power density in dependence to b_{RB} and ASR_{contact}.

Displayed in Figure 7.34 are the predicted stack layer performance results for varied IC flowfield parameters (b_{RB} and b_{CH}) at different ASR_{contact}, while the RPU-width was kept constant at $b_{\mathrm{RPU}} = 1.5$ mm. The individual maxima for each parameter variation marked by black symbols highlight that a different optimum exists at each ASR_{contact}. It is beneficial to increase b_{RB} for ICs with higher contact resistance to minimize the local current density and thus the ohmic loss creation at the cathode/IC interface.

Figure 7.35 shows the 2D-RPU model predicted performance distribution maps for the IC flowfield parameter variations with different ASR_{contact} at $U_{\mathrm{cell}} = 0.8$ V and $T = 800$ °C. Each figure contains ~350 simulation results in order to supply a sufficiently high resolution, even though some of the contour lines do not follow a smooth path. However, it can be seen qualitatively from the color distributions and contour line pathways among the different figures that increasing the contact resistance has a significant impact, not only on the performance itself but on the optimal flowfield parameter composition.

The results displayed in Figure 7.35 are analyzed via MATLAB to identify $b_{\mathrm{RB,opt}}$ at specific b_{RPU}. Furthermore, simulations have also been carried out based on the same model geometry and parameter settings, but for $T = 700/600$ °C (and are analyzed in the same way via MATLAB). The outcome is displayed in Figure 7.36 for the individual temperatures. The highest performance can be achieved using the smallest possible RPU-width, however with the disadvantage of very high rib/channel ratios. As mentioned before, even if it is possible to fabricate such fine structures it is difficult to supply sufficiently high gas flow rates in such narrow gas channels. Similar to what was presented before, $b_{\mathrm{RB,opt}}$ increases with increasing b_{RPU}, but with decreasing performance. At $T = 800$ °C the distribution of $b_{\mathrm{RB,opt}}$ follows an almost linear course for increasing b_{RPU} whereas at lower temperatures and higher ASR_{contact} a non-linear evolution is revealed by the numerical analysis.

The problem for design optimization is that ASR_{contact} is not constant and will increase over time. Depending on the degradation rate and the maximum ASR_{contact}, which has to be expected over time, a certain IC flowfield design optimization is more beneficial. For example, if ASR_{contact} is low from the beginning, but increases quickly, it is advantageous to use a wider rib. Even though initial performance may be lower compared to what would be possible in the long run wider ribs will prevent an extensive performance drop due to increasing ohmic loss. Low contact resistance combined with a low degradation rate of ASR_{contact} can significantly increase the overall power output over time.

With this, the analysis of interconnector parameter influences on SOFC stack performance is completed. The presented numerical results demonstrate that it is crucial to pre-evaluate the stack contacting design in order exploit the full potential of SOFC stack performance. Exactly for this purpose, the 2D-RPU model framework is a perfectly suitable tool. In the next and final Section 7.4 of this thesis, an optimized SOFC stack design is presented based on the gained knowledge from this and the previous section.

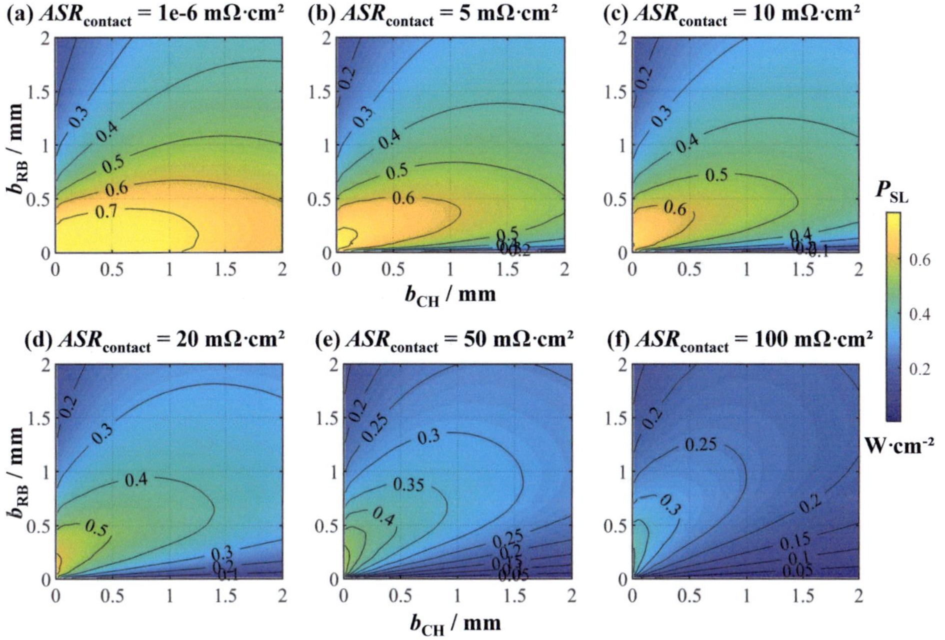

Figure 7.35.: Model predicted power density distribution maps for varied IC geometry at contact resistance ASR_{contact}. $U_{\mathrm{cell}} = 0.8$ V, varied operating temperature, H_2 as fuel with 60 % steam and ambient air as oxidant. ASC Type-A (Table 5.1)

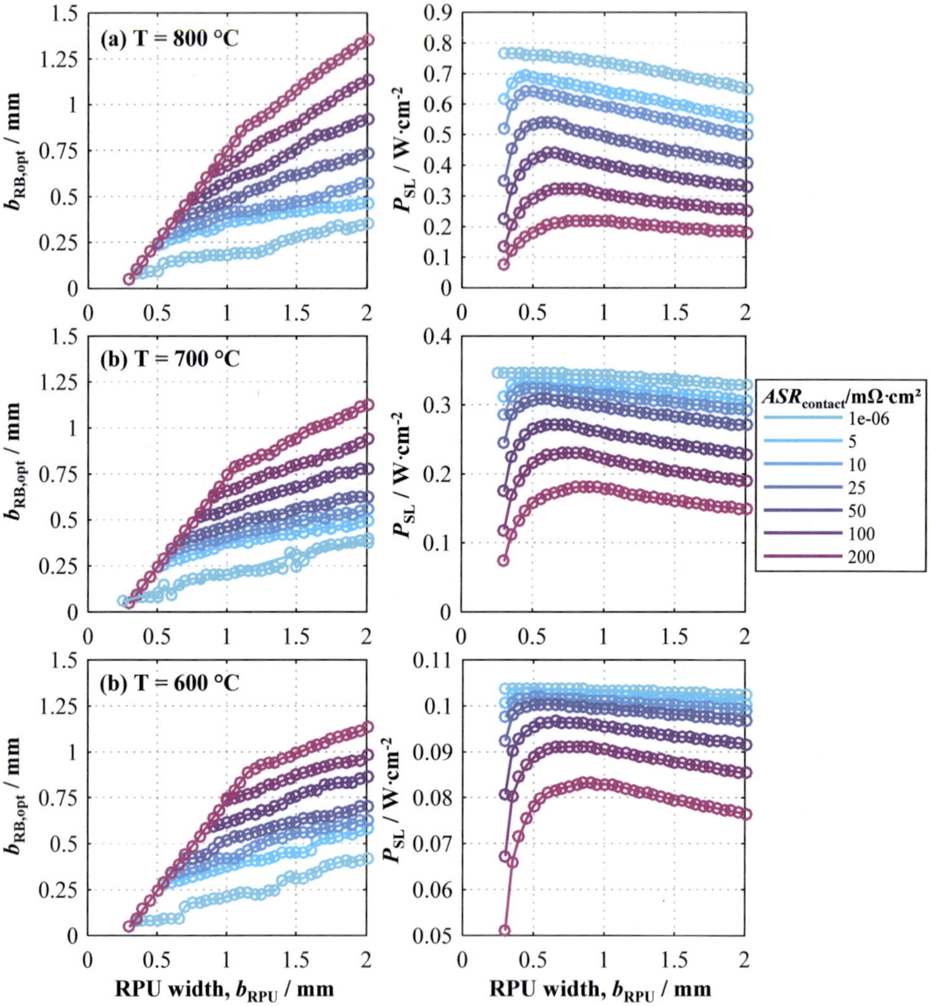

Figure 7.36.: Model based determined optimal IC flowfield contact rib width $b_{\mathrm{RB,opt}}$ in dependence to repeat unit width b_{RPU} and contact resistance ASR_{contact} with corresponding predicted power densities. Minimum gas channel width $b_{\mathrm{CH,min}} = 0.25$ mm. Varied operating temperature, H_2 as fuel with 60 % steam and ambient air as oxidant. ASC Type-A (Table 5.1) with MCF as protective coating (Section 5.6).

7.4. Model-based Performance Optimization

There is clearly more than one area in SOFC stacks with optimization potentials, however the results presented in the previous Sections 7.2 and 7.3 demonstrate that at the cathode side layer and interconnector design offer adequate optimization capacities. First, the application of an advanced current collector (CCL) design is numerically examined using the 2D-RPU model based on the standard Jülich IC design and ASC Type-A with an LSCF cathode. Subsequently, a model-based optimal flowfield design is predicted for different operating temperatures using the optimized CCL design in combination with a planar IC applied with the standard protective coating from Jülich. In the last part, a numerical study is presented using alternative MIEC cathode materials in combination with optimized IC design.

7.4.1. Advanced Current Collector Layer (CCL) Design

In Section 7.2.1 is shown how increasing the cathode layer thickness SOFC stack layer performance increases significantly and in Section 7.2.3, microstructure influences are described. Hence, it is straight forward to simply apply an additional layer on top of the standard, high performance LSCF cathode. The material does not need to be electrochemically active as long as it supplies sufficient pore volume and electronic conductivity. In terms of thermal expansion compatibility, this material could be LSCF itself, which could be screen printed layer-after-layer until the desired thickness is reached. However, this method uses a lot material thus increasing material cost, which is actual one of the most crucial factors to commercial SOFC systems. Furthermore, it is shown in Section 7.2.2 that LSCF supplies enough electronic conductivity in order to keep additional ohmic loss due to in-plane conduction limitations at bay. Thus, an advanced CCL design would be to apply an additional layer only in the area underneath the contact ribs of the IC flowfield. In fact, this method is already installed in state-of-the-art Jülich SOFC stacks (Figure 7.10), using LCC10 as CCL material, however with the mentioned downside of a very low pore volume due to the applied production method. According to researchers in Jülich, it is also possible to use screen printing as application method, for which the following results supply practical information regarding required layer thickness and microstructure properties.

Two 2D-RPU FEM models were set up based with equal IC and cell design as displayed in Figure 7.10, with additional, but different CCL designs. The resulting FEM geometries are displayed in Figure 7.37. The major difference (besides the mentioned layer discontinuity) in both models is the boundary condition setting for the gas inlet ($IF_{\mathrm{GC-cat}}$). In the advanced design, an additional interface between gas channel and CCL allows gas molecules to also enter and leave the porous CCL structure in x direction directly into the area beneath the contact rib while the cathode structure can be entered directly from the gas channel. In the standard CCL design, O_2 molecules have to first bypass the CCL structure before they can enter the cathode layer.

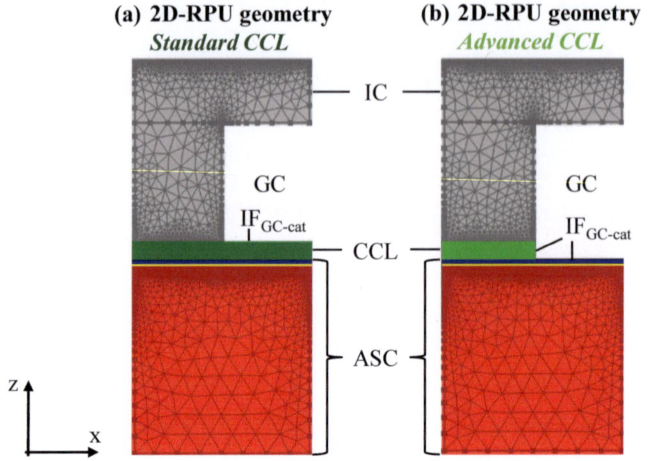

Figure 7.37.: Display of FEM model geometries with current collector layer: (a) Standard layout with full layer coverage over the entire cathode surface area and (b) the advanced layout, where the CCL covers only the area beneath the ICs flowfield contact ribs.

In the models, changes of microstructure properties were accounted for with the help of stochastic LSCF microstructure predictions (Sections 5.1.2 and 5.1.4). The predicted performance results displayed in Figure 7.38a present two conclusions: (i) At high pore volume fraction ($\epsilon = 0.6$, straight lines) performance increases equally for both CCL layouts up to ~ 18 % and converges at $h_{\mathrm{CCL}} = 200$ µm. Apparently, the advanced CCL layout does not offer any advantages besides the lower material proportion. (ii) At low pore volume fraction ($\epsilon = 0.3$, dotted lines) a difference between both CCL layouts is revealed: Power output increases with increasing h_{CCL} for the advanced model (red), however without any tendency to converge and without the same performance gain as with high porosity. In contrast, the model with standard CCL design (black) predicted power decrease after a small initial increase. This is explained by the inverse distribution of material and pore tortuosity (Figure 7.38), coupled with a higher sensitivity of diffusion loss to microstructure changes. Hence, the power loss originates from the increasing impact of cathode diffusion polarization due to the extended transport pathways in the CCL layer, which is demonstrated clearly by the individual loss distributions in Figures 7.38b and 7.38c. $R_{\mathrm{cc,cat}}$ (dotted, dark blue line) increases with growing h_{CCL} in Figure 7.38b while it decreases in Figure 7.38c. The remaining loss contributions show an equal distribution with slightly higher values at lower h_{CCL} for the standard layout, however with minor impact. Only $R_{\Omega,\mathrm{cat}}$ shows an opposing distribution with increasing values using the advanced layout, which is ascribed to the total material fraction inherent in the CCL design, however also with minor impact on the total loss due to the high electronic conductivity of LSCF. In summary, applying a CCL with lower pore fraction, the advanced CCL design clearly makes a difference. Ensuring a pore fraction above 60 %, the advanced design is only advantageous in saving material costs.

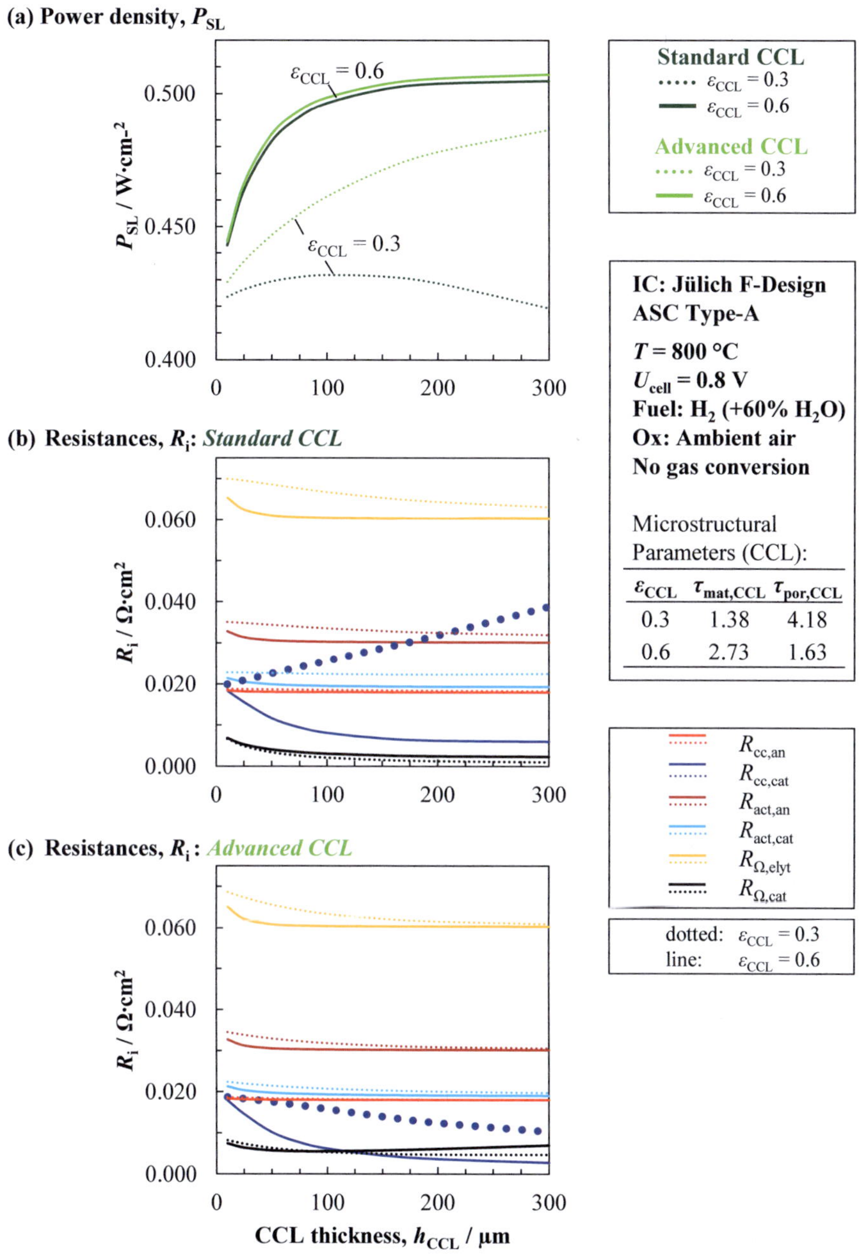

Figure 7.38.: (a) Performance predictions and (b/c) individual loss contributions using the 2D-RPU model with the geometries illustrated in Figure 7.37 for varied CCL pore volume fraction ε_{CCL} and layer thickness h_{CCL}. Constant contact resistance $R_{\Omega,contact} = 58\ \mathrm{m\Omega \cdot cm^2}$, due to the unchanged IC geometry.

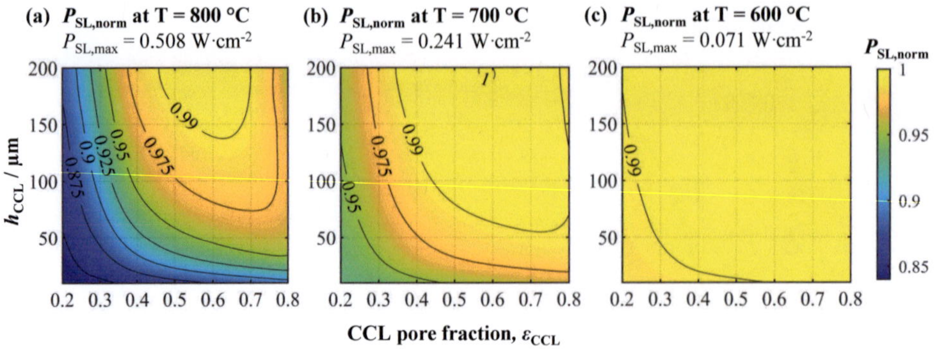

Figure 7.39.: Model predicted performance distribution maps using the 2D-RPU model with advanced CCL FEM geometry for varied CCL pore volume fraction $\varepsilon_{\mathrm{CCL}}$ and layer thickness. h_{CCL}. Operating conditions as displayed in Figure 7.38.

For reasons of completeness, the combined influence of $\varepsilon_{\mathrm{CCL}}$ and h_{CCL} at different temperatures on the advanced CCL layout was analyzed via the 2D-RPU model as well. The results are displayed in Figure 7.39 by distribution maps of the normalized power density $P_{\mathrm{SL,norm}}$, calculated according to the temperature dependent maximum $P_{\mathrm{SL,max}}$. The importance of a well-chosen CCL microstructure and thickness at elevated operating temperatures is clearly emphasized, while the influence becomes less important towards lower T. Due to decreased electrochemical activity, gas and electronic in-plane conduction limitations are less pronounced. However, it is demonstrated that by using an advanced CCL-layout at higher operating temperatures SOFC stack layer performance can be increased up to ~18 %. The pore volume fraction should therefor be $\varepsilon_{\mathrm{CCL}} > 0.5$ and the thickness above $h_{\mathrm{CCL}} > 150$ µm.

7.4.2. Optimal Flowfield Design for LSCF Cathodes

In a final numerical study regarding standard ASCs with LSCF-cathode, but with advanced CCL-layout, the IC flowfield design parameters are varied. Using the 2D-RPU model (Figure 7.37) with standard protective coating (MCF, Section 5.6), the flowfield contact rib (b_{RB}) and gas channel width (b_{CH}) are varied at various operating temperatures with standard fuel and oxidant supply.

Incremental variation of b_{RB} and b_{CH} in the model delivers individual performance distribution maps at a corresponding operating temperature. From these, according maximum values can be determined under selected preconditions. Figure 7.40a displays optimal IC contact rib width, $b_{\mathrm{RB,opt}}$, in dependences of b_{CH} with corresponding power densities, P_{SL}, depicted in Figure 7.40b. The analysis implies that the use of wider channels require increased rib dimensions, however with decreasing performance and with higher impact at high operating temperatures. The underlying mechanism is discussed in Section 7.3.1. At $T = 600$ °C the predicted $b_{\mathrm{RB,opt}}$ values are limited by the chosen parameter range in the simulation. However, the influence on P_{SL} is considered to be low. Assuming a minimum gas channel width of $b_{\mathrm{CH,min}} = 0.25$ mm, Figure 7.40c displays according values for $b_{\mathrm{RB,opt}}$ in dependence to b_{RPU}, with the corresponding values for P_{SL} depicted in Figure 7.40d. The linear increase of $b_{\mathrm{RB,opt}}$ in Figure 7.40c, up to a certain value of b_{RPU}, is caused by the chosen value for $b_{\mathrm{CH,min}}$. The geometrical limitation causes the performance drop for $b_{\mathrm{RPU}} < 1$ mm because wider ribs would be required for a further increased performance. If $b_{\mathrm{CH,min}}$ has to be set wider (due to stack inherent gas distribution requirements), the optimal value for b_{RPU} is shifted to higher values, and the predicted power densities would be lower. If $b_{\mathrm{CH,min}}$ can be even narrower, smaller RPUs can produce even higher power output.

Table 7.3 lists predicted maximum values for P_{SL} at various operating temperatures based on determined optimum IC flowfield parameters. Furthermore, the performance gain, ΔP, compared to what is predicted using the standard Jülich design is given in Table 7.3, thus demonstrating the value of this numerical study.

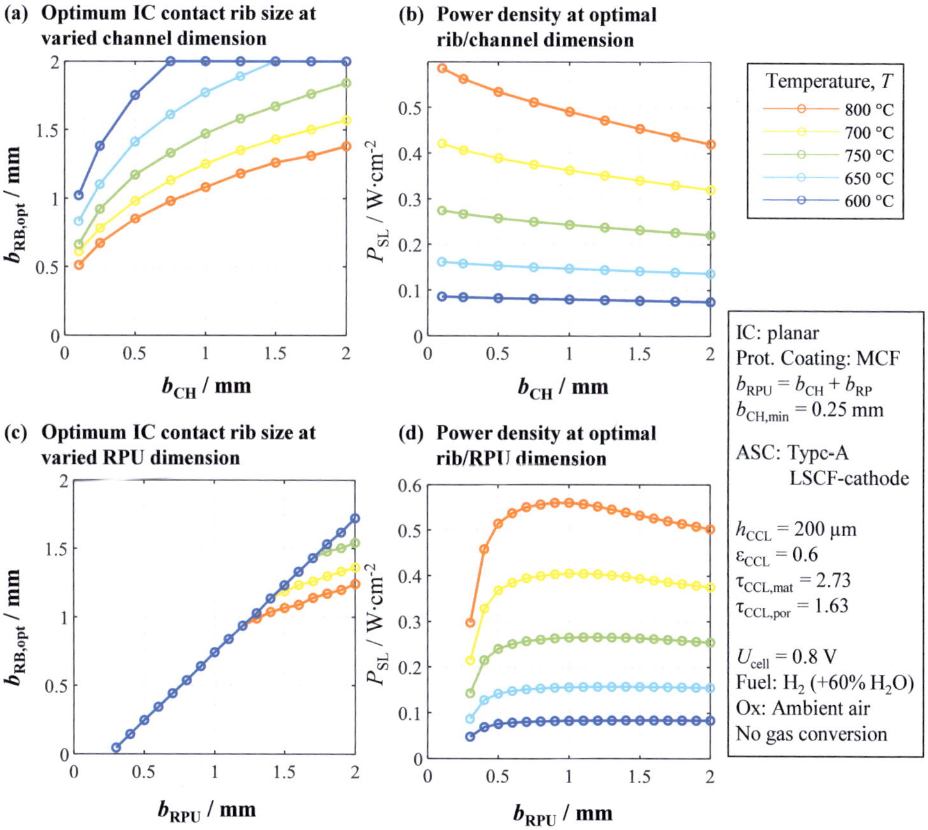

Figure 7.40.: Model predicted performance distribution maps using the 2D-RPU model with advanced CCL FEM geometry for varied CCL pore volume fraction $\varepsilon_{\mathrm{CCL}}$ and layer thickness. h_{CCL}. Operating conditions as displayed in Figure 7.38.

Table 7.3.: Model predicted SOFC stack layer power density P_{SL} and relative power density gain ΔP compared to standard Jülich stack design (Figure 7.10) at various operating temperatures T and $U_{\mathrm{cell}} = 0.8$ V, using a standard ASC Type-A with LSCF-cathode and advanced CCL and IC design with optimized parameters: $h_{\mathrm{CCL}} = 200$ µm, $\epsilon_{\mathrm{CCL}} = 0.6$, $b_{\mathrm{RPU}} = 1$ mm and $b_{\mathrm{RB}} = 0.75$ mm with $b_{\mathrm{CH}} = b_{\mathrm{RPU}} - b_{\mathrm{RB}}$. Hydrogen as fuel with 60 % steam and ambient air as oxidant, both with high gas flow rates. Standard Jülich contact resistance with MCF coating (Section 4.3.1.1).

Temperature, T °C	Power density, P_{SL} $\mathrm{W \cdot cm^{-2}}$	Rel. power density, ΔP %
800	0.560	$+32.5$
750	0.405	$+25.7$
700	0.265	$+18.4$
650	0.156	$+14.6$
600	0.083	$+18.9$

7.4.3. Outlook on SOFC Stack layer Performance

Alternative MIEC-materials with increased electrochemical performance are currently being developed and highly discussed in literature. With LSC and BSCF two promising candidates are already lined up, however with certain drawbacks in terms of stability against CO-containing atmospheres (BSCF) or layer thickness (LSC). J. Hayd has shown that very low cathode losses can be achieved using LSC with microstructure properties at the nano-range [45]. Hence, this last section is dedicated to taking a small look into the future. Using these MIEC-cathode candidates in the 2D-RPU model, the performance is predicted for $T = 600$ °C at varied IC geometries equipped with a metallic protective coating with very low ohmic resistance (Section 5.6). The chosen cathode and CCL parameters are listed in Table 7.4. The BSCF-model is equipped with an LSC-CCL to counter the poor electronic conductivity. The increased active surface area in the nano-LSC model is adopted from the assumptions of J. Hayed, while according $k\&D$ values are assumed from the data of standard LSC. This is certainly a simplification because further performance enhancements due to lattice interface interactions are disregarded [45]. Nevertheless, the study is meant to offer a theoretical view.

Figure 7.41 shows $b_{\mathrm{RB,opt}}$ at specific b_{RPU} on the left and corresponding P_{SL} predictions on the right. For the evaluation $b_{\mathrm{CH,min}} = 0.25$ mm is assumed. While the prediction for LSC and BSCF shows a certain limit for $b_{\mathrm{RB,opt}}$ the nano-LSC prediction for $b_{\mathrm{RB,opt}}$ is only limited by $b_{\mathrm{CH,min}}$. Apparently, the electrochemical conversion rate and oxygen ion bulk diffusion transport are high enough to tolerate very wide ribs. An optimum RPU is therefore not determined for the nano-LSC approach while models with LSC and BSCF show an optimum at $b_{\mathrm{RPU}} \approx 0.6$ mm.

The predicted performance and corresponding cathode loss at $b_{\mathrm{RPU}} \approx 0.6$ mm are depicted in Figure 7.42. The difference between all three models is comparatively low, whereby LSC produces the highest electrochemical loss and BSCF the lowest. Compared to the predictions using the Jülich standard IC design with MCF as protective coating and no CCL, the increase of 60 % promises further improvements off SOFC stack layer performance at low operating temperatures.

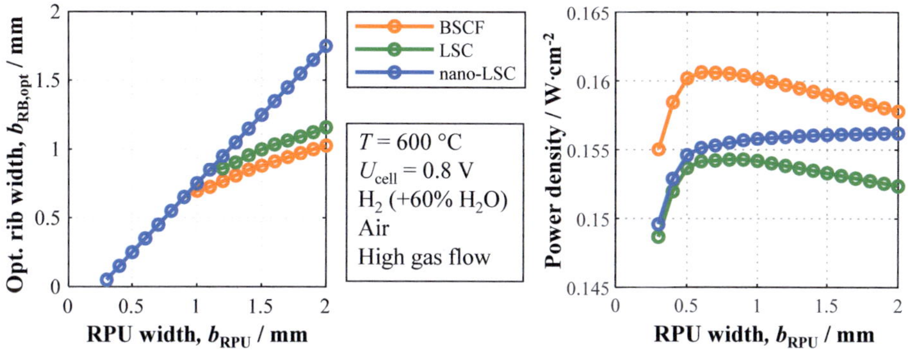

Figure 7.41.: Model based determined optimal IC flowfield contact rib width $b_{\mathrm{RB,opt}}$ in dependence to repeat unit width b_{RPU} using alternative MIEC-cathode material and corresponding predicted power densities. Minimum gas channel width $b_{\mathrm{CH,min}} = 0.25$ mm. $T = 600$ °C, H_2 as fuel with 60 % steam and ambient air as oxidant. Anode and electrolyte/interdiffusion layer of cell Type-A (Table 5.1) and metallic protective coating (Section 5.6).

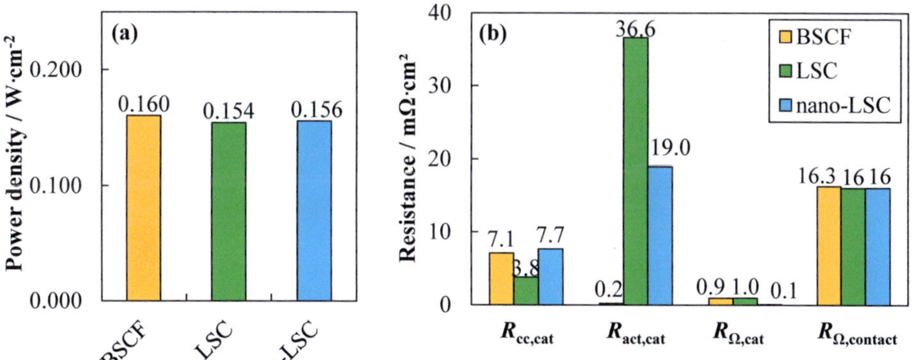

Figure 7.42.: (a) Calculated SOFC stack layer power densities using alternative MIEC-cathode materials and an optimized IC geometry with $b_{\mathrm{RPU}} = 0.6$ mm and $b_{\mathrm{CH}} = 0.25$, $T = 600$ °C, H_2 as fuel with 60 % steam and ambient air as oxidant. Anode and electrolyte/interdiffusion layer of cell Type-A (Table 5.1) and metallic protective coating (Section 5.6).

Table 7.4.: List of parameters used in the 2D-RPU model to calculate possible SOFC performance at $T = 600\ °C$ using varied MIEC-cathode materials.

Layer	Parameter	Unit	MIEC-model		
			LSC	nano-LSC	BSCF
CCL	material	-	LSC	LSC	LSC
	type	-	advanced	standard	standard
	layer thickness, h_{CCL}	µm	200	40	40
	pore fraction, ϵ_{CCL}	-	0.6	0.4	0.6
	pore tortuosity, $\tau_{\mathrm{por,CCL}}$	-	1.6	2.4	1.6
	material tortuosity, $\tau_{\mathrm{mat,CCL}}$	-	2.7	1.6	2.7
	mean pore diameter, $d_{\mathrm{por,CCL}}$	µm	1000	0.2	1000
Cathode	material	-	LSC	LSC [45]	BSCF
	layer thickness, h_{cat}	µm	20	0.2	20
	pore fraction, ϵ_{cat}	-	0.44	0.4	0.44
	pore tortuosity, $\tau_{\mathrm{por,cat}}$	-	2	2.4	2
	material tortuosity, $\tau_{\mathrm{mat,cat}}$	-	1.7	1.6	1.7
	mean pore diameter, $d_{\mathrm{por,cat}}$	µm	556	0.2	556
	active surface area, a_{cat}	μm^{-1}	2.88	1.47	2.88

8. Summary and Conclusions

The main goal of this thesis is to establish a detailed model framework based on the finite element method (FEM) to reliably predict Solid Oxide Fuel Cell (SOFC) performance operated under hydrogen or hydrocarbon-containing gases as fuels. Although many numerical approaches regarding SOFC performance prediction can be found in literature, none provides a comprehensive material parameter set, which has been exclusively determined by in-house experiments, or proves the validity of chosen model equations for sufficient reproduction of the cell-inherent loss mechanisms. In this work, a physically meaningful investigation of cathode and interconnector (IC) parameters was performed for the first time. Model-based parameter investigations are used to derive optimization potentials that can be transferred to commercial SOFC applications and enable an economically attractive alternative in the highly competitive energy market. On this basis, a useful framework is now available for SOFC researchers and manufacturers.

Experimental Methods

The SOFCs used to develop and validate this framework are high performance, state-of-the-art anode supported cells (ASCs), fabricated at Forschungszentrum Jülich. The ASCs are made of Ni/8YSZ cermet anode support and functional layer (AFL), 8YSZ-electrolyte, CGO-interdiffusion layer and LSCF cathode (Figure 3.2). Their 1 cm^2 active electrode area (A_1) and high gas flow rates ensure homogeneously distributed and well defined operating conditions (Section 3.2.1). Ideal gas supply and electronic contacting are ensured by the use of metallic contact meshes placed in between a Al_2O_3-flowfield and porous electrode surface (Figure 2.13). To simulate stack contacting conditions in an experiment, the cathode was contacted by an Au-flowfield (Figure 3.5f). Gas conversion measurements were recorded on ASCs with 16 cm^2 active electrode area (A_{16}) with subsequent gas analysis via gas chromatography (Section 3.2.2). All cells were operated for $\sim$300 h at $T = 800$ °C, with 60 % humidified H_2 as fuel and ambient air as oxidant, and at constant load $j = 1$ A/cm^{-2} before the characterization measurements were started to ensure stable cell behavior. In the measurements temperature, fuel and oxidant gas compositions were varied within an application relevant range.

Model Framework Features

The framework was implemented into COMSOL Multiphysics v4.3b, a commercial FEM software, but can be applied to any other software code that is able to create a numerical mesh for a chosen geometry on which the provided model equations can be solved. In this work, three different *model geometries* are derived based on their intended use (Section 4.2). Due to the symmetrical cell and flowfield design, the geometry and, therefore, the numerical mesh size can be reduced from covering a complete cell or even stack unit down to a repeat unit (RPU). The (i) 3D-RPU geometry can be further divided into a (ii) 2D gas channel (2D-GCh) and a (iii) 2D-RPU geometry (Figure 4.3). In this way, three model geometries are available for predicting SOFC performance, of which the 2D models require less computational effort but are only suitable under certain conditions (Section 6.5).

The relevant physical loss mechanisms covered by the adequate model equations are (i) electronic and ionic charge carrier transport (Section 4.3.1), (ii) gaseous species transport of binary (cathode) and multicomponent mixtures (anode) in open, semi-porous and highly porous media, (iii) electrochemical charge transfer reactions and (iv) fuel gas conversion via chemical reforming reactions (Section 4.3.4). Up to now, heat transport is excluded, out of two reasons: First, the occurring temperature gradients in the test-setup used to record data for validation purposes are < 5 °C under high polarization and their influence on cell-inherent losses is therefore not relevant. Secondly, in SOFC stacks larger temperature gradients are only expected along the gas channel length and can be regarded as constant in the relatively small section of the 2D-RPU model.

In Section 4.3.2, special attention is paid to the *gaseous species transport* description, which occurs in gas channels, semi-porous contact meshes and in the highly porous electrodes. The calculation of individual species fluxes consists of separate convective and diffusive parts, governed by individual momentum and species mass continuity balance equations. It is important to note that this work makes a distinction between mass and molar averaged flux calculation (Section 4.3.2.1), which is rarely mentioned elsewhere in literature and the influence of this on SOFC modeling has not yet been taken into account. However, it is shown in Section 6.3 that the use of model equations from different unit systems (mass and molar based) in the calculation of porous convective and diffusive fluxes in the anode leads to serious errors in the species distribution and thus in the associated calculation of diffusion overpotential if ideal gas law restrictions are taken into account. However, the influence is negligible in the gas channel domains, where the overall transport flux is governed by convection. Nevertheless, a physically correct calculation of individual species gas transport fluxes in porous media can be achieved by a consistent implementation of model equations from the same unit systems. The following porous media diffusion models with versatile implementation and effective transport parameters are presented (Section 4.3.2.3): The Fick Model (FM), the Dusty Gas Model (DGM) and the Mean Pore Transport Model (MPTM). Section 6.5 shows by a comparison of measured and simulated anode gas diffusion loss (R_{1A}), that the MPTM predicts the most accurate results. The DGM with mixed unit system flux terms predicts R_{1A} with a maximum deviation of $\sim 14\%$; it does, however, produce pH_2O_{an}-gradients in the anode substrate which is in opposition to the electrochemical reaction scheme (Figure 6.28). This can be accounted to erroneously predicted convective fluxes. The mass based FM employed under isobar conditions overestimates R_{1A} because its total flux calculation is dependent on the additional convective part (Figure 6.27). It is recommended to use the molar based DGM rather than the MTPM as its implementation is slightly less elaborate. The relative error using the FM-mass is still in the acceptable range and its use is recommended for fuel gas mixtures composed of more than 5 species where the required matrix inversion in the DGM flux computation becomes too expensive (Equation 4.44).

The *electrochemical charge transfer reactions* occurring at anode and cathode are represented in the framework as a reference model by the well-established Butler-Volmer approach (BVM) (Section 4.3.3.1). It is assumed that the electro-oxidation of H_2 in the presence of hydrocarbons in the fuel gas remains the dominant reaction mechanism due to the considerably higher kinetic activity. As an alternative charge transfer model for oxygen reduction at the cathode, an approach proposed by Rüger et al. [65] for mixed ionic-electronic-conductive (MIEC) materials is adopted and extended in this work (Section 4.3.3.2). In this way, a variable penetration depth of the electrochemical reaction into the cathode volume can be derived and the electrochemical kinetics of other MIEC-materials can be easily evaluated in the model framework. This is an important feature for designing cathode layers. A similar approach for the anode by resolution of the anode function layer (AFL) would be interesting for the future.

All occurring loss processes are coupled at the electrode/electrolyte interfaces by the electrochemical activation overpotential calculation (Section 4.4), thus governing the resulting cell voltage and current density prediction. At the interfaces, the local potential distribution of partial pressure dependent half cells and electronic/ionic potentials are correlated according to Kirchhoff's 2nd Law and the relevant current flow direction. This is in fact an advantage over other models, because their disregarding leads to erroneous calculation of the implicit quantity for the corresponding charge transfer reaction equation.

Any operation point can be simulated with the framework and the individual overpotentials of occurring loss processes can be resolved. A special feature of this work is that all differential resistances can be derived for any desired operating point (Section 4.4.2). In this way, simulation results can be directly confirmed by impedance measurements.

Parameterization

In Chapter 5 the acquisition of required material and model equation parameters is described in detail with applied numerical and experimental methods. Where neither was available, analytical approaches were compared with experimental or numerical data or (in selected cases) values were taken from reliable sources. The relevant points in this chapter are summarized below.

The microstructures of the semi-porous meshes and of the porous electrodes are regarded as homogeneously distributed and therefore represented by effective parameters (Section 5.1.5). Required mesh parameters are deduced from the manufacturer's data-sheet (Section 5.4). Characteristic electrode microstructural parameters are determined from 3D reconstructions using high resolution tomography methods (FIB/SEM) and accurate, in-house developed post-processing algorithms (Section 2.6, Section 5.1, Table 5.1). In addition, artificial LSCF cathode structures were created using a stochastic microstructure generator for LSCF-cathodes and 3D-FEM simulations were performed to derive analytical expressions to predict characteristic microstructural properties as a function of pore fraction (Equations 5.3 to 5.5). By comparing numerical results from the artificial cathode simulations and from experimental data, it was found that the Karman-Cozeny approach calculated by surface dependence delivers the best results for prediction of the electrode permeability (Section 5.1.8). Special attention is paid to the calculation of average pore diameters $d_{\mathrm{por,an/cat}}$, the distribution of which was determined by evaluation of 3D reconstructions (Section 5.1.6) and their mean value calculated by averaging the number of occurrences and by weighing the corresponding pore volume.

Furthermore, in a combined numerical and experimental approach the reaction kinetics to calculate gas conversion via the water-gas shift reaction was developed in this work, applicable to any relevant SOFC operation temperature and Ni-catalyst surface area (Section 5.5).

Model Validation

The sensitivity analysis revealed the most critical modeling parameters to be the activation energies for calculating the charge transfer polarization of the electrodes and the effective conductivity of the electrolyte/interdiffusion layer. The sensitivity of these parameters showed a strong temperature dependence, which is why model validation at different temperatures is of great importance. However, not only do individual parameters contribute to the validity of a model, but the implemented loss mechanism approaches do also. Furthermore, a parameter sensitivity may increase significantly under certain operating conditions where the according loss mechanism may rise to be the dominating loss. Hence, it is important to test the framework's ability to reproduce measured data not only at standard operating conditions but also far beyond. This is the only way to ensure that the framework reproduces measurements not by selection or randomness and can also be adapted to different geometries or parameters, while keeping the implemented loss process calculation physically correct.

In summary, the model framework predicts all measured data with high precession ($< 2.5\ \%$ relative deviation, Section 6.2). In particular, the high accuracy in the reproduction of the limiting current density at low $p\mathrm{H}_{2,\mathrm{an}}$-fuel content shows the value of the chosen gas transport model and anode microstructure parameter (Section 6.3). It is interesting that measured data recorded under reverse operation (SOEC-mode) at varied humidification can also be precisely reproduced by the framework (Figure 6.9). A more detailed analysis at various operating conditions and for CO-electrolysis operation with additional hydrocarbons contained in the fuel gas is planned for the near future. A slightly increased deviation is given for C/V predictions for operation at lower temperatures under reformate as fuel for A_1-and for A_{16}-cells in general (Section 6.2). This is accounted to an increased influence of the isothermal approach at lower temperatures and the increased electrode size. At the point of writing, an enhancement of the model is undertaken by implementing the energy balance equation into the framework in order to accordingly regard temperature dependencies on occurring loss processes. It has also been found that the 2D-RPU model geometry in its default configuration with a fixed inlet gas concentration is insufficient to reproduce the measured cathode gas diffusion resistance R_{1C} and recorded C/Vs below $pO_{2,cat} < 0.1$ atm (Section 6.4). Implementing a current-depending function as gas inlet boundary condition to regard gas conversion significantly improved the model accuracy (Section 6.5). Furthermore, it was found that regarding a temperature induced deformation of the contact mesh, resulting from long hours of operation, further improved the $pO_{2,cat}$ dependent loss prediction by the 2D-RPU model. Employing the 3D-RPU geometry delivers the most accurate results, but this increases the required computational time of 3 sec (2D-RPU) by a factor $84 - 1050$, depending on the numerical mesh resolution (Table 6.5). Another work in progress is to apply the model equations to a numerical mesh based on a real electrode microstructure in 3D and compare the calculated performance results with the homogenized approach applied in this thesis.

A detailed analysis of the influences of different microstructure parameters revealed the inapplicability of the mean pore diameters, determined from *number of occurrence* averaging (Section 6.5). Furthermore, a deviation is given for $\tau_{por,an}$, which is accounted to the inhomogeneous pore size distribution in the anode substrate and a certain margin of error in the 3D electrode reconstruction process. It is assumed that the major gas transport flux takes place in the larger pores, which are underrepresented in the reconstructed volume.

Furthermore, it was found that the gas transport model combination of FM-mass + Darcy Equation is not suitable for predicting SOFC stack-layer performance reliably when O_2 depletion is encountered. Due to an incorrect pressure calculation caused by the required boundary condition of the Darcy equation, the balance equations no longer consider the required constraints. The simulation then results in an error; fortunately, the molar based DGM approach predicts a physically plausible solution (Section 7.1.2.1).

2D-RPU Modeling Results

It is one of the goals of this thesis to analyze how SOFC performance is influenced by *stack contacting* (Section 2.5.3). Interconnectors (ICs) are used in the application to combine single SOFCs to a stack unit (Section 2.4.2) and thereby increase the overall power output. Unless mentioned specifically, operating conditions were fixed in measurements and performed simulations to $U_{cell} = 0.8$ V, $T = 800$ °C, $pH_{2,an} = 0.4$ atm, $pH_2O_{an} = 0.6$ atm and $pO_{2,cat} = 0.21$ atm and high gas flow rates of $\dot{v}_{an/cat} = 250$ sccm.

Stack contacting influences on simulated overpotential distributions were revealed by the model, showing that with increasing polarization not only the cathode gas diffusion is affected negatively but all other loss processes increase as well, and that additional ohmic loss is induced in the cathode layer (Figure 7.4). Based on the modeling results (Figures 7.5 and 7.6), it was demonstrated that gas transport limitations cause an electrochemical reaction slowdown in O_2 depleted areas beneath the contact ribs, which results in an inhomogeneous charge transfer current density profile at the interface. Additional ohmic loss is thus induced by in-plane charge carrier transport in the cathode and electrolyte and so the anode activation polarization is affected accordingly. The total overpotential increase for this particular stack contacting setup results in a model predicted performance decrease of $\sim 24\%$.

A comparison of model predicted power density with experimental results obtained from full stack measurements carried out at Forschungszentrum Jülich demonstrated good reproduction (deviation $\sim$ 2 %) of stack performance by the model at $T = 800$ °C, taking into account that the stack cells were fabricated slightly differently.

A 2D-RPU stack-layer model, set up according to the advanced stack design with state-of-the-art planar IC and a Type-A ASC with LSCF-cathode from Jülich (Figure 7.10 and Table 5.1), was employed to systematically investigate relevant cathode parameters for their influence on stack performance (Section 7.2). The following results were found:

- *Layer thickness*: To diminish the influence of O_2 depletion $h_{\mathrm{cat}} > 100 - 300$ µm is required, depending on the applied polarization. The impact decreases at lower operating temperature as increased electrode activation and ohmic losses prevent higher current density and thus gas conversion via the electrochemical reaction. At standard operating conditions the relative performance loss of stack contacting (compared to ideal contacting) is $\sim 14.8\%$ (Section 7.2.1).
- *Electronic conductivity*: In-plane conduction limitations and thereby induced ohmic loss in the cathode layer are insignificant for materials with high $\sigma_{\mathrm{cat,eff}} > 100$ S/m (e.g. LSC). For LSCF a relative power decrease of $\sim 4\%$ is predicted compared to LSC, while for LSM the loss amounts to $\sim 18\%$ at $T = 800$ °C. However, the influence decreases at lower temperatures (Section 7.2.2).
- *Microstructure properties*: An optimal microstructure depends on the operating temperature and polarization. A more open microstructure ($\varepsilon_{\mathrm{cat}} = 0.5$) prevents gas depletion at higher temperatures, while at lower temperatures a denser cathode layer ($\varepsilon_{\mathrm{cat}} = 0.35$) prevents performance limitation by ohmic in-plane conduction loss (Section 7.2.3).
- *MIEC-cathode kinetics*: Application of MIEC-materials with increased charge transfer kinetics significantly influences SOFC stack-layer performance. Especially, an increased bulk diffusion coefficient, D^{δ}, helps to prevent O_2 transport limitations. The analysis of promising alternative MIEC-materials revealed that SOFC stack-layer performance can be increased by $9 - 18\%$ (at 800 °C), when using LSC or BSCF, compared to LSCF and even up to 30% at 600 °C. However, the MIEC-material inherent electronic conductivity has to be taken into account because ohmic in-plane conduction loss ($R_{\Omega,\mathrm{cat}}$) limits the fast kinetics of BSCF, therefore LSC is recommended instead (Section 7.2.4).
- *Current Collector Layer*: An advanced CCL design offers better flexibility in terms of microstructure properties: it functions best for $h_{\mathrm{CCL}} > 200$ µm and $\varepsilon_{\mathrm{CCL}} = 0.6$ ($T = 800$ °C, Section 7.4.1).
- *IC flowfield design*: It was found that narrow RPU-dimensions with thick ribs are beneficial, whereby the optimal rib/channel ratio increases towards thicker ribs at lower temperatures for increased cathode layer thickness (Section 7.3.1). The underlying reason is that an optimal IC flowfield design is mainly governed by ASR_{contact} (Section 7.3.2).

Based on these findings a performance increase of $15 - 32\ \%$ (depending on the operating temperature) is possible using the standard ASC with LSCF cathode combined with an optimized CCL and IC flowfield (Table 7.3).

In a final numerical analysis the optimal IC flowfield design parameters were determined for alternative MIEC-cathode materials in combination with a metallic protective coating, which offers extremely low ASR_{contact}. It was found that with BSCF as active cathode layer combined with a thin CCL composed of LSC a stack-layer power density of $P_{\text{SL}} = 160$ W·cm^{-2} at $U_{\text{cell}} = 0.8$ V and $T = 600$ °C is possible. This would mean a performance increase of more than a factor of 2, compared to the standard. Even higher performance may be achieved by incorporation of a thinner electrolyte layer with increased ionic conductivity (e.g. scandium and cerium-doped zirconia, 10Sc1CeSZ [253, 254]) combined with a highly electrochemical, nano-scaled Ni/8YSZ-anode [255].

It was shown by the above presented results that the FEM model framework of this work can be employed to reliably pre-evaluate SOFC stack-layer performance for variable MIEC-cathode materials and contact conditions. Systematic variation and analysis of numerical results demonstrate the possible optimization potentials of cathode, CCL and IC flowfield design and thus pave the way to increased SOFC stack performance.

Appendix

A. Reformate Fueled SOFC Stack Layer Performance (2D-GCh Model)

As an outlook, the model framework can be employed to predict SOFC stack layer performance under reformate fueled operation. Occurring gas conversion via the watergas-shift reaction is implemented in the framework reliably, demonstrated by the precise reproduction of measured ASRs (Section 6.3) , C/Vs (Sections 6.2 and 6.3) and gas conversion measurements (Section 6.3), recorded under various operating temperatures and fuel gas flow rates. Furthermore, a catalyst-area independent reaction rate expression (Equation 5.31) allows to adapt the framework to different anodes types, if the Ni/pore surface area, $a_{\mathrm{Ni-por}}$, is a known variable (e.g. from 3D-reconstruction). As an example, the 2D-GCh model (Figure 4.3), adapted to a A_{16} flowfield design, was used to predict SOFC performance under varied fuel gas flow rate. Figure A.1 shows the results, simulated for $T = 800$ °C, $U_{\mathrm{cell}} = 0.7$ V and the in the figure stated reformate composition. Depicted in Figure A.1 are the fuel utilization, $f.u.$, and power density, P, in dependence to the set fuel gas flow rate. It is shown how SOFC stack layer performance increases with increasing gas flow rate a lower, but at the cost of lower $f.u.$ and therefore lower system efficiency.

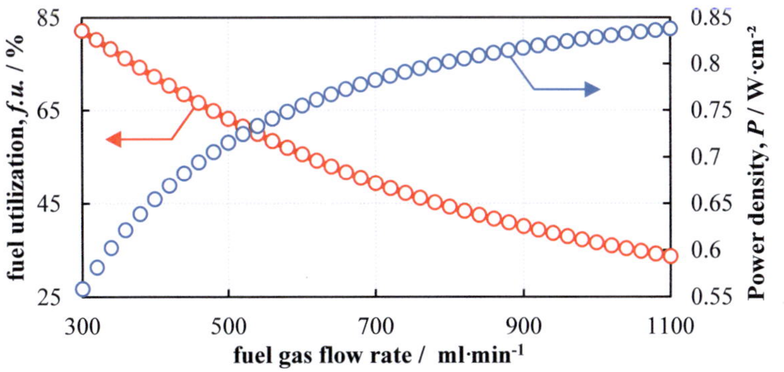

Figure A.1.: Numerical results for predicted fuel utilization, $f.u.$, and stack layer performance density, P, applying the model framework to a 2D-GCh model geometry to an ASC with $A = 16$ cm^2 active electrode area (A_{16}). Operating conditions: $T = 800$ °C, $U_{\mathrm{cell}} = 0.7$ V, $\dot{v}_{\mathrm{an}} = 300 \ldots 1100$ sccm, $pH_{2,\mathrm{an}} = 0.244$ atm, $pHO = 0.246$ atm, $pCO_{\mathrm{an}} = 0.117$ atm, $pCO_{2,\mathrm{an}} = 0.074$ atm, $pN_{2,\mathrm{an}} = 0.319$ atm, air as oxidant with $\dot{v}_{\mathrm{cat}} = 664$ sccm.

Furthermore, the strength of a FEM model approach is to spatially visualize the gas species distribution in the porous electrodes. Figure A.2 depicts predicted H_2 and CO distribution in the anode layer for varied current loads (operating conditions given in the figure caption). It is shown how a H_2 depletion at the anode/electrolyte interface results from high electrochemical fuel consumption and diffusion limitation in the porous electrode with increasing current load [256]. As long as sufficient CO is available, H_2 depletion can be compensated by the water-gas shift reaction (Figure A.2b). However, for increasing current densities, due to a limited reaction rate, also CO depletion occurs in the anode substrate, starting from the anode/electrolyte interface (Figure A.2c). The result is a limited performance, as the measured and predicted limited current density in Figure 6.12 demonstrates. Furthermore, in these areas near the electrolyte very oxidizing atmospheres are encountered due to high H_2O/CO_2 partial pressures and severe damage in the Ni/8YSZ cermet structure will most like occur due to NiO formation [87].

Concluding this section, it shown how the FEM model framework is employed to analyze for SOFC stacks, operated with hydrocarbon containing fuel, the compromise between efficiency and power density. Furthermore, damaging operating conditions can be detected in advance, thus aiding to find optimal operating conditions and providing valuable information for the development of SOFC systems [256].

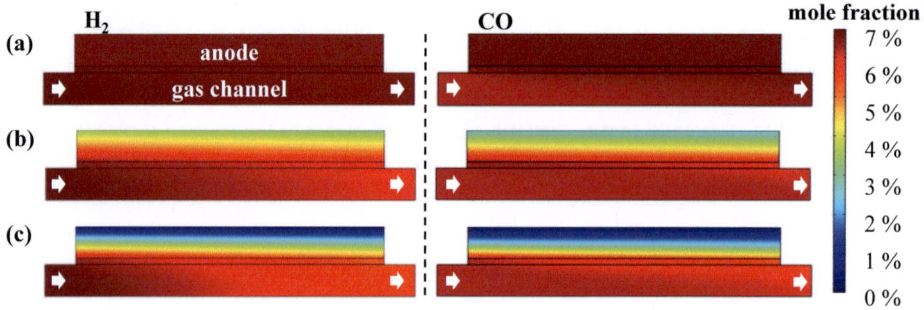

Figure A.2.: Hydrogen (H_2) and carbon monoxide (CO) distribution in the anode substrate (A_1-ASC)c, the Ni mesh and the corresponding gas channel for various current densities at 69 % fuel utilization. a) $j_{\mathrm{cell}} = 0$ A/cm^2, b) $j_{\mathrm{cell}} = 0.45$ A/cm^2 and c) $j_{\mathrm{cell}} = 0.9$ A/cm^2. $T = 780$ °C. Inital fuel gas composition: $pH_{2,\mathrm{an}} = 0.03$ atm, $pH_2O_{\mathrm{an}} = 0.03$ atm, $pCO = 0.22$ atm, $pCO_2 = 0.03$ atm, $pN_{2,\mathrm{an}} = 0.51$ atm, $\dot{v}_{\mathrm{an}} = 187.5$ sccm. Oxidant: air, $\dot{v}_{\mathrm{an}} = 600$ sccm.

B. Parameters for Diffusion Coefficient Calculation

The binary diffusion coefficient, D_{ij}, of species i and j is calculated in this work by the Chapman-Enskog Theory (Equation 4.30). Therein required parameters are the collision integral Ω_D, which can be calculated, based on the following expression [166]:

$$\Omega_D = \frac{A}{(T^*)^B} + \frac{C}{exp(D \cdot T^*)} + \frac{E}{exp(F \cdot T^*)} + \frac{G}{exp(H \cdot T^*)} \tag{B.1}$$

with

$T^* = k_\mathrm{B} \cdot T/\epsilon_{ij}$	A = 1.06036	B = 0.15610
C = 0.19300	D = 0.47635	E = 1.03587
F = 1.52996	G = 1.76474	H = 3.89411

Therein, $\epsilon_{ij} = \sqrt{(\epsilon_i + \epsilon_i)}$ is calculated from the individual Lennard-Jones-Potentials of species i and j. In Table B.1, required parameters from Ref. [166] are listed.

Table B.1.: Required parameters to calculate the binary diffusion coefficient D_{ij} with Equations 4.30 and B.1.

	Unit	CH_4	CO	CO_2	H_2	H_2O_g	N_2	O_2
σ_i	Å	3.758	3.69	3.941	2.827	2.641	3.798	3.467
$(\epsilon_i/k_\mathrm{B})$	K	148.6	91.7	195.2	59.7	809.1	71.4	106.7

C. Calculation of Dynamic Viscosity

In Ref. [257] the following expression is given to calculate the dynamic viscosity, η_i of species i based on experimental data:

$$\frac{\eta_i}{\mathrm{Pa \cdot s}} = A + B \cdot \left(\frac{T}{K}\right) + C \cdot \left(\frac{T}{K}\right)^2 + D \cdot \left(\frac{T}{K}\right)^3 + E \cdot \left(\frac{T}{K}\right)^4 \tag{C.2}$$

The required parameters for Equation C.2 are given in Table C.2 for the relevant species containing in fuel and oxidant gas composition. Due to gas transport, chemical and electrochemical fuel gas conversion, the corresponding total mixture viscosity is a spatial dependent variable, $\eta_e l$. In the framework, $\eta_e l$ is calculated locally by an Wilke approach from the individual gas species concentrations [257]:

$$\eta_i = \sum_{i=1}^{N} \frac{x_i \eta_i}{\sum_{j=1}^{N} x_i F_{\mathrm{ij}}} \tag{C.3}$$

with

$$F_{ij} = \frac{\left[1 + (\eta_i/\eta_j)^{1/2}(M_j/M_i)^{1/4}\right]^2}{\sqrt{8(1 + M_i/M_j)}} \tag{C.4}$$

and

$$F_{ji} = \frac{\eta_j}{\eta_i}\frac{M_i}{M_j}F_{ij} \tag{C.5}$$

Table C.2.: List of anode supported cells and samples used in this work and brief remarks of applied experimental methods.

Substance	A $\cdot 10^5$	B$\cdot 10^7$	C$\cdot 10^{10}$	D$\cdot 10^{12}$	E$\cdot 10^{15}$
CH_4	-0.07759	0.50484	-0.43101	0.03118	-0.00981
CO	0.01384	0.74306	-0.62996	0.03948	-0.01032
CO_2	-0.18024	0.65989	-0.37108	0.01586	-0.003
H_2	0.18024	0.27174	-0.13395	0.00585	-0.00104
H_2O_g	-0.10718	0.35248	0.03575	-	-
N_2	-0.01020	0.74785	-0.59037	0.03230	-0.00673
O_2	-0.10257	0.92625	-0.80657	0.05113	-0.01295

D. Measured Cells

Table D.3.: List of anode supported cells and samples used in this work with brief remarks of applied experimental parameter variation.

IAM-WET ID	Supplier ID	ASC Type	Remarks
Z1_188	FZJ 10539	Type-B (A_1)	Variation of T, $pO_{2,\mathrm{cat}}$, Ref. [9] Sections 6.2 and 6.4 3D-reconstruction (anode, Table 5.1)
Z1_198	FZJ 10651	Type-A (A_1)	3D-reconstruction (cathode, Ref. [120]) Table 5.1
Z1_349	FZJ III-0572-4	Type-C (A_1)	3D-reconstruction anode
Z1_360	FZJ 12798-2	Type-A (A_1)	Variation of $pH_{2,\mathrm{an}}$ and pH_2O_{an} Section 6.3
Z2_190	FZJ 10536	Type-B (A_1)	Variation of reformate composition Sections 6.2 and 6.3
Z2_275	FZJ 12798-8	Type-A (A_1)	Variation of T and $pO_{2,\mathrm{cat}}$-variation Sections 6.2 and 6.4
Z2_276	FZJ 12798-16	Type-A (A_1)	Section 7.1
Z5_218	-	Ni-sheet (A_{16})	Variation of reformate composition Sections 5.5 and 6.3
Z4_191	FZJ 15621	Type-A (A_{16})	Variation of reformate composition Section 6.3
Z8_166	FZJ III-0572-4	Type-C (A_1)	3D-reconstruction (anode) Table 5.1
Z9_072	FZJ 13367	Type-A (A_1)	3D-reconstruction (anode, Ref. [120]) Table 5.1

E. Supervised Diploma, Master and Bachelor Theses

The following lists in chronological order supervised diploma, master and bachelor theses, which have been supervised within the frame of this work.

- Sebastian Hirn: "Transiente Simulation der Stofftransportvorgänge im Brenngas der Hochtemperatur-Brennstoffzelle SOFC", (co-supervised with A. Kromp). Diploma thesis (Dipl.-Ing.), 2012.
- Andreas Kleiser: "Modellierung und Simulation einer 2D-FEM Repeat Unit im SOEC-Betrieb". Bachelor thesis (B.Sc.), 2014.
- Sebastian Dierickx: Leistungssimulation an Hochtemperatur-brennstoffzellen (SOFC) bei Betrieb mit kohlenwasserstoffhaltigen Brenngasen", (co-supervised with A. Kromp). Master thesis (M.Sc.), 2014.
- Franziska Wenz: "3D FEM Modellierung einer Hochtemperaturbrennstoffzelle im Stack-Betrieb". Bachelor thesis (B.Sc.), 2014.
- Fabian Wirtz: "Untersuchung und FEM-Simulation der katalytischen Aktivität von Hochtemperatur-Brennstoffzellen (SOFCs)", (co-supervised with S. Dierickx). Master thesis (M.Sc.), 2015.
- Michael Rößler: "Transiente Simulation der physikalischen und elektrochemischen Prozesse der Hochtemperatur-Brennstoffzelle (SOFC)", (co-supervised with S. Dierickx). Master thesis (M.Sc.), 2015.
- Mike Günthner: "Entwickung eines homogenisierten 1D-Modells für PEM-Brennstoffzellen mit COMSOL Multiphysics", (co-supervised with M. Heinzmann). Master thesis (M.Sc.), 2017.
- Dominik Horny: "Modellierung der viskosen Gasströmung in porösen Elektroden der Hochtemperatur -Brennstoffzelle SOFC", (co-supervised with J. Joos and N. Russner). Master thesis (M.Sc.), 2017.

F. Publications

- A. Kromp, H. Geisler, A. Weber and E. Ivers-Tiffée, "Electrochemical Impedance Modeling of Reformate-Fuelled Anode-Supported SOFC", in F. Lefebvre-Joud (Ed.), Proceedings of the 10th European Solid Oxide Fuel Cell Forum (Lucerne / Switzerland), Chapter 17, pp. 92–101, 26.06.–29.06.2012.
- A. Kromp, H. Geisler, A. Weber and E. Ivers-Tiffée, "Electrochemical Impedance Modeling of Gas Transport and Reforming Kinetics in Reformate Fueled Solid Oxide Fuel Cell Anodes", Electrochimica Acta, 106, pp. 418–424 (2013).
- H. Geisler, M. Kornely, A. Weber, E. Ivers-Tiffée, "Enhancing SOFC-Stack Performance by Model-Based Adaptation of Cathode Gas Transport Conditions", ECS Transaction. 57, pp. 2871–2881 (2013).
- H. Geisler, A. Kromp, S. Hirn, A. Weber, E. Ivers-Tiffée, "Model Based Interpretation of Coupled Gas Conversion and Diffusion in SOFC-Anodes", ECS Trans. 57, pp. 2691–2704 (2013).
- H. Geisler, A. Kromp, A. Weber and E. Ivers-Tiffée, "Stationary FEM Model for Performance Evaluation of Planar Solid Oxide Fuel Cells Connected by Metal Interconnectors – Part I: Model Framework and Validation", Journal of The Electrochemical Society, 161 (6) F778–F788 (2014).
- H. Geisler, A. Kromp, A. Weber and E. Ivers-Tiffée, "Performance of MIEC cathodes in SOFC stacks evaluated by means of FEM modeling", ECS Trans. 61, pp. 191–201 (2014).
- H. Geisler, A. Kromp, A. Weber and E. Ivers-Tiffée, "How SOFC stack performance depends on the interaction of MIC design and cathode material", Proceedings of the 11th European Solid Oxide Fuel Cell Forum (Lucerne / Switzerland), Chapter 05, pp. 24–29, 01.07.–04.07.2014.

  H. Geisler, J. Joos, A. Kromp, A. Weber and E. Ivers-Tiffée "Stationary 2D FEM Model Framework for SOFC Stack Performance Prediction", ECS Trans. 68, pp. 3043–3050 (2015).
- H. Geisler, S. Dierickx, A. Weber and E. Ivers-Tiffée "A 2D Stationary FEM Model for Hydrocarbon Fuelled SOFC Stack Layers", ECS Trans. 68(1), pp. 2151–2158 (2015).
- J. Joos, H. Geisler, A. Weber and E. Ivers-Tiffée, "Designing Porous Cathode Structures for SOFCs", Proceedings of the 12th European Solid Oxide Fuel Cell Forum (Lucerne / Switzerland), 05.07.–08.07.2016.
- H. Geisler, J. Joos, A. Weber and E. Ivers-Tiffée, "Performance Modelling of anode supported cells on a SOFC stack layer level", Proceedings of the 12th European Solid Oxide Fuel Cell Forum (Lucerne / Switzerland), 05.07.–08.07.2016.
- N. Russner, H. Geisler, S. Dierickx, A. Weber and E. Ivers-Tiffée, "A Non-Isothermal 2D Stationary FEM Model for Hydrocarbon Fueled SOFCs Stack Layers", ECS Trans. 78, pp. 2673–2682 (2017).

- J. Joos, T. Carraro, F. Wankmüller, H. Geisler and E. Ivers-Tiffée, "3D-Analyse und Modellierung von gemischtleitenden Kathoden der Hochtemperatur-Brennstoffzelle (SOFC)", in G. Schneider and T. Bernthaler (Eds.), 51. Materialographie-Tagung der DGM, Aalen, Germany, pp. 201–206 (2017).
- N. Russner, H. Geisler, S. Dierickx, A. Weber and E. Ivers-Tiffée, „Performance optimization by means of a non-isothermal FEM model for SOFC stack layers", Proceedings of the 13th European Solid Oxide Fuel Cell Forum (Lucerne / Switzerland), 03.07.–06.07.2018

G. Conference Contributions

- A. Kromp, H. Geisler, A. Weber and E. Ivers-Tiffée, "Electrochemical Impedance Modeling of Reformate-Fuelled Anode-Supported SOFC", in F. Lefebvre-Joud (Ed.), Proceedings of the 10th European Solid Oxide Fuel Cell Forum (Lucerne / Switzerland), Chapter 17, pp. 92–101, 26.06.–29.06.2012.
- A. Kromp, H. Geisler, A. Weber and E. Ivers-Tiffée, Electrochemical Impedance Modeling of Gas Transport and Reforming Kinetics in Ni/YSZ Anodes for SOFCs, 10th Symposium for Fuel Cell and Battery Modeling and Experimental Validation (Bad Boll / Germany), 19.03.–20.03.2013.
- H. Geisler, A. Kromp, A. Weber and E. Ivers-Tiffée, Electrochemical Impedance Modeling of Gas Transport in Ni/YSZ Anodes and associated Gas Channel for SOFCs, 10th Symposium for Fuel Cell and Battery Modeling and Experimental Validation (Bad Boll / Germany), 19.03.–20.03.2013.
- A. Kromp, H. Geisler, A. Weber and E. Ivers-Tiffée, Electrochemical Impedance Modeling of Gas Transport and Reformate-Fueled Ni/YSZ Anodes for SOFCs, 5th International Conference on Fundamentals & Development of Fuel Cells (Karlsruhe / Germany), 16.04.–18.04.2013.
- H. Geisler, M. Kornely, A. Weber and E. Ivers-Tiffée, "Enhancing SOFC-Stack Performance by Model-Based Adaptation of Cathode Gas Transport Conditions", 13th International Symposium on Solid Oxide Fuel Cells, SOFC-XIII (Okinawa / Japan), 6.10.–11.10.2013.
- H. Geisler, A. Kromp, S. Hirn, A. Weber and E. Ivers-Tiffée, "Model Based Interpretation of Coupled Gas Conversion and Diffusion in SOFC-Anodes", 13th International Symposium on Solid Oxide Fuel Cells, SOFC-XIII (Okinawa / Japan), 6.10.–11.10.2013.
- H. Geisler, A. Kromp, A. Weber and E. Ivers-Tiffée, "Prediction of SOFC stack performance by means of FEM modeling: Variation of MIC design and cathode composition", 11th Symposium for Fuel Cell and Battery Modeling and Experimental Validation (Winterthur / Schweiz), 17.03.–19.03.2014.
- H. Geisler, A. Kromp, A. Weber and E. Ivers-Tiffée, "Performance of MIEC cathodes in SOFC stacks evaluated by means of FEM modeling", 225th ECS Meeting (Orlando, Florida / USA), 11.05.–15.05.2015.

- H. Geisler, A. Kromp, A. Weber and E. Ivers-Tiffée, "How SOFC stack performance depends on the interaction of MIC design and cathode material", Proceedings of the 11th European Solid Oxide Fuel Cell Forum (Lucerne / Switzerland), 01.07.–04.07.2014.
- S. Dierickx, H. Geisler, A. Weber, E. Ivers-Tiffée, "FEM Performance Modelling of SOFC stacks operated with hydrocarbon fuels", 12th Symposium for Fuel Cell and Battery Modeling and Experimental Validation (Freiburg / Germany), 26.03.–27.03.2015.
- H. Geisler, A. Weber and E. Ivers-Tiffée, "Stationary 2D FEM model for SOFC stack performance prediction", 12th Symposium for Fuel Cell and Battery Modeling and Experimental Validation (Freiburg / Germany), 26.03.–27.03.2015.
- H. Geisler, J. Joos, A. Kromp, A.Weber, E. Ivers-Tiffée, "Stationary 2D FEM Model Framework for SOFC Stack Performance Prediction", SOFC-XIV (Glasgow, UK), 27.07.–31.07.2015
- H. Geisler, J. Joos, A. Weber, E. Ivers-Tiffée, "FEM Performance Model for SOFC Stack Layer", 13th Symposium for Fuel Cell and Battery Modeling and Experimental Validation (Lausanne, Switzerland), 22.03.–23.03.2016
- J. Joos, H. Geisler, A. Weber and E. Ivers-Tiffée, "Designing Porous Cathode Structures for SOFCs", Proceedings of the 12th European Solid Oxide Fuel Cell Forum (Lucerne / Switzerland), 05.07.–08.07.2016.
- H. Geisler, J. Joos, A. Weber and E. Ivers-Tiffée, "Performance Modelling of anode supported cells on a SOFC stack layer level", Proceedings of the 12th European Solid Oxide Fuel Cell Forum (Lucerne / Switzerland), 05.07.–08.07.2016.
- J. Joos, T. Carraro, H. Geisler, A. Weber, E. Ivers-Tiffée, Microstructure modeling of porous electrodes for solid oxide fuel cells (SOFCs), 8th International Conference on Multiscale Materials Modeling (Dijon, France), 09.10.–14.10.2016
- J. Joos, T. Carraro, H. Geisler, A. Weber, E. Ivers-Tiffée, "Microstructure Modelleing of Porous Cathodes for Solid Oxide Fuel Cells (SOFCs)", ModVal 14 (Karlsruhe, Germany), 02.03.–03.03.2017.
- H. Geisler, J. Joos, A. Weber and E. Ivers-Tiffée, "Multi-Scale Modeling of Cathode Performance in Solid Oxide Fuel Cells (SOFCs)", 231th ECS Meeting (New Orleans, Lousiana/USA), 28.05.–01.06.2017.
- J. Joos, T. Carraro, F. Wankmüller, H. Geisler and E. Ivers-Tiffée, "3D-Analyse und Modellierung von gemischtleitenden Kathoden der Hochtemperatur-Brennstoffzelle (SOFC)", in G. Schneider and T. Bernthaler (Eds.), 51. Materialographie-Tagung der DGM, Aalen, Germany, pp. 201–206 (2017).
- N. Russner, H. Geisler, S. Dierickx, A. Weber and E. Ivers-Tiffée, „Performance optimization by means of a non-isothermal FEM model for SOFC stack layers", Proceedings of the 13th European Solid Oxide Fuel Cell Forum (Lucerne / Switzerland), 03.07.–06.07.2018

Nomenclature

Symbols

a_i	surface area density between two phases	μ^2/μ^3
A_{ch}	single gas channel cross sectional area	mm^2
$A_{k/D}$	pre-exponential factors	m/s
$A_{1/16}$	ASC with active electrode area denoted in subscript	cm^2
$ASR_{\mathrm{ct,cat}}$	area specific cathode charge transfer resistance	$\Omega \cdot cm^2$
ASR_{contact}	area specific contact resistance	$\Omega \cdot cm^2$
b	model width	m
b_{RB}	RPU rib width	m
b_{CH}	RPU channel width	m
b_{RPU}	RPU width	m
$b_{\mathrm{ii/ij}}$	elements of matrix $[\vec{B}]$	-
$[\vec{B}]$	DGM diffusion coefficient matrix	-
c	total concentration of a gas mixture	mol/m^3
c_i	molar concentration of species i	mol/m^3
c_{eq}	oxygen ion equilibrium concentration	mol/m^3
$C_{1/2}$	MIEC-material constants	-
d_{por}	mean pore diamter	m
D_{ij}	binary diffusion coefficient of species i and j	m^2/s
$D_{\mathrm{Kn},i}$	bulk diffusion coefficient of species i	m^2/s
$D_{m,i}$	molar diffusion coefficient of species i	m^2/s
D^{δ}	bulk diffusion coefficient in MIEC material	m^2/s
E_{act}	activation energy	J/mol
$\vec{E}$	electric field	V/m
f	characteristic relaxation frequency	Hz
$f_{\mathrm{ii/ij}}$	elements of matrix $[\vec{F}]$	-
$[\vec{F}]$	MTPM diffusion coefficient matrix	-
F	Faraday constant	C
ΔG	Gibb's free energy	$kg \cdot m^2/s^2$
h	layer/computational domain thickness	m
$i(t)$	sinusoidal current signal	A
I	identity matrix	-

j	current load	A/m^2
j_{cell}	cell/stack-layer current load under operation	A/m^2
$j_{\text{ct},el}$	charge transfer current density	A/m^2
$j_{0,el}$	exchange current density (BVM)	A/m^2
$\vec{j}$	charge carrier flux	A/m^2
$\vec{j}_i$	mass averaged diffusive flux of species i	kg/m^2·s
$\vec{J}_i$	mol averaged diffusive flux of species i	kg/m^2·s
k^δ	surface exchange coefficient of MIEC material	m/s
k_{B}	Boltzmann constant	m^2/kg·s^2·K
$k^+_{\text{sr/sh}}$	SR/WGS forward reaction rate velocity constant	mol/m^3·s·Pa^2
$\hat{k}^+_{\text{sh,ASC}}$	modified WGS forward reaction rate velocity constant	mol/m^2·s·Pa^2
K	equlibrium constant	-
$K_{\text{eq,sr/sh}}$	equlibrium constant of SR/WGS-reation	-
K_{n}	Knudsen number	-
l	model length	m
L	representative length variable	m
M_i	molecular mass of species i	kg/mol
n	number of species	-
n_{ch}	number of flowfiel gas channels	-
$\vec{n}$	normal vector	-
$\vec{n}_i$	total mass averaged flux of species i	kg/m^2·s
$\vec{N}_i^{\text{c}}$	mol averaged convective flux of species i	mol/m^2·s
$\vec{N}_i$	total mol averaged flux of species i	mol/m^2·s
N	number of species	-
p	absolute pressure	atm
p_i	partial pressure of gas species i	atm
P_i	loss process i	-
Q_i	current source	A/m^3
R	universal gas constant	kg·m^2/s^2·mol·K
Re	Reynold's number	-
R_0	ohmic resistance	Ω·cm^2
R_{pol}	polarization resistance	Ω·cm^2
R_i	resistance of loss process i	Ω·cm^2
R_i^{s}	surface dependent MIEC reaction rate	mol/m^3·s
Re	Reynolds-number	-
s_i	mass source term of species i	kg/m^2·s
S_i	molar source term of species i	mol/m^2·s
t	time	s
T	temperature	°C
T_{k}	absolute temperature	K

$u(t)$	sinusoidal voltage response signal	V
U_{cell}	cell operating voltage	V
U_{N}	Nernst voltage	V
U_{th}	theoretical Nernst voltage	V
$v_{0,el}$	inlet mass flow rate	sccm
$\vec{v}$	mass averaged total velocity	m/s
$\vec{v}_i$	mass averaged velocity of species i	m/s
$\vec{v}_a$	average mixture velocity	m/s
$\dot{v}_i$	total mass flow rate	sccm
$\dot{v}_i$	total mass flow rate	sccm
V	total electrode volume	m^3
V_i	volume of phase i	m^3
$\vec{V}$	mol averaged total velocity	m/s
$\vec{V}_i$	mol averaged velocity of species i	m/s
x_i	mole fraction of species i	-
z	number of per mole transfered electrons	-
$\underline{Z}_{\text{cell}}$	complex cell impedance	$\Omega\cdot cm^2$
Z'	real part of complex impedance	$\Omega\cdot cm^2$
Z''	imaginary part of complex impedance	$\Omega\cdot cm^2$
α_i	MTPM diffusion matrix component	-
α_{el}	charge transfer coefficient	
$\alpha_{k/D}$	oxygen partial pressure dependency coefficients	-
γ_n	distribution function	-
$\Delta\Phi_{el}$	half-cell potential of electrode el	V
ε_0	permittivity in vacuum	F/m
ε_τ	relative material permittivity	F/m
ε_{el}	pore volume fraction of layer el	-
η_i	overpotential of process i	V
$\kappa_e l$	permeability of layer el	m^2
κ_i	individual permeability factor of species i	m^2
λ	air to fuel ratio	-
μ_i	dynamic viscosity of species i	Pa·s
μ_{el}	dynamic viscosity of gas mixture el	Pa·s
$\nu_{\text{sr/sh,i}}$	stoichiometric coefficient of reforming reaction	-
ρ_{el}	density of fluid/gas in domain el	kg/m^3
σ_{el}	electronic (ec) or ionic (io) conductivity of layer el	S/m
σ_{ij}	collision diameter of species i and j	Å
τ_0	characteristics relaxation time	-
$\tau_{\text{por/mat},el}$	pore/material phase tortuosity of electrode el	-
Φ_{el}	electronic / ionic potential in layer el	V
Ψ_{el}	microstructure parameter of layer el	-
ω	angular frequency	Hz
ω_i	mass fraction of species i	-

Acronyms

AFC	alkaline fuel cell
AFL	anode functional layer
APS	atmospheric plasma spraying
APU	axillary power unit
ASC	anode supported cell
BFM	Binary friction Model
BVM	Butler-Volmer Model
BSCF	barium strontium cobalt ferrite
CCL	current collector layer (or cathode contact layer)
CGO	ceria doped gadolinium
CHP	combined heat and power
CNLS	complex non-linear least-squares
CPIM	Cylindrical Pore Interpolation Model
C/V	current voltage (measurement / curve / characteristic)
DFO	degrees of freedom
DGM	Dusty-Gas Model
DIR	direct internal reforming
DRT	distribution of relaxation times
ECM	equivalent circuit model
EDT	Euclidean Distance Transform
ETD	Everhart-Thornley detector
EIS	electrochemical impedance spectroscopy
EMF	electromotive force
ESC	electrode supported cell
FDM	finite differential method
FEM	finite element method
FIB	focused ion beam
FM	Fick Model
FVM	finite volume method
FRA	frequency response analyzer
f.u.	fuel utilization
GC	gas conversion
GCh	gas channel
GDC	gadolium doped ceria
GT	gas transport
KIT	Karlsruhe Institute of Technology
IAM-WET	Institute of Applied Materials Materials for Electrical and Electronic Engineering
IF	interface
IIR	indirect internal reforming
IC	interconnector

LCC	lanthanum manganese copper cobaltite
LSC	lanthanum strontium cobaltite
LSCF	lanthanum strontium cobalt ferrite
LSM	strontium doped lanthanum manganite
MCF	manganese cobalt ferrite oxide
MCFC	molton carbonate fuel cell
MEA	membrane electrode assembly
MIEC	mixed-ionic-electrionic conductive
MSC	metall supported cell
MSR	methane steam reforming
MTPM	Mean Transport Pore Model
OCC	open circuit conditions
OCV	open circuit voltage
PAFC	phosphoric acid fuel cell
PDE	partial differential equation
PSD	particle/pore size distribution
PtG	power to gas
PtL	power to liquid
PVD	physical vapor deposition
ROI	region of interest
RPU	repeat unit
R&D	research and developement
RVE	representative volume element
SEM	scanning electrode microscopy
SL	stack layer
SMM	Stefan-Maxwell Model
SNR	signal to noise ratio
SOE	solid oxide electrolysis
SOEC	solid oxide electrolysis cell
SOFC	solid oxide fuel cell
SR	steam reforming
SrZr	strontium zirconite
TPB	triple phase boundary
WGS	watergas-shift reaction
WPS	wet powder spraying
YSZ	ytrium stabilized zircon oxide
μGC	gas chromatograph
μCT	μ computer tomography

Super and subscripts

act	activation
an	anode
bulk	solid material
cat	cathode
cc	concentration
ct	charge transfer
contact	contact between IC and cathode
CH	channel
ec	electronic
eff	effective
el	layer
elyt	electrolyte
eq	equilibrium
GC	gas channel
mat	material
mh	mesh
n	number based averaging
i	species
io	ionic
KC	Kozeny-Carman
por	pore
RB	rib
s	surface
sh	watergas-shift reaction
sr	steam reforming reaction
v	volume based averaging

Bibliography

[1] T. F. Stocker, D. Qin, G. K. Plattner, M. M. Tignor, S. K. Allen, J. Boschung, A. Nauels, Y. Xia, V. Bex, and P. M. Midgley, "Climate change 2013 the physical science basis: Working Group I contribution to the fifth assessment report of the intergovernmental panel on climate change," pp. 1–1535, 2013.

[2] A. Palzer and H. M. Henning, "A comprehensive model for the German electricity and heat sector in a future energy system with a dominant contribution from renewable energy technologies - Part II: Results," *Renew. Sustain. Energy Rev.*, vol. 30, pp. 1019–1034, 2014.

[3] G. Richter, "Widerstand gegen Windkraft: Die Energiewende als Soziallabor," https://www.deutschlandfunkkultur.de/widerstand-gegen-windkraft-die-energiewende-als-soziallabor.976.de.html?dram:article_id=354904, 2016.

[4] G. B. Volk, "Das Gasnetz als universeller Speicher - Chancen und Grenzen," in *1. Energiespeichertagung*, 2013, pp. 1–22.

[5] C. Schenuit, R. Heuke, and J. Paschke, "Potenzialatlas Power to Gas - Eine Studie der Deutschen Energie-Agentur (dena)," *Dtsch. Energie-Agentur GmbH*, p. 111, 2016.

[6] J. Lehmann and T. Luschtinetz, "Wasserstoff und Brennstoffzellen," *BWK*, vol. 69, no. 5, 2014.

[7] H. Ammermann, P. Hoff, M. Atanasiu, J. Ayllor, M. Kaufmann, and O. Tisler, "Advancing Europe's energy systems: Stationary fuel cells in distributed generation," *FCH JU Publ.*, p. 185, 2015.

[8] P. Schmidt, W. Weindorf, A. Roth, V. Batteiger, and F. Riegel, *Power-to-Liquids – Potentials and Perspectives for the Future Supply of Renewable Aviation Fuel.* German Environment Agency, 2016.

[9] A. Leonide, "SOFC Modelling and Parameter Identification by means of Impedance Spectroscopy," Dissertation, Karlsruhe Institute of Technology (KIT), 2010.

[10] W. Vielstich, A. Lamm, and H. A. Gasteiger, *Handbook of Fuel Cells: Fundamentals, Technology, Applications.* New York: Wiley, 2003.

[11] E. Ivers-Tiffée and A. V. Virkar, *Electrode Polarisation - High Temperature Solid Oxide Fuel Cells : Fundamentals, Design, and Applicatons.* Oxford: Elsevier Advanced Technology, 2003.

[12] M. Heneka, "Alterung der Festelektrolyt-Brennstoffzelle unter thermischen und elektrischen Lastwechseln," Dissertation, Universität Karlsruhe (TH), 2006.

[13] W. Nernst, “Die elektromotorische Wirksamkeit der Ionen,” *Zeitschrift für Phys. Chemie*, vol. 4, no. 2, pp. 129–181, 1889.

[14] A. Weber and E. Ivers-Tiffée, “Materials and concepts for solid oxide fuel cells (SOFCs) in stationary and mobile applications,” *J. Power Sources*, vol. 127, no. 1-2, pp. 273–283, 2004.

[15] A. Leonide, Y. Apel, and E. Ivers-Tiffée, “SOFC Modeling and Parameter Identification by means of Impedance Spectroscopy,” *ECS Trans.*, vol. 19, no. (20), pp. 81–109, 2009.

[16] A. Leonide, S. Hansmann, A. Weber, and E. Ivers-Tiffée, “Performance simulation of current/voltage-characteristics for SOFC single cell by means of detailed impedance analysis,” *J. Power Sources*, vol. 196, no. 17, pp. 7343–7346, sep 2011.

[17] A. Kromp, “Model-Based Interpretation of the Performance and Degradation of Reformate Fueled Solid Oxide Fuel Cells,” Dissertation, Karlsruher Institut für Technologie (KIT), 2013.

[18] J. Macdonald, *Impedance spectroscopy: emphasizing solid materials and systems*. Wiley-Interscience Publication, 1987.

[19] B. Y. B.-Y. Chang, S.-M. S. M. Park, M. E. Orazem, B. Tribollet, B. Y. B.-Y. Chang, and S.-M. S. M. Park, *Electrochemical Impedance Spectroscopy*. John Wiley & Sons, 2008, vol. 3.

[20] V. Lvovich, *Impedance spectroscopy application to electrochemical and dielectric phenomena*, 1st ed. New York: John Wiley & Sons, Ltd, 2012.

[21] E. Ivers-Tiffée, A. Weber, and H. Schichlein, “Electrochemical Impedance Spectroscopy,” in *Analysis*, 2nd ed., W. Vielstich, H. A. Gasteiger, and A. Lamm, Eds. John Wiley & Sons, Ltd, 2003, pp. 220 – 235.

[22] E. Ivers-Tiffée, *Brennstoffzellen und Batterien*. lecture notes, Institute for Applied Materials - Materials for Electrical and Electronic Engineering (IAM-WET), Karlsruhe Institute of Technology (KIT), 2017.

[23] D. Klotz, “Characterization and Modeling of Electrochemical Energy Conversion Systems by Impedance Techniques,” Dissertation, Karlsruher Institut für Technologie (KIT), 2014.

[24] A. Müller, “Mehrschicht-Anode für die Hochtemperatur-Brennstoffzelle (SOFC),” Dissertation, Universität Karlsruhe (TH), 2004.

[25] H. Schichlein, “Experimentelle Modellbildung für die Hochtemperatur-Brennstoffzelle SOFC,” Dissertation, Universität Karlsruhe (TH), 2003.

[26] A. Kromp, A. Leonide, H. Timmermann, A. Weber, and E. Ivers-Tiffée, “Internal reforming kinetics in SOFC-anodes,” *ECS Trans.*, vol. 28, no. 11, pp. 205–215, 2010.

[27] A. Leonide, V. Sonn, A. Weber, and E. Ivers-Tiffée, “Evaluation and Modeling of the Cell Resistance in Anode-Supported Solid Oxide Fuel Cells,” *J. Electrochem. Soc.*, vol. 155, no. 1, pp. B36–B41, 2008.

[28] A. Kromp, A. Leonide, A. Weber, and E. Ivers-Tiffée, "Electrochemical Analysis of Reformate-Fuelled Anode Supported SOFC," *J. Electrochem. Soc.*, vol. 158, no. 8, p. B980, 2011.

[29] H. Geisler, "Simulation der Stofftransportvorgänge im Anodensubstrat der SOFC," Diploma Thesis, Karlsruher Institut für Technologie (KIT), 2011.

[30] S. B. Adler, "Mechanism and kinetics of oxygen reduction on porous La1-xSrxCoO3-d electrodes," *Solid State Ionics*, vol. 111, pp. 125–134, 1998.

[31] C. Endler-Schuck, J. Joos, C. Niedrig, A. Weber, and E. Ivers-Tiffée, "The chemical oxygen surface exchange and bulk diffusion coefficient determined by impedance spectroscopy of porous $La_{0.58}Sr_{0.4}Co_{0.2}Fe_{0.8}O_{3-\delta}$ (LSCF) cathodes," *Solid State Ionics*, vol. 269, pp. 67–79, 2015.

[32] H. W. Engl, M. Hanke-Bourgeois, and A. Neubauer, "Regularization of inverse problems," in *Math. its Appl.* Kluwer Academic Publishing, 2000, vol. 375.

[33] H. W. Engl, "Inverse Problems and Their Regularisation," in *Comput. Math. Driven by Ind. Probl.*, H. W. E. R. E. Burkard, A. Jameson, G. Strang, P. Deuflhard, J.-L. Lions, V. Capasso, J. Periaux, Ed. Heidelberg: Springer, 2000, vol. 173, pp. 127–150.

[34] A. Tikhonov, A. Goncharsky, V. Stepanov, and A. Yagola, *Numerical Methods for the Solution of Ill-Posed Problems.* Dordrecht: Kluwer Academic Publishing, 1995.

[35] J. B. Bell, A. N. Tikhonov, and V. Y. Arsenin, *Solutions of Ill-Posed Problems.* Winston, Washington, D.C.: Scripta series in mathematics. A Halsted Press book, 1978, vol. 32.

[36] V. Sonn, A. Leonide, and E. Ivers-Tiffée, "Combined Deconvolution and CNLS Fitting Approach Applied on the Impedance Response of Technical Ni-8YSZ Cermet Electrodes," *J. Electrochem. Soc.*, vol. 155, no. 7, p. B675, 2008.

[37] M. Kornely, "Elektrische Charakterisierung und Modellierung von metallischen Interkonnektoren (MIC) des SOFC-Stacks," Dissertation, Karlsruhe Institute of Technology (KIT), 2012.

[38] S. P. S. Badwal, "Zirconia-based solid electrolytes: microstructure, stability and ionic conductivity," *Solid State Ionics*, vol. 52, no. 1-3, pp. 23–32, 1992.

[39] N. Q. Minh, "Ceramic Fuel Cells," *J. Am. Ceram. Soc.*, vol. 76, no. 3, pp. 563–588, 1993.

[40] O. Yamamoto, "Solid oxide fuel cells: fundamental aspects and prospects," *Electrochim. Acta*, vol. 45, no. 15-16, pp. 2423–2435, 2000.

[41] F. Tietz, H. P. Buchkremer, and D. Stöver, "Components manufacturing for solid oxide fuel cells," *Solid State Ionics*, vol. 152-153, no. December 2002, pp. 373–381, 2002.

[42] S. M. Haile, "Fuel cell materials and components," *Acta Mater.*, vol. 51, no. 19, pp. 5981–6000, 2003.

[43] N. Brandon, S. Skinner, and B. Steele, "Recent Advances in Materials for Fuel Cells," *Annu. Rev. Mater. Res.*, vol. 33, no. 1, pp. 183–213, 2003.

[44] E. Ivers-Tiffée and W. V. Münch, *Werkstoffe der Elektrotechnik*. Springer-Verlag, 2007.

[45] J. Hayd, "Nanoskalige Kathoden für den Einsatz in Festelektrolyt-Brennstoffzellen bei abgesenkten Betriebstemperaturen," Dissertation, Karlsruher Institut für Technologie (KIT), 2012.

[46] M. Mori, "Thermal Expansion of Nickel-Zirconia Anodes in Solid Oxide Fuel Cells during Fabrication and Operation," *J. Electrochem. Soc.*, vol. 145, no. 4, p. 1374, 1998.

[47] P. Costamagna, P. Costa, and V. Antonucci, "Micro-modelling of solid oxide fuel cell electrodes," *Electrochim. Acta*, vol. 43, no. 3-4, pp. 375–394, 1998.

[48] W. Z. Zhu and S. C. Deevi, "A review on the status of anode materials for solid oxide fuel cells," *Mater. Sci. Eng. A*, vol. 362, no. 1-2, pp. 228–239, 2003.

[49] S. Jiang and S. Chan, "A review of anode materials development in solid oxide fuel cells," *J. Mater. Sci.*, vol. 39, no. 14, pp. 4405–4439, 2004.

[50] A. Atkinson, S. Barnett, R. J. Gorte, J. T. S. Irvine, a. J. McEvoy, M. Mogensen, S. C. Singhal, and J. Vohs, "Advanced anodes for high-temperature fuel cells." *Nat. Mater.*, vol. 3, no. 1, pp. 17–27, 2004.

[51] C. Sun and U. Stimming, "Recent anode advances in solid oxide fuel cells," *J. Power Sources*, vol. 171, no. 2, pp. 247–260, 2007.

[52] F. Tietz, "Evaluation of commercial nickel oxide powders for components in solid oxide fuel cells," *J. Eur. Ceram. Soc.*, vol. 20, no. 8, pp. 1023–1034, 2000.

[53] R. J. Kee, H. Zhu, and D. G. Goodwin, "Solid-oxide fuel cells with hydrocarbon fuels," *Proc. Combust. Inst.*, vol. 30 II, no. 2, pp. 2379–2404, 2005.

[54] D. Fouquet, A. C. Müller, A. Weber, and E. Ivers-Tiffée, "Kinetics of oxidation and reduction of Ni/YSZ cermets," *Ionics (Kiel).*, vol. 9, no. 1-2, pp. 103–108, 2003.

[55] D. Sarantaridis and A. Atkinson, "Redox cycling of Ni-based solid oxide fuel cell anodes: A review," *Fuel Cells*, vol. 7, no. 3, pp. 246–258, 2007.

[56] Q. Fu, F. Tietz, D. Sebold, S. Tao, and J. T. S. Irvine, "An efficient ceramic-based anode for solid oxide fuel cells," *J. Power Sources*, vol. 171, no. 2, pp. 663–669, 2007.

[57] Q. Ma, F. Tietz, A. Leonide, and E. Ivers-Tiffée, "Anode-supported planar SOFC with high performance and redox stability," *Electrochem. commun.*, vol. 12, no. 10, pp. 1326–1328, 2010.

[58] Q. Ma and F. Tietz, "Comparison of Y- and La-substituted SrTiO3as the anode materials for SOFCs," *Solid State Ionics*, vol. 225, pp. 108–112, 2012.

[59] M. C. Tucker, "Progress in Metal-Supported Solid Oxide Fuel Cells," *J. Power Sources*, vol. 195, no. 15, pp. 4570–4582, 2010.

[60] S. R. Hui, D. Yang, Z. Wang, S. Yick, C. Decès-Petit, W. Qu, A. Tuck, R. Maric, and D. Ghosh, "Metal-supported solid oxide fuel cell operated at 400 - 600 °C," *J. Power Sources*, vol. 167, no. 2, pp. 336–339, 2007.

[61] G. Schiller, A. Ansar, M. Lang, and O. Patz, "High temperature water electrolysis using metal supported solid oxide electrolyser cells (SOEC)," *J. Appl. Electrochem.*, vol. 39, no. 2, pp. 293–301, 2009.

[62] P. Blennow, J. Hjelm, T. Klemensø, S. Ramousse, A. Kromp, A. Leonide, and A. Weber, "Manufacturing and characterization of metal-supported solid oxide fuel cells," *J. Power Sources*, vol. 196, no. 17, pp. 7117–7125, 2011.

[63] F. Tietz, A. Mai, and D. Stöver, "From powder properties to fuel cell performance - A holistic approach for SOFC cathode development," *Solid State Ionics*, vol. 179, no. 27-32, pp. 1509–1515, 2008.

[64] J. Richter, P. Holtappels, T. Graule, T. Nakamura, and L. J. Gauckler, "Materials design for perovskite SOFC cathodes," *Monatshefte für Chemie*, vol. 140, no. 9, pp. 985–999, 2012.

[65] B. Rueger, "Mikrostrukturmodellierung von Elektroden für die Festelektrolytbrennstoffzelle," Dissertation, Universität Karlsruhe (TH), 2009.

[66] A. Dany, "Development of lanthanum strontium manganite perovskite cathode materials of solid oxide fuel cells: A review," *J. Mater. Sci.*, vol. 43, no. 21, pp. 6799–6833, 2008.

[67] J. Mizusaki, "Electronic conductivity, Seebeck coefficient, defect and electronic structure of nonstoichiometric $La_{1-x}Sr_xMnO_{3-\delta}$," *Solid State Ionics*, vol. 132, no. 3-4, pp. 167–180, 2000.

[68] M. Kornely, A. Leonide, A. Weber, and E. Ivers-Tiffée, "Performance limiting factors in anode-supported cells originating from metallic interconnector design," *J. Power Sources*, vol. 196, no. 17, pp. 7209–7216, sep 2011.

[69] C. Endler-Schuck, "Alterungsverhalten mischleitender LSCF kathoden für Hochtemperatur Festoxid-Brennstoffzellen (SOFCs)," Dissertation, Universität Karlsruhe (TH), 2012.

[70] O. Yamamoto, "Perovskite-type oxides as oxygen electrodes for high temperature oxide fuel cells," *Solid State Ionics*, vol. 22, no. 2-3, pp. 241–246, 1987.

[71] S. Uhlenbruck, N. Jordan, D. Sebold, H. P. Buchkremer, V. A. C. Haanappel, and D. Stöver, "Thin film coating technologies of $(Ce,Gd)O_{2-\delta}$ interlayers for application in ceramic high-temperature fuel cells," *Thin Solid Films*, vol. 515, no. 7-8, pp. 4053–4060, 2007.

[72] J. Szász, F. Wankmüller, J. Joos, V. Wilde, H. Störmer, D. Gerthsen, and E. Ivers-Tiffée, "Correlating cathode/electrolyte interface characteristics to SOFC performance," *ECS Trans.*, vol. 77, no. 10, pp. 27–34, 2017.

[73] A. Mai, V. A. C. Haanappel, S. Uhlenbruck, F. Tietz, and D. Stöver, "Ferrite-based perovskites as cathode materials for anode-supported solid oxide fuel cells: Part I. Variation of composition," *Solid State Ionics*, vol. 176, no. 15-16, pp. 1341–1350, 2005.

[74] A. Mai, V. A. C. Haanappel, F. Tietz, and D. Stöver, "Ferrite-based perovskites as cathode materials for anode-supported solid oxide fuel cells. Part II. Influence of the CGO interlayer," *Solid State Ionics*, vol. 177, no. 19-25 SPEC. ISS., pp. 2103–2107, 2006.

[75] F. Tietz, V. A. C. Haanappel, A. Mai, J. Mertens, and D. Stöver, "Performance of LSCF cathodes in cell tests," *J. Power Sources*, vol. 156, no. 1 SPEC. ISS., pp. 20–22, 2006.

[76] A. Mai, M. Becker, W. Assenmacher, F. Tietz, D. Hathiramani, E. Ivers-Tiffée, D. Stöver, and W. Mader, "Time-dependent performance of mixed-conducting SOFC cathodes," *Solid State Ionics*, vol. 177, no. 19-25 SPEC. ISS., pp. 1965–1968, 2006.

[77] C. Endler, A. Leonide, A. Weber, F. Tietz, and E. Ivers-Tiffée, "Time-Dependent Electrode Performance Changes in Intermediate Temperature Solid Oxide Fuel Cells," *J. Electrochem. Soc.*, vol. 157, no. 2, pp. B292–B298, 2010.

[78] L. Blum, H. P. Buchkremer, S. Gross, A. Gubner, L. G. De Haart, H. Nabielek, W. J. Quadakkers, U. Reisgen, M. J. Smith, R. Steinberger-Wilckens, R. W. Steinbrech, F. Tietz, and I. C. Vinke, "Solid oxide fuel cell development at Forschungszentrum Juelich," *Fuel Cells*, vol. 7, no. 3, pp. 204–210, 2007.

[79] Q. Fang, L. Blum, R. Peters, M. Peksen, P. Batfalsky, and D. Stolten, "SOFC stack performance under high fuel utilization," *Int. J. Hydrogen Energy*, vol. 40, no. 2, pp. 1128–1136, 2015.

[80] S. B. Beale, H.-W. Choi, J. G. Pharoah, H. K. Roth, H. Jasak, and D. H. Jeon, "Open-source computational model of a solid oxide fuel cell," *Comput. Phys. Commun.*, vol. 200, pp. 15–26, 2015.

[81] S. Singhal, "Solid oxide fuel cells for stationary, mobile, and military applications," *Solid State Ionics*, vol. 152, pp. 405–410, 2002.

[82] K. Huang and J. B. Goodenough, "Solid Oxide Fuel Cell Technology: Principles, Performance and Operations," *Solid Oxide Fuel Cell Technol. Princ. Perform. Oper.*, pp. 1–328, 2009.

[83] Thyssen. Krupp, *Crofer 22 APU, Werkstoffdatenblatt Nr. 4143*. Thyssen Krupp VDM, 2005.

[84] A. Neumann, *Chrom-bezogene Degradation von Festoxid-Brennstoffzellen.* Schriften des Forschungszentrums Jülich, 2011.

[85] N. Grigorev, K. Sick, N. H. Menzler, O. Guillom, R. Telle, "Development of a $MnCo_{1.9}Fe_{0.1}O_4$ protection layer for SOFC interconnects applied by WPS and densified by reactive sintering," *Proc. 13th Eur. Solid Oxide Fuel Cell Forum*, Lucerne, 2018.

[86] B. C. H. Steele, "Running on natural gas," *Nature*, vol. 400, no. August, pp. 619–621, 1999.

[87] J. Van herle, D. Larrain, N. Autissier, Z. Wuillemin, M. Molinelli, and D. Favrat, "Modeling and experimental validation of solid oxide fuel cell materials and stacks," *J. Eur. Ceram. Soc.*, vol. 25, no. 12, pp. 2627–2632, 2005.

[88] Z. Zhan, J. Liu, and S. A. Barnett, "Operation of anode-supported solid oxide fuel cells on propane-air fuel mixtures," *Appl. Catal. A Gen.*, vol. 262, no. 2, pp. 255–259, 2004.

[89] A. Lindermeir, S. Kah, S. Kavurucu, and M. Mühlner, "On-board diesel fuel processing for an SOFC-APU-Technical challenges for catalysis and reactor design," *Appl. Catal. B Environ.*, vol. 70, no. 1-4, pp. 488–497, 2007.

[90] P. K. Cheekatamarla, C. M. Finnerty, C. R. Robinson, S. M. Andrews, J. A. Brodie, Y. Lu, and P. G. DeWald, "Design, integration and demonstration of a 50 W JP8/kerosene fueled portable SOFC power generator," *J. Power Sources*, vol. 193, no. 2, pp. 797–803, 2009.

[91] M. Santin, A. Traverso, and L. Magistri, "Liquid fuel utilization in SOFC hybrid systems," *Appl. Energy*, vol. 86, no. 10, pp. 2204–2212, 2009.

[92] O. A. Marina, C. A. Coyle, E. C. Thomsen, D. J. Edwards, G. W. Coffey, and L. R. Pederson, "Degradation mechanisms of SOFC anodes in coal gas containing phosphorus," *Solid State Ionics*, vol. 181, no. 8-10, pp. 430–440, 2010.

[93] A. L. Dicks, "Hydrogen generation from natural gas for the fuel cell systems of tomorrow," *J. Power Sources*, vol. 61, p. 113, 1996.

[94] S. H. Clarke, A. L. Dicks, K. Pointon, T. A. Smith, and A. Swann, "Catalytic Aspects of the Steam Reforming of Hydrocarbons in Internal Reforming Fuel Cells," *Catal. Today*, vol. 38, no. 4, pp. 411–423, 1997.

[95] A. L. Dicks, "Advances in catalysts for internal reforming in high temperature fuel cells," *J. Power Sources*, vol. 71, no. 1-2, pp. 111–122, 1998.

[96] F. Joensen and J. R. Rostrup-Nielsen, "Conversion of hydrocarbons and alcohols for fuel cells," *J. Power Sources*, vol. 105, no. 2, pp. 195–201, 2002.

[97] M. B. Mogensen and K. Kammer Hansen, "Conversion of hydrocarbons in solid oxide fuel cells," *Annu. Rev. Mater. Res.*, vol. 33, pp. 321–331, 2003.

[98] D. Mogensen, J. D. Grunwaldt, P. V. Hendriksen, K. Dam-Johansen, and J. U. Nielsen, "Internal steam reforming in solid oxide fuel cells: Status and opportunities of kinetic studies and their impact on modelling," *J. Power Sources*, vol. 196, no. 1, pp. 25–38, 2011.

[99] W. Lehnert, J. Meusinger, and F. Thom, "Modelling of gas transport phenomena in SOFC anodes," *J. Power Sources*, vol. 87, no. 1-2, pp. 57–63, apr 2000.

[100] P. Aguiar, C. S. Adjiman, and N. P. Brandon, "Anode-supported intermediate temperature direct internal reforming solid oxide fuel cell. I: Model-based steady-state performance," *J. Power Sources*, vol. 138, no. 1-2, pp. 120–136, 2004.

[101] E. S. Hecht, G. K. Gupta, H. Zhu, A. M. Dean, R. J. Kee, L. Maier, and O. Deutschmann, "Methane reforming kinetics within a Ni-YSZ SOFC anode support," *Appl. Catal. A Gen.*, vol. 295, no. 1, pp. 40–51, 2005.

[102] V. M. Janardhanan and O. Deutschmann, "CFD analysis of a solid oxide fuel cell with internal reforming: Coupled interactions of transport, heterogeneous catalysis and electrochemical processes," *J. Power Sources*, vol. 162, no. 2, pp. 1192–1202, 2006.

[103] H. Timmermann, W. Sawady, R. Reimert, and E. Ivers-Tiffée, "Kinetics of (reversible) internal reforming of methane in solid oxide fuel cells under stationary and APU conditions," *J. Power Sources*, vol. 195, no. 1, pp. 214–222, 2010.

[104] J. P. Van Hook, "Methane-Steam Reforming," *Catal. Rev.*, vol. 21, no. 1, pp. 1–51, 1980.

[105] J. R. Rostrup-Nielsen, *Catalytic Steam Reforming*. Berlin, Heidelberg: Springer-Verlag, 1984, vol. 5.

[106] M. Twigg, *Catalyst Handbook*, 2nd ed. CRC Press, 1989.

[107] F. Fischer and H. Tropsch, "Conversion of methane into hydrogen and carbon monoxide," *Brennst. Chem.*, vol. 3, no. 9, 1922.

[108] S. S. Bharadwaj and L. D. Schmidt, "Catalytic partial oxidation of natural gas to syngas," *Fuel Process. Technol.*, vol. 42, no. 2, pp. 109–127, 1995.

[109] B. C. Enger, R. Loedeng, and A. Holmen, "A review of catalytic partial oxidation of methane to synthesis gas with emphasis on reaction mechanisms over transition metal catalysts." *Appl. Catal. A Gen.*, vol. 346, no. 1-2, pp. 1–27, 2008.

[110] A. Weber, "Entwicklung von Kathodenstrukturen für die Hochtemperatur-Brenstoffzelle SOFC," Dissertation, Universität Karlsruhe (TH), 2002.

[111] J. Liu and S. A. Barnett, "Operation of anode-supported solid oxide fuel cells on methane and natural gas," *Solid State Ionics*, vol. 158, no. 1-2, pp. 11–16, 2003.

[112] Y. Lin, Z. Zhan, J. Liu, and S. A. Barnett, "Direct operation of solid oxide fuel cells with methane fuel," *Solid State Ionics*, vol. 176, no. 23-24, pp. 1827–1835, 2005.

[113] C. M. Finnerty, N. J. Coe, R. H. Cunningham, and R. Ormerod, "Carbon formation on and deactivation of nickel-based/zirconia anodes in solid oxide fuel cells running on methane," *Catal. Today*, vol. 46, no. 2-3, pp. 137–145, 1998.

[114] H. Timmermann, W. Sawady, D. Campbell, A. Weber, R. Reimert, and E. Ivers-Tiffée, "Coke Formation and Degradation in SOFC Operation with a Model Reformate from Liquid Hydrocarbons," *J. Electrochem. Soc.*, vol. 155, no. 4, p. B356, 2008.

[115] S. Zha, Z. Cheng, and M. Liu, "Sulfur Poisoning and Regeneration of Ni-Based Anodes in Solid Oxide Fuel Cells," *J. Electrochem. Soc.*, vol. 154, no. 2, p. B201, 2007.

[116] M. Gong, X. Liu, J. Trembly, and C. Johnson, "Sulfur-tolerant anode materials for solid oxide fuel cell application," *J. Power Sources*, vol. 168, no. 2, pp. 289–298, 2007.

[117] Z. Cheng, J. H. Wang, Y. M. Choi, L. Yang, M. C. Lin, and M. Liu, "From Ni-YSZ to sulfur-tolerant anode materials for SOFCs," *Energy Environ. Sci.*, no. 4, pp. 4380–4409, 2011.

[118] M. Kornely, A. Neumann, N. H. Menzler, A. Leonide, A. Weber, and E. Ivers-Tiffée, "Degradation of anode supported cell (ASC) performance by Cr-poisoning," *J. Power Sources*, vol. 196, no. 17, pp. 7203–7208, sep 2011.

[119] D. Udomsilp, D. Roehrens, N. H. Menzler, R. Conradt, and O. Guillon, "Characterization of the Contact Resistance of Cathodic SOFC Contacting," *ECS Trans.*, vol. 68, no. 1, pp. 751–756, 2015.

[120] J. Joos, "Modelling and Simulation of Solid Oxide Fuel Cell Cathodes," Dissertation, Karlsruhe Institute of Technology (KIT), 2015.

[121] M. Ender, "Mikrostrukturelle Charakterisierung, Modellentwicklung und Simulation poröser Elektroden für Lithiumionenzellen," Dissertation, Karlsruhe Institute of Technology (KIT), 2014.

[122] J. Ohser and K. Schladitz, *3D Images of Materials Structures: Processing and Analysis*. Wiley Online Library, 2010.

[123] L. O'Gorman, M. J. Sammon, M. Seul, and M. Seul, *Practical algorithms for image analysis: description, examples, programs, and projects*, 2nd ed. Cambridge University Press, 2008, vol. 2.

[124] N. Otsu, "A threshold selection method from gray-level histograms," *Automatica*, vol. 11, pp. 23–27, 1975.

[125] COMSOL, "COMSOL Multiphysics® User's Guide," Stockholm, Sweden, p. 1292, 2012. [Online]. Available:

[126] R. L. T. O. C. Zienkiewicz, *The Finite Element Method Set*, 6th ed. Butterworth-Heinemann, 2006, vol. 2.

[127] M. Jung and U. Langer, *Methode der finiten Elemente für Ingenieure*, 2nd ed. Wiesbaden: Springer Vieweg, 2013. [Online]. Available:

[128] E. Hairer, *Solving ordinary differential equations I: Nonstiff problems, Springer series in computational mathematics*, 8th ed. Berlin: Springer, 2000.

[129] E. Hairer and G. Wanner, *Solving Ordinary Differential Equations II: Stiff and Differential-Algebraic Problems*. Springer Series, 2010.

[130] H. R. Schwarz, *Numerische Mathematik, Lehrbuch Mathematik*, 6th ed. Wiesbaden: Teubner, 2006.

[131] R. Steinberger-Wilckens, L. Blum, H. P. Buchkremer, S. Gross, L. G. J. d. Haart, W. J. Quadakkers, U. Reisgen, R. W. Steinbrech, and F. Tietz, "Overview of the Development of Solid Oxide Fuel Cells at Forschungszentrum Jülich," in *Proc. Electrochem. Soc.* Quebec, Canada: Electrochemical Society, 2005, pp. 39–47.

[132] D. Stöver, H. P. Buchkremer, F. Tietz, and N. H. Menzler, "Trends in Processsing of SOFC Components," in *Proc. 5th Eur. Solid Oxide Fuel Cell Forum*, Lucerne, 2002, pp. 1–9.

[133] D. Stöver, H. Buchkremer, and J. Huijsmans, "Fuel Cell Technology and Applications Part 2," in *Handb. Fuell Cells*, W. Vielstich, A. Lamm, and H. Gasteiger, Eds. Chichester, UK: John Wiley and Sons Ltd., 2003, vol. 5, pp. 1013–1031.

[134] R. Mücke, N. H. Menzler, H. P. Buchkremer, and D. Stöver, "Cofiring of thin zirconia films during SOFC manufacturing," *J. Am. Ceram. Soc.*, vol. 92, no. SUPPL. 1, 2009.

[135] N. H. Menzler, W. Schafbauer, F. Han, O. Büchler, R. Mücke, H. P. Buchkremer, and D. Stöver, "Development of High Power Density Solid Oxide Fuel Cells (SOFCs) for Long-Term Operation," *Mater. Sci. Forum*, vol. 654-656, pp. 2875–2878, 2010.

[136] D. Klotz, A. Weber, and E. Ivers-Tiffée, "Practical Guidelines for Reliable Electrochemical Characterization of Solid Oxide Fuel Cells," *Electrochim. Acta*, vol. 227, pp. 110–126, 2017.

[137] M. Becker, A. Mai, E. Ivers-Tiffée, and F. Tietz, "Long-term mearsurements of anode-supported solid oxide fuel cells with LSCF cathodes under various operating conditions," *Proc. - Electrochem. Soc.*, vol. 2005-07, pp. 514–523, 2005.

[138] E. Dannhäuser, "Bewertung verschiedener quantitativer Multikomponenten-Gasanalysemethoden anhand von Produktgasen der internen Methanreformierung in der SOFC-Brennstoffzelle," Dissertation, Universität Karlsruhe (TH), 2003.

[139] P. König, "Modellgestützte Analyse und Simulation von stationären Brennstoffzellensystemen," Dissertation, Univeristät Karlsruhe (TH), 2007.

[140] M. Peksen, R. Peters, L. Blum, and D. Stolten, "Numerical modelling and experimental validation of a planar type pre-reformer in SOFC technology," *Int. J. Hydrogen Energy*, vol. 34, no. 15, pp. 6425–6436, 2009.

[141] M. Peksen, "A coupled 3D thermofluid–thermomechanical analysis of a planar type production scale SOFC stack," *Int. J. Hydrogen Energy*, vol. 36, no. 18, pp. 11 914–11 928, sep 2011.

[142] M. Peksen, R. Peters, L. Blum, and D. Stolten, "Hierarchical 3D multiphysics modelling in the design and optimisation of SOFC system components," *Int. J. Hydrogen Energy*, vol. 36, no. 7, pp. 4400–4408, 2011.

[143] M. Peksen, "3D CFD/FEM analysis of thermomechanical long-term behaviour in SOFCs: Furnace operation with different fuel gases," *Int. J. Hydrogen Energy*, vol. 40, no. 36, pp. 12 362–12 369, 2015.

[144] M. Andersson, H. Paradis, J. Yuan, and B. Sundén, "Three dimensional modeling of an solid oxide fuel cell coupling charge transfer phenomena with transport processes and heat generation," *Electrochim. Acta*, vol. 109, pp. 881–893, oct 2013.

[145] S. Liu, W. Kong, and Z. Lin, "Three-dimensional modeling of planar solid oxide fuel cells and the rib design optimization," *J. Power Sources*, vol. 194, no. 2, pp. 854–863, dec 2009.

[146] W. G. Bessler, S. Gewies, and M. Vogler, "A new framework for physically based modeling of solid oxide fuel cells," *Electrochim. Acta*, vol. 53, no. 4, pp. 1782–1800, dec 2007.

[147] A. Bertei, J. Mertens, and C. Nicolella, "Electrochemical Simulation of Planar Solid Oxide Fuel Cells with Detailed Microstructural Modeling," *Electrochim. Acta*, vol. 146, pp. 151–163, nov 2014.

[148] K. Tseronis, I. Kookos, and C. Theodoropoulos, "Modelling mass transport in solid oxide fuel cell anodes: a case for a multidimensional dusty gas-based model," *Chem. Eng. Sci.*, vol. 63, no. 23, pp. 5626–5638, dec 2008.

[149] A. Bertei, B. Nucci, and C. Nicolella, "Microstructural modeling for prediction of transport properties and electrochemical performance in SOFC composite electrodes," *Chem. Eng. Sci.*, vol. 101, pp. 175–190, 2013.

[150] W. G. Bessler and S. Gewies, "Gas Concentration Impedance of Solid Oxide Fuel Cell Anodes," *J. Electrochem. Soc.*, vol. 154, no. 6, p. B548, 2007.

[151] H. Geisler, A. Kromp, S. Hirn, A. Weber, and E. Ivers-Tiffée, "Model Based Interpretation of Coupled Gas Conversion and Diffusion in SOFC-Anodes," *ECS Trans.*, vol. 57, no. 1, pp. 2691–2704, 2013.

[152] S. Hirn, "Transiente Simulation der Stofftransportvorgänge im Brenngas der Hochtemperatur-Brennstoffzelle," Diploma Thesis, Karlsruhe Institut of Technology (KIT), 2012.

[153] C. W. Tanner and A. V. Virkar, "A simple model for interconnect design of planar solid oxide fuel cells," *J. Power Sources*, vol. 113, no. 1, pp. 44–56, 2003.

[154] S. Liu, C. Song, and Z. Lin, "The effects of the interconnect rib contact resistance on the performance of planar solid oxide fuel cell stack and the rib design optimization," *J. Power Sources*, vol. 183, no. 1, pp. 214–225, aug 2008.

[155] W. Kong, J. Li, S. Liu, and Z. Lin, "The influence of interconnect ribs on the performance of planar solid oxide fuel cell and formulae for optimal rib sizes," *J. Power Sources*, vol. 204, pp. 106–115, apr 2012.

[156] J. Joos, T. Carraro, A. Weber, and E. Ivers-Tiffée, "Reconstruction of porous electrodes by FIB/SEM for detailed microstructure modeling," *J. Power Sources*, vol. 196, no. 17, pp. 7302–7307, 2011.

[157] J. Joos, M. Ender, T. Carraro, A. Weber, and E. Ivers-Tiffée, "Representative volume element size for accurate solid oxide fuel cell cathode reconstructions from focused ion beam tomography data," *Electrochim. Acta*, vol. 82, pp. 268–276, nov 2012.

[158] J. Joos, M. Ender, I. Rotscholl, N. H. Menzler, and E. Ivers-Tiffée, "Quantification of double-layer Ni/YSZ fuel cell anodes from focused ion beam tomography data," *J. Power Sources*, vol. 246, pp. 819–830, jan 2014.

[159] T. Carraro, J. Joos, B. Rueger, A. Weber, and E. Ivers-Tiffée, "3D finite element model for reconstructed mixed-conducting cathodes: I. Performance quantification," *Electrochim. Acta*, vol. 77, pp. 315–323, 2012.

[160] T. Carraro, J. Joos, B. Rüger, A. Weber, and E. Ivers-tiffée, "3D finite element model for reconstructed mixed-conducting cathodes : II . Parameter sensitivity analysis," *Electrochim. Acta*, vol. 77, pp. 309–314, 2012.

[161] A. Bieberle and L. J. Gauckler, "State-space modeling of the anodic SOFC system Ni, H2-H2O|YSZ," *Solid State Ionics*, vol. 146, no. 1-2, pp. 23–41, 2002.

[162] W. G. Bessler, "A new computational approach for SOFC impedance from detailed electrochemical reaction-diffusion models," *Solid State Ionics*, vol. 176, no. 11-12, pp. 997–1011, 2005.

[163] W. G. Bessler, M. Vogler, H. Störmer, D. Gerthsen, A. Utz, A. Weber, and E. Ivers-Tiffée, "Model anodes and anode models for understanding the mechanism of hydrogen oxidation in solid oxide fuel cells," *Phys. Chem. Chem. Phys.*, vol. 12, no. 42, pp. 13 888–13 903, 2010.

[164] D. G. Goodwin, H. Zhu, A. M. Colclasure, and R. J. Kee, "Modeling Electrochemical Oxidation of Hydrogen on Ni–YSZ Pattern Anodes," *J. Electrochem. Soc.*, vol. 156, no. 9, p. B1004, 2009.

[165] M. Vogler, A. Bieberle-Huetter, L. Gauckler, J. Warnatz, and W. G. Bessler, "Modelling Study of Surface Reactions, Diffusion, and Spillover at a Ni/YSZ Patterned Anode," *J. Electrochem. Soc.*, vol. 156, no. 5, p. B663, 2009.

[166] B. E. Poling, M. Prausnitz, and J. P. O'Connell, *The Properties of Gases and Liquids*, 5th ed. New York: McGraw-Hill Education, 2006, vol. 23.

[167] S. Megel and A. Michaelis, "Kathodische Kontaktierung in planaren Hochtemperatur-Brennstoffzellen," Dissertation, Technische Universität Dresden, 2009.

[168] H. Zhu and R. J. Kee, "A general mathematical model for analyzing the performance of fuel-cell membrane-electrode assemblies," *J. Power Sources*, vol. 117, no. 1-2, pp. 61–74, may 2003.

[169] R. Suwanwarangkul, E. Croiset, M. Fowler, P. Douglas, E. Entchev, and M. Douglas, "Performance comparison of Fick's, dusty-gas and Stefan–Maxwell models to predict the concentration overpotential of a SOFC anode," *J. Power Sources*, vol. 122, no. 1, pp. 9–18, jul 2003.

[170] A. Bertei and C. Nicolella, "Common inconsistencies in modeling gas transport in porous electrodes: The dusty-gas model and the Fick law," *J. Power Sources*, vol. 279, pp. 133–137, 2015.

[171] A. Kromp, H. Geisler, A. Weber, and E. Ivers-Tiffée, "Electrochemical impedance modeling of gas transport and reforming kinetics in reformate fueled solid oxide fuel cell anodes," *Electrochim. Acta*, vol. 106, pp. 418–424, sep 2013.

[172] R. B. Bird, W. E. Stewart, E. N. Lig, E. N. Lightfoot, and J. Wiley, *Transport Phenomena*, 2nd ed. John Wiley & Sons, 2007.

[173] B. Todd and J. Young, "Thermodynamic and transport properties of gases for use in solid oxide fuel cell modelling," *J. Power Sources*, vol. 110, no. 1, pp. 186–200, jul 2002.

[174] D. Arnošt and P. Schneider, "Dynamic transport of multicomponent mixtures of gases in porous solids," *Chem. Eng. J. Biochem. Eng. J.*, vol. 57, no. 2, pp. 91–99, 1995.

[175] E. L. Cussler, *Diffusion : Mass Transfer in Fluid Systems*. Cambridge: Cambridge University Press, 1984.

[176] E. A. Mason and A. Malinauskas, "Gas transport in porous media: the dusty-gas model," *Elsevier Sci. Ltd.*, vol. 17, 1983.

[177] H. Zhu, R. J. Kee, V. M. Janardhanan, O. Deutschmann, and D. G. Goodwin, "Modeling Elementary Heterogeneous Chemistry and Electrochemistry in Solid-Oxide Fuel Cells," *J. Electrochem. Soc.*, vol. 152, no. 12, p. A2427, 2005.

[178] J. Young and B. Todd, "Modelling of multi-component gas flows in capillaries and porous solids," *Int. J. Heat Mass Transf.*, vol. 48, no. 25-26, pp. 5338–5353, dec 2005.

[179] P. J. Kerkhof, "A modified Maxwell-Stefan model for transport through inert membranes: the binary friction model," *Chem. Eng. J. Biochem. Eng. J.*, vol. 64, no. 3, pp. 319–343, dec 1996.

[180] S. Wang, W. M. Worek, and W. Minkowycz, "Performance comparison of the mass transfer models with internal reforming for solid oxide fuel cell anodes," *Int. J. Heat Mass Transf.*, vol. 55, no. 15-16, pp. 3933–3945, jul 2012.

[181] R. Krishna and J. A. Wesselingh, "The Maxwell-Stefan approach to mass tranfer," *Chem. Eng. Sci.*, vol. 52, no. 6, pp. 861–911, 1997.

[182] D. F. Fairbanks and C. R. Wilke, "Diffusion Coefficients in Multicomponent Gas Mixtures," *Ind. Eng. Chem.*, vol. 42, no. 3, pp. 471–475, 1950.

[183] J. M. Zalc, S. C. Reyes, and E. Iglesia, "The effects of diffusion mechanism and void structure on transport rates and tortuosity factors in complex porous structures," *Chem. Eng. Sci.*, vol. 59, no. 14, pp. 2947–2960, 2004.

[184] A. V. Akkaya, B. Sahin, and H. Huseyin Erdem, "Exergetic performance coefficient analysis of a simple fuel cell system," *Int. J. Hydrogen Energy*, vol. 32, no. 17, pp. 4600–4609, 2007.

[185] R. Jackson, *Transport in porous catalysts*. Elsevier North-Holland, 1977.

[186] R. Taylor and R. Krishna, *Chemical Engineering*. Wiley Series, 1983.

[187] K. Tseronis, I. K. Kookos, and C. Theodoropoulos, "Modelling mass transport in solid oxide fuel cell anodes: a case for a multidimensional dusty gas-based model," *Chem. Eng. Sci.*, vol. 63, no. 23, pp. 5626–5638, 2008.

[188] Y. Vural, L. Ma, D. B. Ingham, and M. Pourkashanian, "Comparison of the multicomponent mass transfer models for the prediction of the concentration overpotential for solid oxide fuel cell anodes," *J. Power Sources*, vol. 195, no. 15, pp. 4893–4904, aug 2010.

[189] I. K. Kookos, "On the diffusion in porous electrodes of SOFCs," *Chem. Eng. Sci.*, vol. 69, no. 1, pp. 571–577, feb 2012.

[190] H. A. Jakobsen, *Chemical reactor modeling: Multiphase reactive flows:*, 2nd ed. Springer International Publishing, 2014, vol. 9783319050.

[191] R. Krishna, "A simplified procedure for the solution of the dusty gas model equations for steady-state transport in non-reacting systems," *Chem. Eng. J.*, vol. 35, no. 2, pp. 75–81, jun 1987.

[192] L. B. Rothfeld, "Gaseous counterdiffusion in catalyst pellets," *AIChE J.*, vol. 9, no. 1, pp. 19–24, 1963.

[193] P. Schneider, "Characterization of catalyst pore structure by parameters describing mass transport," *Stud. Surf. Sci. Catal.*, vol. 39, no. C, pp. 333–343, 1988.

[194] M. Novák, K. Ehrhardt, K. Klusáček, and P. Schneider, "Dynamics of non-isobaric diffusion in porous catalysts," *Chem. Eng. Sci.*, vol. 43, no. 2, pp. 185–193, 1988.

[195] A. J. L. R. F. Bard, *Electrochemical Methods*, 2nd ed. Wiley, 2000.

[196] A. V. Virkar, J. Chen, C. W. Tanner, and J. W. Kim, "Role of electrode microstructure on activation and concentration polarizations in solid oxide fuel cells," *Solid State Ionics*, vol. 131, no. 1, pp. 189–198, 2000.

[197] G. Wang, Y. Yang, H. Zhang, and W. Xia, "3-D model of thermo-fluid and electrochemical for planar SOFC," *J. Power Sources*, vol. 167, no. 2, pp. 398–405, 2007.

[198] H. Geisler, A. Kromp, A. Weber, and E. Ivers-Tiffee, "Stationary FEM Model for Performance Evaluation of Planar Solid Oxide Fuel Cells Connected by Metal Interconnectors: I. Model Framework and Validation," *J. Electrochem. Soc.*, vol. 161, no. 6, pp. F778–F788, may 2014.

[199] M. Andersson, J. Yuan, and B. Sundén, "SOFC modeling considering electrochemical reactions at the active three phase boundaries," *Int. J. Heat Mass Transf.*, vol. 55, no. 4, pp. 773–788, 2012.

[200] D. Sánchez, R. Chacartegui, a. Muñoz, and T. Sánchez, "Thermal and electrochemical model of internal reforming solid oxide fuel cells with tubular geometry," *J. Power Sources*, vol. 160, no. 2, pp. 1074–1087, oct 2006.

[201] R. C. Alkire, *Electrochemical engineering.* Berlin: Springer-Verlag, 1983, vol. 60.

[202] C. Niedrig, *Electrochemical Performance and Stability of* $Ba_{0.5}Sr_{0.5}Co_{0.8}Fe_{0.2}O_{3-\delta}$ *for Oxygen Transport Membranes.* Dissertation, Karlsruher Institute of Technology (KIT), 2015.

[203] M. Søgaard, P. V. Hendriksen, T. Jacobsen, and M. Mogensen, "Modelling of the Polarization Resistance from Surface Exchange and Diffusion Coefficient Data," in *Proc. 7th Eur. Solid Oxide Fuel Cell Forum*, Lucerne, pp. 1–17, 2006.

[204] H. Timmermann, "Untersuchungen zum Einsatz von Reformat aus flüssigen Kohlenwasserstoffen in der Hochtemperaturbrennstoffzelle SOFC," Dissertation, (TH) Karlsruhe, 2009.

[205] H. Yokokawa, S. Yamauchi, and T. Matsumoto, "Thermodynamic database MALT for windows with gem and CHD," *Calphad Comput. Coupling Phase Diagrams Thermochem.*, vol. 26, no. 2, pp. 155–166, 2002.

[206] A. Kromp, "Modellbildung und Simulation der internen Reformierung an SOFC-Anoden bei Betrieb mit kohlenwasserstoffhaltigen Brenngasen," Diploma thesis, Universität Karlsruhe (TH), 2008.

[207] S. Primdahl, "Gas Diffusion Impedance in Characterization of Solid Oxide Fuel Cell Anodes," *J. Electrochem. Soc.*, vol. 146, no. 8, p. 2827, 1999.

[208] A. Leonide, V. Sonn, A. Weber, and E. Ivers-Tiffée, "Evaluation and Modeling of the Cell Resistance in Anode-Supported Solid Oxide Fuel Cells," *J. Electrochem. Soc.*, vol. 155, no. 1, pp. B36–B41, 2008.

[209] D. A. G. Bruggeman, "Calculation of various physical constants of heterogeneous substances. I. Dielectric," *Ann. Phys.*, vol. 24, no. 8, pp. 636–679, 1935.

[210] L. Shen and Z. Chen, "Critical review of the impact of tortuosity on diffusion," *Chem. Eng. Sci.*, vol. 62, no. 14, pp. 3748–3755, 2007.

[211] T. Carraro, J. Joos, B. Rueger, A. Weber, and E. Ivers-Tiffée, "3D finite element model for reconstructed mixed-conducting cathodes: II. Parameter sensitivity analysis," *Electrochim. Acta*, vol. 77, pp. 309–314, 2012.

[212] J. Joos, H. Geisler, A. Weber, and E. Ivers-Tiffée, "Designing Porous Cathode Structures for SOFCs," in *Proc. 12th Eur. SOFC SOE Forum*, vol. 8, Lucerne, 2016, p. B0803.

[213] W. E. Lorensen and H. E. Cline, "Marching Cubes: A high resolution 3D surface construction algorithm," *Comput. Graph. Vol. 21, Nr. 4*, 1987.

[214] J. Joos, T. Carraro, M. Ender, B. Rueger, A. Weber, and E. Ivers-Tiffée, "Detailed Microstructure Analysis and 3D Simulations of Porous Electrodes," *ECS Trans.*, vol. 35, no. 1, pp. 2357–2368, 2011.

[215] U. Hornung, *Homogenization and Porous Media.* New York: Springer-Verlag, 1991, vol. 98.

[216] M. Ender, J. Joos, T. Carraro, and E. Ivers-Tiffee, "Quantitative Characterization of $LiFePO_4$ Cathodes Reconstructed by FIB/SEM Tomography," *J. Electrochem. Soc.*, vol. 159, no. 7, pp. A972–A980, 2012.

[217] M. Ender, J. Joos, A. Weber, and E. Ivers-Tiffée, "Anode microstructures from high-energy and high-power lithium-ion cylindrical cells obtained by X-ray nanotomography," *J. Power Sources*, vol. 269, pp. 912–919, 2014.

[218] S. Gewies, "Modellgestützte Interpretation der elektrochemischen Charakteristik von Festoxid-Brennstoffzellen mit Ni/YSZ-Cermetanoden," Dissertation, Ruprecht-Karls-Universität Heidelberg, 2009.

[219] F. Zhao and A. V. Virkar, "Dependence of polarization in anode-supported solid oxide fuel cells on various cell parameters," *J. Power Sources*, vol. 141, no. 1, pp. 79–95, 2005.

[220] D. Simwonis, A. Naoumidis, F. Dias, J. Linke, and A. Moropoulou, "Material characterization in support of the development of an anode substrate for solid oxide fuel cells," *J. Mater. Res.*, vol. 12, no. 6, pp. 1508–1518, 1997.

[221] D. Simwonis, H. Thülen, F. J. Dias, a. Naoumidis, and D. Stöver, "Properties of Ni/YSZ porous cermets for SOFC anode substrates prepared by tape casting and coat-mix process," *J. Mater. Process. Technol.*, vol. 92-93, pp. 107–111, 1999.

[222] W. Schafbauer and D. Stöver, "Entwicklung und Herstellung von foliengegossenen, anodengestützten Festoxidbrennstoffzellen," Dissertation, Universität Bochum, 2009.

[223] J. Kozeny, "Ueber kapillare Leitung des Wassers im Boden," *Akad. Wiss.*, vol. 136(2a), pp. 271–306, 1927.

[224] S. P. Sutera and R. Skalak, "The History of Poiseuille's Law," *Annu. Rev. Fluid Mech.*, vol. 25, no. 1, pp. 1–20, 1993.

[225] M. S. Paterson, "The Equivalent Channel Model for Permiability and Resistivity in Fluid-Saturated Rock - A Re-Appraisal," *Mech. Mater.*, vol. 2, no. 4, pp. 345–352, 1983.

[226] M. N. Panda and L. W. Lake, "Estimation of single-phase permeability from parameters of particle-size distribution," *Am. Assoc. Pet. Geol. Bull.*, vol. 78, no. 7, pp. 1028–1039, 1994.

[227] A. Costa, "Permeability-porosity relationship: A reexamination of the Kozeny-Carman equation based on a fractal pore-space geometry assumption," *Geophys. Res. Lett.*, vol. 33, no. 2, pp. 1–5, 2006.

[228] B. P. C. Carman and P. D. Graduate, "Fluid Flow Through Granular Beds *," *Chem. Eng. Res. Des.*, vol. 75, pp. S32–S48, 1932.

[229] D. Horny, "Modellierung der viskosen Gasströmung in porösen Elektroden der Hochtemperatur - Brennstoffzelle SOFC," Master thesis, Karlsruhe Institute of Technology (KIT), 2017.

[230] E. Bucher, M. Yang, and W. Sitte, "In Situ Investigations of the Chromium-Induced Degradation of the Oxygen Surface Exchange Kinetics of IT-SOFC Cathode Materials," *J. Electrochem. Soc.*, vol. 159, no. 5, pp. B592–B596, 2012.

[231] L. Almar, J. Szász, A. Weber, and E. Ivers-Tiffée, "Oxygen Transport Kinetics of Mixed Ionic-Electronic Conductors by Coupling Focused Ion Beam Tomography and Electrochemical Impedance Spectroscopy," *J. Electrochem. Soc.*, vol. 164, no. 4, pp. F289–F297, 2017.

[232] J. Szász, "Charakterisierung und Modellentwicklung von Natur und Funktionalität der Kathoden / Elektrolyt-Grenzfläche von Hochtemperatur-Brennstoffzellen (SOFC)," Dissertation, Karlsruhe Institute of Technology (KIT), 2018.

[233] J. Druce, T. Ishihara, and J. Kilner, "Surface composition of perovskite-type materials studied by Low Energy Ion Scattering (LEIS)," *Solid State Ionics*, vol. 262, pp. 893–896, 2014.

[234] J. Druce, H. Téllez, N. Simrick, T. Ishihara, and J. Kilner, "Surface composition of solid oxide electrode structures by laterally resolved low energy ion scattering (LEIS)," *Int. J. Hydrogen Energy*, vol. 39, no. 35, pp. 20 850–20 855, 2014.

[235] F. Wang, K. Yamaji, D. H. Cho, T. Shimonosono, M. Nishi, H. Kishimoto, M. E. Brito, T. Horita, and H. Yokokawa, "Evaluation of sulfur dioxide poisoning for LSCF cathodes," *Fuel Cells*, vol. 13, no. 4, pp. 520–525, 2013.

[236] S. Y. Lai, D. Ding, M. Liu, M. Liu, and F. M. Alamgir, "Operando and in situ X-ray spectroscopies of degradation in $La_{0.6}Sr_{0.4}Co_{0.2}Fe_{0.8}O_{3-\delta}$ thin film cathodes in fuel cells," *ChemSusChem*, vol. 7, no. 11, pp. 3078–3087, 2014.

[237] A. Egger, E. Bucher, M. Yang, and W. Sitte, "Comparison of oxygen exchange kinetics of the IT-SOFC cathode materials," *Solid State Ionics*, vol. 225, pp. 55–60, 2012.

[238] H. Jung, WC, Tuller, "Investigation of cathode behavior of model thin-film $SrTi_{1-x}Fe_xO_{3-\delta}$) (x=0.35 and 0.5) mixed ionic-electronic conducting electrodes," *J. Electrochem. Soc.*, vol. 155, no. 11, pp. B1194 – B1201, 2008.

[239] T. S. Stefanik, "Electrical Properties and Defect Structure of Praseodymium–Cerium Oxide Solid Solutions," Dissertation, Massachusetts Institute of Technology, 2004.

[240] H. Tuller, S. Bishop, D. Chen, Y. Kuru, J.-J. Kim, and T. Stefanik, "Praseodymium doped ceria: Model mixed ionic electronic conductor with coupled electrical, optical, mechanical and chemical properties," *Solid State Ionics*, vol. 225, pp. 194–197, oct 2012.

[241] F. Wirtz, "Untersuchung und FEM-Simulation der katalytischen Aktivität von Hochtemperatur-Brennstoffzellen (SOFCs)," Master thesis, Karlsruhe Institute of Technology (KIT), 2015.

[242] W. Z. Zhu and S. C. Deevi, "Opportunity of metallic interconnects for solid oxide fuel cells: A status on contact resistance," *Mater. Res. Bull.*, vol. 38, no. 6, pp. 957–972, 2003.

[243] J. W. Fergus, "Metallic interconnects for solid oxide fuel cells," *Mater. Sci. Eng. A*, vol. 397, no. 1-2, pp. 271–283, 2005.

[244] J. G. Grolig, J. Froitzheim, and J. E. Svensson, "Coated stainless steel 441 as interconnect material for solid oxide fuel cells: Oxidation performance and chromium evaporation," *J. Power Sources*, vol. 248, pp. 1007–1013, 2014.

[245] J. Malzbender, P. Batfalsky, R. Vaßen, V. Shemet, and F. Tietz, "Component interactions after long-term operation of an SOFC stack with LSM cathode," *J. Power Sources*, vol. 201, pp. 196–203, 2012.

[246] M.-J. Garcia-Vargas, M. Zahid, F. Tietz, T. Kiefer, and P. Stevens, "Protection of SOFC Interconnects Against Cr-Evaporation Using Atmospheric Plasma-Sprayed Spinel Layers," *7th Eur. SOFC Forum*, vol. 1, no. B081, p. File No. B081, 2006.

[247] K. Sick, N. Grigorev, N. Menzler, and O. Guillon, "Development of Cathode Contacting for SOFC Stacks," in *42nd Int. Conf. Expo Adv. Ceram. Compos. - Ceram. Trans.*, vol. 39, Daytona Beach, FL United States, 2018.

[248] C. Goebel, A. G. Fefekos, J. E. Svensson, and J. Froitzheim, "Does the conductivity of interconnect coatings matter for solid oxide fuel cell applications?" *J. Power Sources*, vol. 383, pp. 110–114, 2018.

[249] M. Andersson, J. Yuan, and B. Sundén, "SOFC modeling considering hydrogen and carbon monoxide as electrochemical reactants," *J. Power Sources*, vol. 232, pp. 42–54, 2013.

[250] J.-C. Njodzefon, D. Klotz, A. Kromp, A. Weber, and E. Ivers-Tiffee, "Electrochemical Modeling of the Current-Voltage Characteristics of an SOFC in Fuel Cell and Electrolyzer Operation Modes," *J. Electrochem. Soc.*, vol. 160, no. 4, pp. F313–F323, 2013.

[251] W. Jiang, R. Fang, J. A. Khan, and R. A. Dougal, "Parameter setting and analysis of a dynamic tubular SOFC model," *J. Power Sources*, vol. 162, no. 1, pp. 316–326, 2006.

[252] L. Blum, P. Batfalsky, Q. Fang, L. J. de Haart, J. Malzbender, N. Margaritis, N. Menzler, and R. Peters, "Solid Oxide Fuel Cell, Stack and System Development Status at Forschungszentrum Jülich L. Blum," *ECS Trans.*, vol. 68, no. 1, pp. 157–169, 2015.

[253] S. Omar, A. Belda, A. Escardino, and N. Bonanos, "Ionic conductivity ageing investigation of 1Ce10ScSZ in different partial pressures of oxygen," *Solid State Ionics*, vol. 184, no. 1, pp. 2–5, 2011.

[254] M. A. Laguna-Bercero and V. M. Orera, "Micro-spectroscopic study of the degradation of scandia and ceria stabilized zirconia electrolytes in solid oxide electrolysis cells," *Int. J. Hydrogen Energy*, vol. 36, no. 20, pp. 13 051–13 058, 2011.

[255] J. Szász, S. Seils, D. Klotz, H. Störmer, M. Heilmaier, D. Gerthsen, H. Yokokawa, and E. Ivers-Tiffée, "High-Resolution Studies on Nanoscaled Ni/YSZ Anodes," *Chem. Mater.*, vol. 29, no. 12, pp. 5113–5123, 2017.

[256] H. Geisler, S. Dierickx, A. Weber, and E. Ivers-Tiffee, "A 2D Stationary FEM Model for Hydrocarbon Fuelled SOFC Stack Layers," *ECS Trans.*, vol. 68, no. 1, pp. 2151–2158, 2015.

[257] Gesellschaft, VDI, "5. Stoffwerte und Zustandsgrößen," in *VDI-Wärmeatlas*. Düsseldorf: Springer Vieweg, p.1760, 2006.

Werkstoffwissenschaft @ Elektrotechnik /
Universität Karlsruhe, Institut für Werkstoffe der Elektrotechnik

---

Band 1 Helge Schichlein
Experimentelle Modellbildung für die Hochtemperatur-Brennstoffzelle SOFC. 2003
ISBN 3-86130-229-2

Band 2 Dirk Herbstritt
Entwicklung und Optimierung einer leistungsfähigen Kathodenstruktur für die Hochtemperatur-Brennstoffzelle SOFC. 2003
ISBN 3-86130-230-6

Band 3 Frédéric Zimmermann
Steuerbare Mikrowellendielektrika aus ferroelektrischen Dickschichten. 2003
ISBN 3-86130-231-4

Band 4 Barbara Hippauf
Kinetik von selbsttragenden, offenporösen Sauerstoffsensoren auf der Basis von $Sr(Ti,Fe)O_3$. 2005
ISBN 3-86130-232-2

Band 5 Daniel Fouquet
Einsatz von Kohlenwasserstoffen in der Hochtemperatur-Brennstoffzelle SOFC. 2005
ISBN 3-86130-233-0

Band 6 Volker Fischer
Nanoskalige Nioboxidschichten für den Einsatz in hochkapazitiven Niob-Elektrolytkondensatoren. 2005
ISBN 3-86130-234-9

Band 7 Thomas Schneider
Strontiumtitanferrit-Abgassensoren. Stabilitätsgrenzen / Betriebsfelder. 2005
ISBN 3-86130-235-7

Band 8 Markus J. Heneka
Alterung der Festelektrolyt-Brennstoffzelle unter thermischen und elektrischen Lastwechseln. 2006
ISBN 3-86130-236-5

Die Bände sind im Verlagshaus Mainz (Aachen) erschienen.

Band 9 Thilo Hilpert
Elektrische Charakterisierung von Wärmedämmschichten mittels Impedanzspektroskopie. 2007
ISBN 3-86130-237-3

Band 10 Michael Becker
Parameterstudie zur Langzeitbeständigkeit von Hochtemperaturbrennstoffzellen (SOFC). 2007
ISBN 3-86130-239-X

Band 11 Jin Xu
Nonlinear Dielectric Thin Films for Tunable Microwave Applications. 2007
ISBN 3-86130-238-1

Band 12 Patrick König
Modellgestützte Analyse und Simulation von stationären Brennstoffzellensystemen. 2007
ISBN 3-86130-241-1

Band 13 Steffen Eccarius
Approaches to Passive Operation of a Direct Methanol Fuel Cell. 2007
ISBN 3-86130-242-X

Die Bände sind im Verlagshaus Mainz (Aachen) erschienen.

Fortführung als

Schriften des Instituts für Werkstoffe der Elektrotechnik, Karlsruher Institut für Technologie (ISSN 1868-1603)

bei KIT Scientific Publishing

Band 14 Stefan F. Wagner
Untersuchungen zur Kinetik des Sauerstoffaustauschs an modifizierten Perowskitgrenzflächen. 2009
ISBN 978-3-86644-362-4

Band 15 Christoph Peters
Grain-Size Effects in Nanoscaled Electrolyte and Cathode Thin Films for Solid Oxide Fuel Cells (SOFC). 2009
ISBN 978-3-86644-336-5

Band 16 Bernd Rüger
Mikrostrukturmodellierung von Elektroden für die Festelektrolytbrennstoffzelle. 2009
ISBN 978-3-86644-409-6

Band 17 Henrik Timmermann
Untersuchungen zum Einsatz von Reformat aus flüssigen Kohlenwasserstoffen in der Hochtemperaturbrennstoffzelle SOFC. 2010
ISBN 978-3-86644-478-2

Band 18 André Leonide
SOFC Modelling and Parameter Identification by Means of Impedance Spectroscopy. 2010
ISBN 978-3-86644-538-3

Band 19 Cornelia Endler-Schuck
Alterungsverhalten mischleitender LSCF Kathoden für Hochtemperatur-Festoxid-Brennstoffzellen (SOFCs). 2011
ISBN 978-3-86644-652-6

Band 20 Annika Utz
The Electrochemical Oxidation of H_2 and CO at Patterned Ni Anodes of SOFCs. 2011
ISBN 978-3-86644-686-1

Die Bände sind unter www.ksp.kit.edu als PDF frei verfügbar oder als Druckausgabe bestellbar.

Band 21 Jan Hayd
Nanoskalige Kathoden für den Einsatz in Festelektrolyt-Brennstoffzellen bei abgesenkten Betriebstemperaturen. 2012
ISBN 978-3-86644-838-4

Band 22 Michael Kornely
Elektrische Charakterisierung und Modellierung von metallischen Interkonnektoren (MIC) des SOFC-Stacks. 2012
ISBN 978-3-86644-833-9

Band 23 Dino Klotz
Characterization and Modeling of Electrochemical Energy Conversion Systems by Impedance Techniques. 2012
ISBN 978-3-86644-903-9

Band 24 Alexander Kromp
Model-based Interpretation of the Performance and Degradation of Reformate Fueled Solid Oxide Fuel Cells. 2013
ISBN 978-3-7315-0006-3

Band 25 Jan Philipp Schmidt
Verfahren zur Charakterisierung und Modellierung von Lithium-Ionen Zellen. 2013
ISBN 978-3-7315-0115-2

Band 26 Moses Ender
Mikrostrukturelle Charakterisierung, Modellentwicklung und Simulation poröser Elektroden für Lithiumionenzellen. 2014
ISBN 978-3-7315-0205-0

Band 27 Jörg Illig
Physically based Impedance Modelling of Lithium-Ion Cells. 2014
ISBN 978-3-7315-0246-3

Fortführung als

Schriften des Instituts für Angewandte Materialien – Werkstoffe der Elektrotechnik Karlsruher Institut für Technologie (ISSN 2365-8029)

bei KIT Scientific Publishing

Band 28 Christian Niedrig
Electrochemical Performance and Stability of $Ba_{0.5}Sr_{0.5}Co_{0.8}Fe_{0.2}O_{3-\delta}$ for Oxygen Transport Membranes. 2015
ISBN 978-3-7315-0437-5

Band 29 Daniel Manka
Elektrochemisch-optische Impedanzspektroskopie. 2016
ISBN 978-3-7315-0547-1

Band 30 Jochen Joos
Microstructural Characterisation, Modelling and Simulation of Solid Oxide Fuel Cell Cathodes. 2017
ISBN 978-3-7315-0625-6

Band 31 Christian Uhlmann
Charakterisierung und Modellierung der wässrigen Lithium-Luft Batterie. 2017
ISBN 978-3-7315-0671-3

Band 32 Michael Schönleber
Verfahren zur Charakterisierung des Niederfrequenzverhaltens von Lithium-Ionen Batterien. 2017
ISBN 978-3-7315-0685-0

Band 33 Janina Costard
Einfluss von Mikrostruktur und Materialparametern auf die Leistungsfähigkeit poröser Elektroden für Lithium-Ionen Batterien. 2018
ISBN 978-3-7315-0821-2

Band 34 Lana-Simone Unger
Phasenstabilisierung und Oberflächenaktivierung von Sauerstoffseparationsmembranen aus dotiertem $Ba_{0,5}Sr_{0,5}Co_{0,8}Fe_{0,2}O_{3-\delta}$. 2019
ISBN 978-3-7315-0847-2

Die Bände sind unter www.ksp.kit.edu als PDF frei verfügbar oder als Druckausgabe bestellbar.

Band 35 Julian Tibor Szász
Charakterisierung und Modellentwicklung von Natur und Funktionalität der Kathoden/Elektrolyt-Grenzfläche von Hochtemperatur-Brennstoffzellen (SOFC). 2019
ISBN 978-3-7315-0861-8

Band 36 Helge Ingolf Geisler
Finite Element Method (FEM) Model and Performance Analysis of Solid Oxide Fuel Cells. 2019
ISBN 978-3-7315-0895-3